Retinoids, differentiation and disease

Retinoids, differentiation and disease

Ciba Foundation Symposium 113

1985

Pitman
London

© Ciba Foundation 1985

ISBN 0 272 79813 4

Published in June 1985 by Pitman Publishing Ltd., 128 Long Acre, London WC2E 9AN, UK.
Distributed in North America by CIBA Pharmaceutical Company (Medical Education Division),
Post Office Box 18060, Newark, NJ 07101, USA

Suggested series entry for library catalogues:
Ciba Foundation symposia

Ciba Foundation symposium 113
x + 286 pages, 59 figures, 34 tables

British Library cataloguing in publication data:
Retinoids, differentiation and disease.—(Ciba Foundation symposium; 113)
   1. Retinoids—Therapeutic use
   I. Series
   615′.328   RM666.R4/

Typeset and printed in Great Britain at The Bath Press, Avon

Contents

Participants

J. S. Bertram   Cancer Center of Hawaii, 1236 Lauhala Street, Honolulu, Hawaii 96813, USA

T. R. Breitman   Laboratory of Medicinal Chemistry & Pharmacology, DTP, DCT, National Cancer Institute, Bldg 37, Rm 5B14, National Institutes of Health, Bethesda, Maryland 20205, USA

C. E. Brinckerhoff   Department of Medicine, Connective Tissue Disease Section, Dartmouth Medical School, Hanover, New Hampshire 03756, USA

M. I. Dawson   Bio-Organic Chemistry Laboratory, SRI International, 333 Ravenswood Avenue, Menlo Park, California 94025, USA

G. Dennert   University of Southern California Comprehensive Cancer Center, 2025 Zonal Avenue, Los Angeles, California 90033, USA

S. A. Eccles   Division of Tumour Immunology, Block X, Institute of Cancer Research, Clifton Avenue, Belmont, Sutton, Surrey SM2 5PX, UK

F. Frickel   BASF Aktiengesellschaft, Main Laboratory A.30, D-6700 Ludwigshafen, Federal Republic of Germany

D. Hartmann   Hoffmann-La Roche & Co Ltd, Basel, Switzerland

R. M. Hicks   The School of Pathology, Middlesex Hospital Medical School, Riding House Street, London W1P 7LD, UK

A. M. Jetten   Cell Biology Group, Laboratory of Pulmonary Function & Toxicology, National Institute of Environmental Health Sciences, National Institutes of Health, PO Box 12233, Research Triangle Park, North Carolina 27709, USA

I. A. King *(Ciba Foundation Bursar)*   Dermatology Research Group, Division of Clinical Sciences, MRC Clinical Research Centre, Watford Road, Harrow, Middlesex HA1 3UJ, UK

H. P. Koeffler Department of Medicine, 32–139 CHS, Division of Hematology/Oncology, Factor Bldg, 11–240, University of California School of Medicine, 405 Hilgard Avenue, Los Angeles, California 90024, USA

R. Lotan Department of Tumor Biology, The University of Texas System Cancer Center, MD Anderson Hospital and Tumor Institute, 6723 Bertner Avenue, Houston, Texas 77030, USA

M. Maden Division of Developmental Biology, MRC National Institute for Medical Research, The Ridgeway, Mill Hill, London NW7 1AA, UK

M. Malkovský Division of Immunological Medicine, MRC Clinical Research Centre, Watford Road, Harrow, Middlesex HA1 3UJ, UK

R. C. Moon Laboratory of Pathophysiology, Life Sciences Research, IIT Research Institute, 10 West 35th Street, Chicago, Illinois 60616, USA

G. L. Peck Dermatology Branch, National Cancer Institute, Building 10, Room 12N 238, National Institutes of Health, Bethesda, Maryland 20205, USA

G. A. J. Pitt Department of Biochemistry, University of Liverpool, PO Box 147, Liverpool L69 3BX, UK

M. I. Sherman Department of Cell Biology, Roche Institute of Molecular Biology, Nutley, New Jersey 07110, USA

M. B. Sporn *(Chairman)* Laboratory of Chemoprevention, Division of Cancer Etiology, National Cancer Institute, Building 41, Room D.201, National Institutes of Health, Bethesda, Maryland 20205, USA

S. Strickland Department of Pharmacology, Health Science Center, State University of New York, Stony Brook, New York 11794, USA

C. Tickle Department of Anatomy and Biology as Applied to Medicine, The Middlesex Hospital Medical School, Cleveland Street, London W1P 6DP, UK

D. Tsambaos Department of Dermatology, University of Essen, Hufelandstraße 55, D-4300 Essen 1, Federal Republic of Germany

J. A. Turton   The School of Pathology, Middlesex Hospital Medical School, Riding House Street, London W1P 7LD, UK

N. J. Wald   Department of Environmental and Preventive Medicine, The Medical College of St Bartholomew's Hospital, Charterhouse Square, London EC1M 6BQ, UK

F. H. White   Department of Anatomy and Cell Biology, University of Sheffield, Sheffield S10 2TN, UK

S. H. Yuspa   Laboratory of Cellular Carcinogenesis and Tumor Promotion, National Cancer Institute, Bldg 37, Rm 3A21, National Institutes of Health, Bethesda, Maryland 20205, USA

Nomenclature note

(1) The names of retinoic acid derivatives are based on the retinoid skeleton:

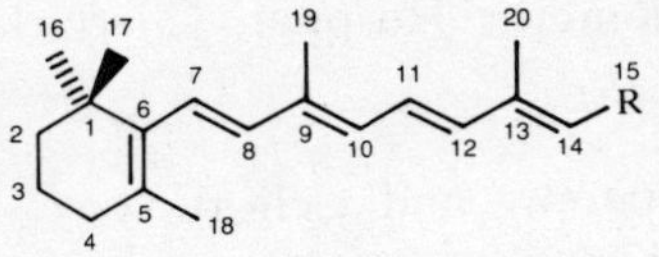

(IUPAC–IUB)

(2) The aromatic benzoic acid derivative:

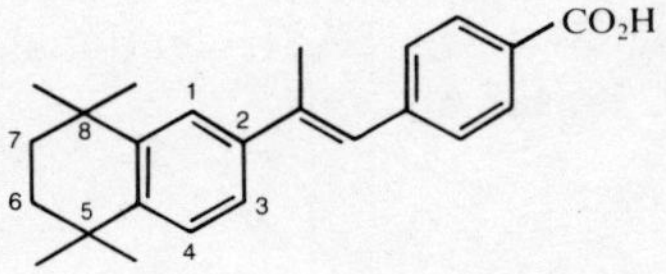

is usually referred to in this book as (*E*)-4-[2-(5,6,7,8-tetrahydro-5,5,8,8-tetramethyl-2-naphthalenyl)-1-propenyl]benzoic acid, abbreviated to TTNPB. This compound is also known as arotinoid or Ro 13-7410. TTNPB ethyl ester is also known as arotinoid ethyl ester or Ro 13-6298.

The numbering system shown for TTNPB is also used here for other aromatic retinoic acid analogues.

Introduction: what is a retinoid?

MICHAEL B. SPORN and ANITA B. ROBERTS

Laboratory of Chemoprevention, National Cancer Institute, Bethesda, Maryland 20205, USA

1985 Retinoids, differentiation and disease. Pitman, London (Ciba Foundation Symposium 113) p 1–5

This symposium deals with the fundamental role of retinoids in the control of cell differentiation and cell proliferation, and the application of this know- ledge to the understanding and treatment of human disease. It is clear that research on retinoids has expanded far beyond its original domain of studies of nutrition and vision. This expansion is the result of the creative efforts of organic chemists who have made a very large set of new substances which now have applications in many areas of both basic science and clinical medicine. More than a thousand new retinoids have been synthesized (Frickel 1984), and many of them are far more potent than retinol, retinyl palmitate, retinaldehyde or retinoic acid (Sporn & Roberts 1984). The greater potency of the most active new retinoids has been measured in a broad spectrum of assays, as diverse as the regulation of specific gene activity (Eckert & Green 1984), the control of either cell differentiation or cell proliferation in tissue culture or organ culture (Lotan 1980, Sporn & Roberts 1984), the prevention of cancer in experimental animals (Moon & Itri 1984), and even the classical assay of support of growth in vitamin A-deficient rats (Stephens-Jarnagin et al 1985).

The most fundamental question that may be asked of this set of substances is: 'What is a retinoid?' The orthodox response to this question would be to state that retinoids are substances related to vitamin A (retinol), and then to define them (as IUPAC–IUB has done) as 'those diterpenoids derived from a monocyclic parent compound containing five carbon–carbon double bonds and a functional group at the terminus of the acyclic portion'. While this is an entirely reasonable definition, it unfortunately suffers from being totally at variance with a large set of recent experimental data. The problem with the IUPAC–IUB definition is that it is fixated on retinol as a reference substance and does not take into account the fact that tricyclic or tetracyclic retinoidal benzoic acid derivatives (which are *not* diterpenoids and are *not* derived from a monocyclic parent compound containing five carbon–carbon double bonds) such as (*E*)-4-[2-(5,6,7,8-tetrahydro-5,5,8,8-tetramethyl-2-naphthalenyl)-

1

1-propenyl]benzoic acid (TTNPB, Fig. 1) and 5′,6′,7′,8′-tetra-hydro-5′,5′,8′,8′-tetramethyl-[2,2′-binaphthalene]-6-carboxylic acid (TTNN, Fig. 1) are more than a thousand times more potent than retinol or its closely related derivatives in many biological assays (Loeliger et al 1980, Newton et al 1980, Strickland et al 1983, Dawson et al 1983).

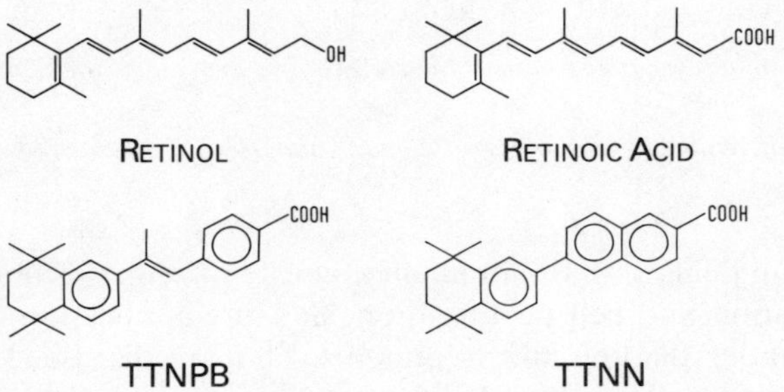

FIG. 1. Structures of retinol, retinoic acid, (E)-4-[2-(5,6,7,8-tetrahydro-5,5,8,8-tetra-methyl-2-naphthalenyl)-1-propenyl]benzoic acid (TTNPB) and 5′,6′,7′,8′-tetrahydro-5′,5′,8′,8′-tetramethyl-[2,2′-binaphthalene]-6-carboxylic acid (TTNN).

Ultimately, the definition of a substance as a retinoid must rest on its ability to elicit a specific biological response, just as the original definition of vitamin A ('fat-soluble A') was based on the ability of certain substances to maintain growth in animals fed specific diets. When the tricyclic retinoidal benzoic acid derivatives were first synthesized and shown to be unusually potent (thousands of times more active than retinoic acid) in causing regression of skin papillomas in mice (Loeliger et al 1980), it was suggested by sceptics that this did not represent an effect of a true retinoid, that there was something peculiar about the skin cancer assay. Soon, however, TTNPB was shown to be much more active than any naturally occurring retinoid in a variety of *in vitro* assays that measured effects of retinol or retinoic acid on cell differentiation and cell proliferation (Newton et al 1980, Strickland et al 1983). Recently, the experimental data have come full circle with detailed reports of the extreme potency of TTNPB in supporting growth in vitamin A-deficient rats (Stephens-Jarnagin et al 1985), i.e. in the original assay in which retinol and retinyl esters were discovered almost 75 years ago. Thus, we now have the paradox that a group of substances which do *not* fit the IUPAC–IUB chemical definition of a retinoid are indeed the most potent of *all* known substances in the original, classic biological assay for defining retinoids.

If the use of a biological definition of the word 'retinoid' has made the IUPAC–IUB chemical definition of this term essentially worthless, then what is left? Are we merely playing with words, or is there something more meaningful and constructive from a scientific viewpoint to be gained from this semantic exercise? The answer is clearly that we are not playing with words, and that an entirely new definition of the term 'retinoid' is now needed. By 'constructive', we imply that a new definition of the term 'retinoid' should eventually lead to further synthetic activity for the organic chemists, to the isolation and characterization of the true cellular receptors for retinoids, and to the elucidation of the molecular mechanism of action of retinoids in controlling cell growth, differentiation and proliferation.

The problem with the orthodox chemical definition of a retinoid is that it fails to take into account the basic mechanisms whereby cellular information is transferred and the biological activity of cells is regulated. The IUPAC–IUB definition is fixated on the chemical structure of the first effector substances (namely retinol and retinoic acid) that were found to elicit a specific biological response: growth stimulation in rats. However, it is now clear that the cellular programmes for many similar biological growth responses do not reside intrinsically in a particular effector (the ligand), but rather in the specific receptor that is activated by the effector (Roth & Taylor 1983). For many cellular systems that are under hormonal control, the full programme for activating the cell is within the receptor, and the role of the ligand (effector) is largely or exclusively to cause the receptor to express its programme. This is particularly striking for many peptide hormones, whose activities may be mimicked by specific monoclonal antibodies or anti-idiotype antibodies. As Roth (1981) has elegantly expressed this concept, 'the receptor has the message'.

What, then, is a retinoid? We propose the following definition: a retinoid is a substance that can elicit specific biological responses by binding to and activating a specific receptor or set of receptors. The classical ligands for these receptors are retinol and retinoic acid, but it is clear that synthetic ligands can have a better molecular fit to these receptors than retinol or retinoic acid. The programme for the biological response of the target cell resides in the retinoid receptor rather than in the retinoid itself.

The implications of this new definition are clear. First, it liberates one from the restrictive orthodox definition of retinoids as diterpenoid, polyene substances, and thus allows the newly synthesized retinoidal benzoic acid derivatives to be classified as retinoids. Secondly, the new definition focuses on a receptor system as the locus for future research to define the molecular mechanism of action of retinoids. At present, we have little if any knowledge of such receptor molecules. The extensive studies that have been performed on cellular retinoid-binding proteins (for a review see Chytil & Ong 1984) have done little to clarify this situation, and it is unlikely that any of the binding proteins that have been identified so far are true receptors (Roberts & Sporn 1984). At best, they may fulfil some type of cellular transport role;

certainly there is little if any evidence that they are capable of activating any specific cellular response, as receptors for steroids or peptide hormones are known to do.

Retinoids affect the differentiation and proliferation of many types of cells, whether they are of ectodermal, endodermal or mesodermal origin; whether they are epithelial, fibroblastic or mesenchymal; or whether they are neoplastic, preneoplastic or non-neoplastic. This universality of effects suggests that there may be a common molecular mechanism of action. At present, it appears most likely that retinoids exert some fundamental control on gene expression. The synthesis of potent new retinoidal benzoic acid derivatives of high chemical stability should now facilitate the identification, affinity labelling, isolation and molecular cloning of true, functional retinoid receptors that are not merely binding proteins. Once this has been accomplished, it should be possible to address the basic problem of how retinoids selectively regulate gene transcription and thus control cell differentiation and proliferation.

REFERENCES

Chytil F, Ong DE 1984 Cellular retinoid-binding proteins. In: Sporn MB et al (eds) The retinoids. Academic Press, Orlando, vol 2:89-123

Dawson MI, Chan RL, Derdzinski K, Hobbs PD, Chao W, Schiff LJ 1983 Synthesis and pharmacological activity of 6-[(E)-2-(2,6,6-trimethyl-1-cyclohexen-1-yl)ethen-1-yl]- and 6-(1,2,3,4-tetrahydro-1,1,4,4-tetramethyl-6-naphthyl)-2-naphthalenecarboxylic acids. J Med Chem 26:1653-1656

Eckert RL, Green H 1984 Cloning of cDNAs specifying vitamin A-responsive human keratins. Proc Natl Acad Sci USA 81:4321-4325

Frickel F 1984 Chemistry and physical properties of retinoids. In: Sporn MB et al (eds) The retinoids. Academic Press, Orlando, vol 1:7-145

Loeliger P, Bollag W, Mayer H 1980 Arotinoids, a new class of highly active retinoids. Eur J Med Chem Chim Ther 15:9-15

Lotan R 1980 Effects of vitamin A and its analogs (retinoids) on normal and neoplastic cells. Biochim Biophys Acta 605:33-91

Moon RC, Itri LM 1984 Retinoids and cancer. In: Sporn MB et al (eds) The retinoids. Academic Press, Orlando, vol 2:327-371

Newton DL, Henderson WR, Sporn MB 1980 Structure–activity relationship of retinoids in hamster tracheal organ culture. Cancer Res 40:3413-3425

Roberts AB, Sporn MB 1984 Cellular biology and biochemistry of the retinoids. In: Sporn MB et al (eds) The retinoids. Academic Press, Orlando, vol 2:209-286

Roth J 1981 Insulin binding to its receptor: is the receptor more important than the hormone? Diabetes Care 4:27-32

Roth J, Taylor SI 1983 Information transfer, cell regulation, and disease mechanisms: insights from studies of cell surface receptors. In: Harvey Lect Ser 77. Academic Press, New York, p 81-127

Sporn MB, Roberts AB 1984 Biological methods of analysis and assay of retinoids. In: Sporn MB et al (eds) The retinoids. Academic Press, Orlando, vol 1:235-279

Stephens-Jarnagin A, Miller DA, DeLuca HF 1985 Growth supporting activity of a retinoidal benzoic acid derivative and 4,4-difluororetinoic acid. Arch Biochem Biophys 237:11-16

Strickland S, Breitman TR, Frickel F, Nurrenbach A, Hadicke E, Sporn MB 1983 Structure–activity relationships of a new series of retinoidal benzoic acid derivatives as measured by induction of differentiation of murine F9 teratocarcinoma cells and human HL-60 promyelocytic leukemia cells. Cancer Res 43:5268-5272

Conformational restrictions of the retinoid skeleton

M. I. DAWSON, P. D. HOBBS, R. CHAN, K. DERDZINSKI, C. T. HELMES, W. CHAO, E. MEIERHENRY and L. J. SCHIFF*

*Life Sciences Division, SRI International, Menlo Park, California 94025, and *Life Sciences Division, IIT Research Institute, Chicago, Illinois 60616, USA*

Abstract. A series of conformationally restricted retinoids has been synthesized and assayed for biological activity. These compounds have aromatic rings in place of selected double bonds of the tetraene side-chain of retinoic acid and could be considered as analogues of retinoic acid in which some of the double bonds possess s-*cis* topology. Thus far, analogues in which the bonds corresponding to the (5,7*E*)-, (7,9*E*)-, (9,11,13*E*)- and (11,13*E*)-double bond systems of retinoic acid are restricted to a cisoid conformation have been studied. Analogues were screened for their ability to reverse keratinization in hamster tracheal organ culture and to inhibit the induction of ornithine decarboxylase in mouse epidermis. Selected compounds were also screened in the antipapilloma assay in mice. The toxicity of some analogues on intraperitoneal injection in mice was determined.

1985 Retinoids, differentiation and disease. Pitman, London (Ciba Foundation Symposium 113) p 6–28

Evidence that retinoic acid and other retinoids are able to regulate the process of cellular differentiation and reverse the transformation of cells from the normal to the neoplastic state (Breitman et al 1980, Davies 1967, Lasnitzki 1955, Roberts & Sporn 1984, Sporn & Roberts 1984, Strickland & Mahdavi 1978) has stimulated interest in the synthesis of new, more effective retinoids that might have reduced toxic side-effects. These new retinoids would have potential value as therapeutic agents for the treatment and prophylaxis of proliferative diseases (Moon & Itri 1984, Sporn et al 1977). To develop these new drugs and to determine how structural modifications of the retinoic acid skeleton affect biological activity and toxicity, we have undertaken the synthesis and screening of new analogues of retinoic acid for the past seven years (Dawson et al 1980, 1981a,b, 1983a,b, 1984). Although modifications have been made in the β-cyclogeranylidene ring and polar terminus regions of the retinoic acid skeleton, a major portion of our research

6

effort has been devoted to the design of analogues in which the tetraene chain (carbons 7 to 14) is modified.

The conformation of retinoic acid is generally depicted as structure **1** of Fig. 1. The olefinic bonds of the side-chain are transoid and planar with a 6-s-*cis*, 8-s-*trans*, 10-s-*trans*, 12-s-*trans* topology. However, this conformation is not necessarily the one that the side-chain adopts upon binding to an active site

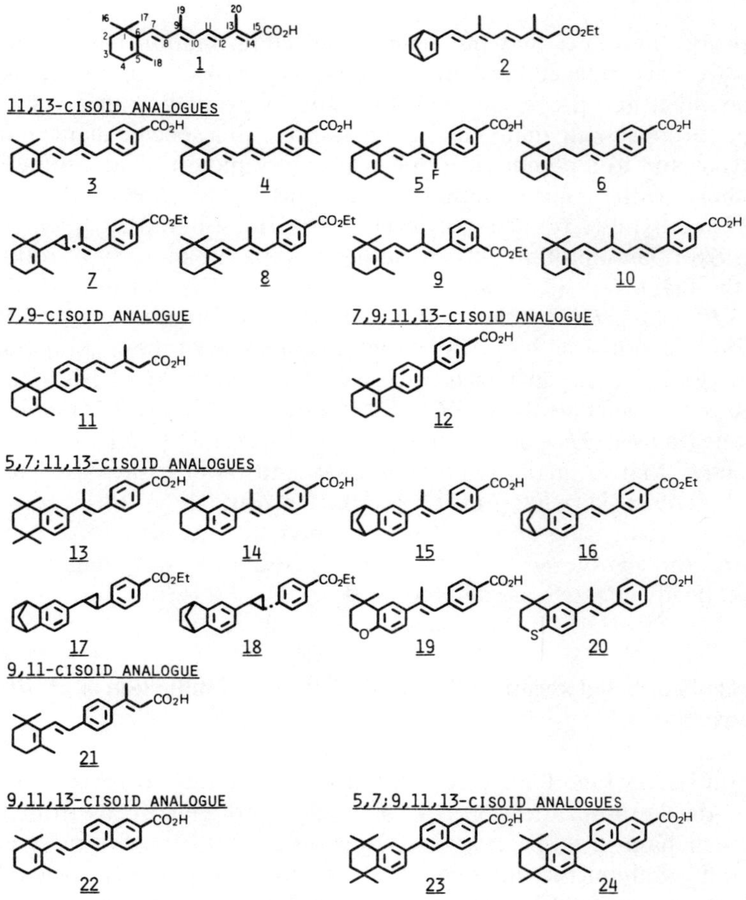

FIG. 1. Structures of retinoic acid (**1**) and other retinoids.

and exerting a biological effect. Planar cisoid or intermediate conformations derived from **1** by rotation about the single bonds joining the double bonds may be the optimal ones for activity.

TRANSOID   CISOID   AROMATIC ANALOGUE
                                                      OF CISOID BOND

To probe the effect that the cisoid bond conformation has on biological activity we have replaced specific diene (C=C—C=C) bond systems of the tetraene chain and β-cyclogeranylidene ring by aromatic rings. As a result, the four carbons in an analogue that correspond to a specific dienic moiety of **1** are restricted to a planar cisoid or s-*cis* conformation by an aromatic ring. Beginning with our synthesis of the (11,13*E*)-cisoid analogue (*E*)-4-[2-methyl-4-(2,6,6-trimethylcyclohexenyl)-2,4-butadienyl]benzoic acid (**3**) in 1978 (Dawson et al 1981a), we have prepared an array of retinoids in which the (11,13*E*)-, (7,9*E*)-, (7,9*E*; 11,13*E*)-, (5,7*E*; 11,13*E*)-, (9,11,13*E*)- and (5,7*E*; 9,11,13*E*)-double bond systems of **1** are replaced by aromatic rings (Dawson et al 1983b, 1984). The structures of some of these compounds are shown in Fig. 1. Other groups have also undertaken the synthesis of aromatic analogues of **1**, such as the (5,7*E*; 11,13*E*)-cisoid analogue **13** (Loeliger et al 1980) and the (9,11*E*)-cisoid analogue **21** (Frickel 1984). Retinoid **13** was more active than **1** in the papilloma assay and other assays used to assess retinoid activity (Loeliger et al 1980). Unfortunately, this retinoid has appreciable toxic side-effects that may preclude its use as a drug, and therefore the synthesis and screening of retinoids that might be more therapeutically effective are still required.

Reversal of epithelial keratinization and inhibition of induction of ornithine decarboxylase

The retinoids in Fig. 1 were screened for their ability to reverse, in organ culture, the keratinization process of tracheal epithelial cells from vitamin A-deficient hamsters (TOC assay) (Newton et al 1980b, Sporn et al 1975, 1976) and to inhibit the induction of the enzyme ornithine decarboxylase by the tumour promoter 12-*O*-tetradecanoylphorbol-13-acetate (TPA) in mouse epidermis (ODC assay) (Verma et al 1978). Activity in these two assays correlates with the ability of retinoids to prevent the transformation of normal epithelial cells to preneoplastic ones (Sporn & Roberts 1984, Verma et al 1979). The results of these assays are given in Table 1. As Table 1 illustrates, the analogues did not necessarily have similar activities in the two test systems. The variation in activities may be the result of differences in the sensitivities of the test systems or in the pharmacokinetic behaviour of the retinoids screened (Sporn & Roberts 1984). However, those retinoids with a carboxylic acid polar terminus that exhibited high activity in the TOC assay generally also had high activity in the ODC assay. The cyclopropane analogues were screened as the ethyl esters (compounds **7**, **8**, **17** and **18**) to ensure compound stability during testing. The parent carboxylic acid of ethyl ester **2** was too labile for testing. Because the ethyl esters **9** and **16** had such low activity in the two assays, the parent acids were not tested.

Replacement of the (11,13E)-dienic bond system of **1** by a benzene ring (analogue **3**) decreased activity in the TOC assay. The concentration of retinoid **3** required for reversal of keratinization in half the cultures (ED$_{50}$) was 20-fold that of **1**. Activity in the ODC assay also decreased. Placement of a methyl group *ortho* to the carboxylic acid group of **3** (analogue **4**) increased the ED$_{50}$ by a factor of 10 in the TOC assay and essentially eliminated activity at the 17-nmol dose level in the ODC assay. This methyl group would correspond to the 20-methyl group of **1**.

Replacement of a vinylic hydrogen on the tetraene chain with a fluorine has been reported to enhance activity from sixfold to 20-fold in the papilloma assay (Pawson et al 1979). Retinoid **5**, which has a fluoro group at the chain position corresponding to the C-10 hydrogen of **1**, was as active as **3** in the TOC assay and was about twice as active at the 1.7-nmol dose in the ODC assay as **3** was.

Saturation of the double bond of **3** corresponding to the 7,8-double bond of **1** (analogue **6**) decreased activity in the TOC assay by 50% but did not decrease activity in the ODC assay. In contrast, 7,8-dihydroretinoic acid had an ED$_{50}$ of 1×10^{-8}M in the TOC assay, whereas the ED$_{50}$ of **1** was 1×10^{-11}M (Sporn & Roberts 1984). Replacement of this bond by a cyclopropane ring (analogue **7**) decreased TOC activity by 50% but enhanced ODC activity approximately twofold at the 1.7-nmol dose. Replacement of the 5,6-double bond of **3** by a cyclopropane ring (compound **8**) almost tripled the ED$_{50}$ in the TOC assay but did not affect ODC activity at the 17-nmol dose.

The activity of these analogues is retained on conversion of the polar terminus from a carboxyl group to a carbethoxy group (L.J. Schiff & M.I. Dawson, unpublished work 1982). For example, the ethyl ester of **3** had an ED$_{50}$ in the TOC assay of 2×10^{-10}M. However, shifting the polar terminus of **3** from a position *para* to the side-chain to one that was *meta* (analogue **9**) eliminated activity. Lengthening the side-chain by one double bond (analogue **10**) only moderately reduced activity in both assays. A similar vinylic homologation of **1** increased the ED$_{50}$ in the TOC assay to 3×10^{-9}M (Sporn & Roberts 1984).

TABLE 1 Activity of retinoids in assays for reversal of keratinization (TOC assay) and for inhibition of induction of ornithine decarboxylase (ODC assay)

Retinoid	TOC assay[a]			ODC assay[b]	
	Concn (M)	Active/total cultures (%)	ED_{50}[c] (M)	Dose (nmol)	% Inhibition[d] ±SE
1	10^{-10}	18/19 (95)	1×10^{-11}	17.0	89 ± 2
	10^{-11}	10/20 (50)		1.7	84 ± 2
	10^{-12}	3/19 (16)		0.17	48 ± 10
2	10^{-9}	5/19 (26)	$>1 \times 10^{-9}$	17	54 ± 10
3			2×10^{-10}[e]	17	77 ± 6
				1.7	34 ± 13
4	10^{-8}	7/7 (100)	2×10^{-9}	17	9 ± 1
	10^{-9}	8/19 (42)			
	10^{-10}	2/6 (33)			
5	10^{-8}	13/13 (100)	2×10^{-10}	17	80 ± 1
	10^{-9}	12/13 (92)		1.7	70 ± 1
	10^{-10}	5/13 (38)			
6	10^{-8}	7/7 (100)	3×10^{-10}	17	83 ± 4
	10^{-9}	14/15 (93)		1.7	45 ± 12
	10^{-10}	4/15 (27)			
	10^{-11}	1/8 (13)			
7	10^{-8}	7/7 (100)	3×10^{-10}	17	80 ± 1
	10^{-9}	7/7 (100)		1.7	77 ± 0
	10^{-10}	0/6 (0)			
8	10^{-8}	7/7 (100)	5×10^{-10}	17	80 ± 16
	10^{-9}	10/13 (77)			
	10^{-10}	4/14 (29)			
	10^{-11}	2/7 (29)			
9	10^{-8}	1/7 (14)	$>1 \times 10^{-8}$	17	6 ± 4
	10^{-9}	1/6 (17)		1.7	13 ± 3
	10^{-10}	2/7 (29)			
10	10^{-8}	6/7 (86)	4×10^{-10}	17	56 ± 4
	10^{-9}	4/7 (57)		1.7	33 ± 8
	10^{-10}	3/7 (43)			
11	10^{-8}	2/7 (29)	$>1 \times 10^{-8}$	17	23 ± 12
	10^{-9}	1/7 (14)		1.7	9 ± 6
	10^{-10}	1/7 (14)			
12	10^{-8}	7/7 (100)	3×10^{-9}	17	55 ± 7
	10^{-9}	2/12 (17)		1.7	34 ± 6
	10^{-10}	2/12 (17)			
13	10^{-10}	15/15 (100)	1×10^{-12}	17	90 ± 1
	10^{-11}	15/15 (100)		1.7	89 ± 3
	10^{-12}	6/15 (40)		0.17	79 ± 3
				0.017	34 ± 8
14	10^{-9}	7/7 (100)	3×10^{-11}	17	86 ± 3
	10^{-10}	9/11 (82)		1.7	62 ± 6
	10^{-11}	2/7 (29)			
	10^{-12}	1/7 (14)			

TABLE 1 (*ctd*)

Retinoid	TOC assay[a] Concn (M)	Active/total cultures (%)	ED_{50}[c] (M)	ODC assay[b] Dose (nmol)	% Inhibition[d] ±SE
15	10^{-8}	8/8 (100)	6×10^{-10}	17	69 ± 2
	10^{-9}	8/14 (57)		1.7	33 ± 8
	10^{-10}	3/14 (21)			
16	10^{-8}	4/7 (57)	3×10^{-9}	17	39 ± 8
	10^{-9}	3/7 (43)		1.7	15 ± 10
	10^{-10}	1/6 (17)			
17	10^{-8}	4/7 (57)	5×10^{-9}	17	0 ± 5
	10^{-9}	2/6 (33)		1.7	12 ± 6
	10^{-10}	1/7 (14)			
18	10^{-8}	3/7 (43)	$>1 \times 10^{-8}$	17	4 ± 5
	10^{-9}	2/7 (29)		1.7	0 ± 3
	10^{-10}	1/7 (14)			
19	10^{-8}	7/7 (100)	2×10^{-10}	17	81 ± 1
	10^{-9}	7/7 (100)		1.7	42 ± 5
	10^{-10}	4/13 (31)			
	10^{-11}	1/7 (14)			
20	10^{-8}	7/7 (100)	5×10^{-11}	17	85 ± 2
	10^{-9}	7/7 (100)		1.7	68 ± 6
	10^{-10}	12/14 (86)			
	10^{-11}	1/7 (14)			
21	10^{-8}	1/5 (20)	$>1 \times 10^{-8}$[f]		
	10^{-9}	0/5 (0)			
22	10^{-8}	7/7 (100)	1×10^{-10}	17	48 ± 1
	10^{-9}	11/12 (92)		1.7	30 ± 2
	10^{-10}	6/13 (46)			
	10^{-11}	1/5 (20)			
23	10^{-9}	7/7 (100)	3×10^{-12}	17	80 ± 3
	10^{-10}	21/22 (95)		1.7	56 ± 1
	10^{-11}	17/22 (77)			
	10^{-12}	4/15 (27)			
24	10^{-9}	7/7 (100)	2×10^{-11}	17	70 ± 4
	10^{-10}	12/14 (86)		1.7	66 ± 5
	10^{-11}	3/7 (43)			
	10^{-12}	1/7 (14)			

[a]This assay measures the ability of retinoids to reverse the process of keratinization in tracheal organ cultures obtained from hamsters that are in the early stages of vitamin A deficiency. Retinoids are considered 'active' if, after treatment, tracheae contain no keratohyaline granules.
[b]Studies on the mechanism of tumour promotion suggest that the induction of ornithine decarboxylase (ODC) is one of the essential events in carcinogenesis. The ODC assay measures the ability of topically applied retinoids to inhibit the induction of ornithine decarboxylase by a tumour-promoting phorbol ester (12-O-tetradecanoylphorbol-13-acetate) applied to the dorsal skin of female CD-1 mice. Results reported are for groups of three mice assayed in triplicate.
[c]ED_{50} is the concentration of retinoid required to effect reversal of keratinization in 50% of the cultures. Interpolation of ED_{50} was by cubic regression analysis.
[d]Compared to control; $n = 9$.　　　[e]Newton et al (1980b).　　　[f]Newton et al (1980a).

Replacement of the 9-methyl-$(7,9E)$-dienic bond system of 1 by a benzene ring with a methyl group *ortho* to the side-chain (analogue 11) eliminated activity. Replacement of the $(7,9E)$- and $(11,13E)$-dienic bond systems of 1 with a biphenyl group (analogue 12) decreased activity in both assays. The ED_{50} of 12 in the TOC assay was approximately sevenfold that of 3, whereas the activity of 12 in the ODC assay at the 1.7-nmol dose was the same as that found for 3. Evidently the methyl group on the benzene ring adjacent to the β-cyclogeranylidene ring has a greater effect on reducing activity than the ring itself.

Replacement of the $(5,7E)$- and $(11,13E)$-dienic bond systems of 1 with aromatic rings and substitution of the hydrogens at the C-4 position of 1 with methyl groups (retinoid 13, Ro 13-7410) increased activity in the TOC assay 10-fold. Removal of the geminal methyl groups at the C-5 position of the tetrahydronaphthalene ring of 13 (retinoid 14) increased the ED_{50} in the TOC assay by a factor of 30 and also reduced activity in the ODC assay when compared to that of 13. However, 14 still had one-third the activity of 1 in the TOC assay.

Analogue 2, which has a norbornene ring in place of the β-cyclogeranylidene ring of 1, was an extremely unstable compound and therefore gave variable results on screening (Dawson et al 1980). To assess more accurately the effect that this type of substitution has on activity, the benzonorbornenyl analogue 15 was synthesized (Dawson et al 1984). This compound was less active than either 13 or 14. Evidently lipophilic bulk in the β-cyclogeranylidene ring region increases activity. Analogues of 15 were prepared to probe the effect on activity of modifications of the propenyl side-chain. Removal of the methyl group from the side-chain (analogue 16) increased the ED_{50} fivefold in the TOC assay and decreased activity by half in the ODC assay. Evidently, a methyl group at a position corresponding to the 19-methyl group of 1 augments activity. Substitution of the propenyl group with a cyclopropane ring (compounds 17 and 18) also decreased activity. From these results it appears that planarity of the 8,9,10,11-bond system enhances activity.

Two isosteres with a heterocyclic atom at the C-1 position of the tetrahydronaphthalene ring were also synthesized. The dihydrobenzopyran 19 was less active in both assays than 14 was; however, its activity in the ODC assay was still appreciable. The dihydrobenzothiopyran 20 had about half the activity of 14 in the TOC assay and comparable activity in the ODC assay. This finding is of interest because the 4-thia analogue of retinyl acetate is reported to lack activity in a growth promotion assay (Baas et al 1966).

Replacement of the $(9,11E)$-dienic bond system of 1 with a benzene ring (compound 21) eliminated activity in the TOC assay (Newton et al 1980a), whereas replacement of the $(9,11,13E)$-trienic bond system of 1 with a naphthalene ring gave an analogue (22) that had an ED_{50} 10 times that of 1 in the TOC assay and low activity in the ODC assay. Activity was appreciably increased on incorporation of the (5,7E)-dienic bond system of 22 into an aromatic ring. The resultant retinoid, 6-(5,6,7,8-tetrahydro-5,5,8,8-tetra-methyl-2-naphthalenyl)-2-naphthalenecarboxylic acid (23), had an ED_{50} three times that of 13 and approximately one-third that of 1 in the TOC assay. However, it was less active than 1 or 13 in the ODC assay. Substitution of the hydrogen at the C-3 position of the tetrahydronaphthalene ring of 23 by a methyl group (analogue 24) increased the ED_{50} by a factor of seven in the TOC assay but did not have a significant effect on activity in the ODC assay.

Antipapilloma assay

Compounds 1, 5, 6, 13, 19, 20, 23 and 24 were screened in the antipapilloma assay in mice (Bollag 1975, 1979, Verma et al 1979). This assay measures the ability of topically applied retinoids to inhibit papilloma tumour formation in mouse epidermis that has been treated initially with the carcinogen 7,12-dimethylbenz[a]anthracene and then treated twice weekly over the course of the 20-week experiment with the tumour promoter TPA. The decrease in the average number of tumours formed per mouse is reported in Table 2. The (11,13E)-cisoid analogues 5 and 6 at both the 17- and 170-nmol doses were less effective in decreasing the average number of tumours per mouse than 1 was. The fluoro analogue 5 at the 170-nmol dose (83% inhibition compared to the control) was far more effective than the 7,8-dihydro analogue 6 at this dose (57% inhibition). The (5,7E; 11,13E)-cisoid analogue 13 at the 1.7-nmol dose inhibited the number of papillomas per mouse by 92%. Higher doses caused the death of the test animals. The heterocyclic analogues 19 and 20 at the 170-nmol dosage were as effective at suppressing tumour formation as 1 was. At the 17-nmol dose, the dihydro-benzopyran 19 had inhibitory activity comparable to that of 1, whereas the dihydrobenzothiopyran 20 (93% inhibition) was more active than 1 (74% inhibition). The naphthalenecarboxylic acid 23 at the 170-nmol dose (95% inhibition) was as effective as 1. However, at the 17-nmol dose, 23 (46% inhibition) was less effective than 1 (74% inhibition). The 3-methyl analogue of 23 (24) was toxic at the 170-nmol dose and had activity comparable to that of 23 at the 17-nmol dose.

Toxicity in mice

Compounds 1, 3, 6, 12, 13, 14, 19, 20, 23 and 24 were screened for toxicity in mice. Compounds were administered by intraperitoneal injection daily on

TABLE 2 Effect of retinoids on the prevention of papilloma tumour formation in mice treated with a carcinogen and a tumour promoter

Retinoid	Dose[a] (nmol)	Mice with papillomas (%)	% Decrease in average no. of papillomas per mouse ± SE[b]
Control[c]	0	93	0
1	170	40	90 ± 4
	17	47	74 ± 5
	1.7	69	56 ± 10
5	170	41	83 ± 8
	17	70	45 ± 14
6	170	70	57 ± 9
	17	68	65 ± 7
13	170[d]	—[e]	
	17[d]	—[f]	
	1.7[d]	50	92 ± 2
	0.17	73	63 ± 9
19	170	38	94 ± 2
	17	53	75 ± 12
20	170	33	94 ± 3
	17	40	93 ± 2
23	170[d]	28	95 ± 3
	17	82	46 ± 12
24	170	—[g]	
	17	63	53 ± 11

[a]Female Charles River CD-1 mice (29–30 animals/group) were treated with 7,12-dimethylbenz-[a]anthracene, followed by twice weekly applications of 12-O-tetradecanoylphorbol-13-acetate and a retinoid.
[b]Decrease in average number of papillomas per retinoid-treated mouse compared to control mouse at 20 weeks.
[c]Average number of papillomas per mouse in control group was 17.3 ± 2.6 (SE).
[d]Toxic symptoms of the hypervitaminosis A syndrome were observed: skin redness and scaling and hair loss.
[e]Animals died at 1–2 weeks.
[f]Animals died at 3–7 weeks.
[g]Animals died at 6–13 weeks.

weekdays over a period of two weeks (Hixson & Denine 1979). The results of this preliminary study are shown in Table 3. The (11,13E)-cisoid analogue **3** was more toxic than **1**, whereas its 7,8-dihydro analogue, compound **6**, was less toxic. The (7,9E; 11,13E)-cisoid analogue **12** was also less toxic than **1**. The (5,7E; 11,13E)-cisoid analogue **13** was the most toxic retinoid examined. Its 5,5-desmethyl analogue (**14**) was slightly less toxic. The hetero analogues **19** and **20** were less toxic than **1**. Evidently substitution of the methylene group at the C-1 position of the tetrahydronaphthalene ring by a more polar oxygen or sulphur group decreases toxic effects. The (5,7E; 9,11,13E)-cisoid analogue **23** was more toxic than **1** but less toxic than **13**. Replacement of the hydrogen at the C-3 position of the tetrahydronaphthalene ring of **23** by a methyl group (analogue **24**) increased toxicity.

TABLE 3 Toxicity of retinoids in mice

Retinoid	Dose[a] ($\mu mol\,kg^{-1}\,day^{-1}$)	% Survivors Day 8	Day 15	Mortality range (days)	Total animals
Control[b]	0	100	100	—	60
1	600	97	0	7–13	30
	300	100	0	10–14	20
	100	100	100	—	40
	30	100	100	—	30
3	600	0	0	2–6	10
	300	100	100	—	10
	100	100	100	—	10
	30	100	100	—	10
6	600	100	40	8–13	10
	300	100	100	—	10
	100	100	100	—	10
	30	100	100	—	10
12	600	27	27	2–6	15
	300	100	100	—	15
	100	100	100	—	15
	30	100	100	—	15
13	100	0	0	4–8	10
	30	50	0	6–8	30
	10	65	0	5–10	40
	3.3	97	0	7–11	30
	1.0	94	18	7–15	50
14	600	10	0	6–8	10
	300	0	0	5–7	10
	100	20	0	5–8	10
	30	80	0	7–8	10
	10	90	0	7–9	10
	1.0	100	80	12–14	10
19	600	70	0	7–10	10
	300	100	50	12–15	10
	100	100	100	—	10
	30	100	100	—	10
20	600	100	0	9–10	10
	300	100	80	14–15	10
	100	100	100	—	10
	30	100	100	—	10

continued on p. 16

TABLE 3 (*ctd*)

Retinoid	Dose[a] ($\mu mol\,kg^{-1}day^{-1}$)	% Survivors Day 8	Day 15	Mortality range (days)	Total animals
23	100	100	0	8	10
	30	100	0	9–12	10
	10	100	68	10–15	30
	3.3	100	100	—	20
	1.0	100	100	—	10
24	30	100	0	8–9	10
	10	100	0	8–10	10
	3.3	100	10	11–15	10

[a] Retinoid was administered by i.p. injection of a suspension in 8% Cremophor EL and 10% propylene glycol in water. Female Swiss mice (22–25 g) were injected daily on weekdays for two weeks, according to the procedure of Hixson & Denine (1979).
[b] Animals were injected with vehicle only.

Thus far, compounds **19** and **20**, which are less toxic than **1** but which still have significant activity in the *in vivo* antipapilloma and ODC assays and in the *in vitro* TOC assay, appear to be the most promising leads for the design of new retinoids. Our research efforts are now directed to the synthesis of analogues of these compounds.

Acknowledgements

Support of this work by United States Public Health Service Grants CA30512 and CA32428 and Contracts NO1-CP-05600 and NO1-CP-05610, which were awarded by the Division of Cancer Cause and Prevention, National Cancer Institute, DHHS, is gratefully acknowledged. We also wish to thank Ms Saundra Smith, Ms Janice Schindler and Mr Randall Newbury for their technical assistance.

REFERENCES

Baas JL, Davies-Fidder A, Visser FR, Huisman HO 1966 Vitamin A analogues. II. Synthesis of 4-thia vitamin A. Tetrahedron 22:265-275
Bollag W 1975 Therapy of epithelial tumors with an aromatic retinoic acid analog. Chemotherapy 21:236-247
Bollag W 1979 Retinoids and cancer. Cancer Chemother Pharmacol 3:207-215
Breitman TR, Selonick SE, Collins SJ 1980 Induction of differentiation of the human promyelocytic leukemia cell line (HL-60) by retinoic acid. Proc Natl Acad Sci USA 77:2936-2940
Davies RE 1967 Effect of vitamin A on 7,12-dimethylbenz[a]anthracene-induced papillomas in rhino mouse skin. Cancer Res 27:237-241
Dawson MI, Hobbs PD, Kuhlmann K, Fung VA, Helmes CT, Chao W 1980 Retinoic acid analogues. Synthesis and potential as cancer chemopreventive agents. J Med Chem 23:1013-1022

Dawson MI, Hobbs PD, Chan RL, Chao W, Fung VA 1981a Aromatic retinoic acid analogues. Synthesis and pharmacological activity. J Med Chem 24:583-592

Dawson MI, Hobbs PD, Chan RL, Chao W 1981b Retinoic acid analogues with ring modifications. Synthesis and pharmacological activity. J Med Chem 24:1214-1223

Dawson MI, Chan R, Hobbs PD, Chao W, Schiff LJ 1983a Aromatic retinoic acid analogues. 2. Synthesis and pharmacological activity. J Med Chem 26:1282-1293

Dawson MI, Chan RL, Derdzinski K, Hobbs PD, Chao W, Schiff LJ 1983b Synthesis and pharmacological activity of 6-[(E)-2-(2,6,6-trimethyl-1-cyclohexen-1-yl)ethen-1-yl]-2-naphthalenecarboxylic acid and 6-(1,2,3,4-tetrahydro-1,1,4,4-tetramethyl-6-naphthyl)-2-naphthalenecarboxylic acid. J Med Chem 26:1653-1656

Dawson MI, Hobbs PD, Derdzinski K et al 1984 Conformationally restricted retinoids. J Med Chem 27:1516-1531

Frickel F 1984 Chemistry and physical properties of retinoids. In: Sporn MB et al (eds) The retinoids. Academic Press, Orlando, vol 1:7-145

Hixson EJ, Denine EP 1979 Comparative subacute toxicity of retinyl acetate and three synthetic retinamides in Swiss mice. J Natl Cancer Inst 63:1359-1364

Lasnitzki I 1955 The influence of A hypervitaminosis on the effect of 20-methylcholanthrene on mouse prostate glands in vitro. Br J Cancer 9:434-441

Loeliger P, Bollag W, Mayer H 1980 Arotinoids, a new class of highly active retinoids. Eur J Med Chem Chim Ther 15:9-15

Moon RC, Itri LM 1984 Retinoids and cancer. In: Sporn MB et al (eds) The retinoids. Academic Press, Orlando, vol 2:327-371

Newton DL, Henderson WR, Sporn MB 1980a Structure–activity relationships of retinoids. Laboratory of Chemoprevention, Division of Cancer Cause and Prevention, National Cancer Institute, Bethesda, Maryland, C-3

Newton DL, Henderson WR, Sporn MB 1980b Structure–activity relationships of retinoids in hamster tracheal organ culture. Cancer Res 40:3413-3425

Pawson BA, Chan K, DeNoble J et al 1979 Fluorinated retinoic acids and their analogues. 1. Synthesis and biological activity of (4-methoxy-2,3,6-trimethylphenyl)nonatetraenoic acid analogues. J Med Chem 22:1059-1067

Roberts AB, Sporn MB 1984 Cellular biology and biochemistry of the retinoids. In: Sporn MB et al (eds) The retinoids. Academic Press, Orlando, vol 2:209-286

Sporn MB, Clamon GH, Dunlop NM, Newton DL, Smith JM, Saffiotti U 1975 Activity of vitamin A analogues in cell cultures of mouse epidermis and organ cultures of hamster trachea. Nature (Lond) 253:47–50

Sporn MB, Dunlop NM, Newton DL, Henderson WR 1976 Relationships between structure and activity of retinoids. Nature (Lond) 263:110-113

Sporn MB, Squire RA, Brown CC, Smith JM, Wenk ML, Springer S 1977 13-cis-Retinoic acid: inhibition of bladder carcinogenesis in the rat. Science (Wash DC) 195:487-489

Sporn MB, Roberts AB 1984 Biological methods for analysis and assay of retinoids. In: Sporn MB et al (eds) The retinoids. Academic Press, Orlando, vol 1:236-279

Strickland S, Mahdavi V 1978 The induction of differentiation in teratocarcinoma stem cells by retinoic acid. Cell 15:393-403

Verma AK, Rice HM, Shapas BG, Boutwell RK 1978 Inhibition of 12-O-tetradecanoyl-phorbol-13-acetate-induced ornithine decarboxylase activity in mouse epidermis by vitamin A analogs (retinoids). Cancer Res 38:793-801

Verma AK, Shapas BG, Rice HM, Boutwell RK 1979 Correlation of the inhibition by retinoids of tumor promoter-induced mouse epidermal ornithine decarboxylase activity and of skin tumor promotion. Cancer Res 39:419-425

DISCUSSION

Yuspa: One of the things that would be very helpful in determining the mechanisms of action of retinoids, in particular whether there are single or multiple sites of action, would be to have competitors or inhibitors. Have you looked for any inhibitors of, for example, retinoic acid actions?

Dawson: No. The major goal of the present work has been the design of new retinoids.

Sporn: When we started working with analogues the thought of developing some sort of retinoid antagonist was very appealing, but no work has ever been done on this.

Yuspa: Antagonists may already be available; it's just that no one has ever tested for them. I think it would be more useful to search for antagonists than to keep synthesizing new agents looking for greater activity. Many analogues with agonist activity have already been synthesized and although some advances have been made, they have not been tremendous.

Sporn: It depends on what your interests are; for many pharmaceutical companies there is no real motivation to do that.

Yuspa: That is very short-sighted. There is always a trade off between toxicity and therapeutic activity. If retinoids have a common mechanism of action for both effects there may be no incentive to look for antagonists. However, if there are separate mechanisms of action, the incentives are substantial. It would probably be a very fruitful area of research. The same assays currently used to look for biological potency could also be applied to the search for antagonists.

Sporn: We were up against a similar problem with insulin antagonists. There have been hundreds of synthetic modifications of the insulin molecule and all the analogues have been tested to see whether they act like insulin. However, there have been no reports that any of these substances are insulin antagonists.

Yuspa: Can you draw any general conclusions about the chemical properties of your analogues and their biological potency, in terms of stability, partition coefficients or that sort of thing?

Dawson: We have not investigated the partition coefficients of these compounds. The naphthalenecarboxylic acid analogues are the most stable retinoids that we have designed. They are stable to oxygen, heat and light, unlike other retinoids that have olefinic double bonds.

Yuspa: And what is their potency relative to agents that are less stable?

Dawson: It depends on the cell system in which the compounds are screened. In the tracheal organ culture system the naphthalenecarboxylic acids are more active than retinoic acid, but in the *in vivo* ornithine decarboxylase assay they are less potent, and that may be because they are less soluble and are not penetrating the skin as well.

Yuspa: Have you ever tried the ornithine decarboxylase assay in cell culture?

Dawson: No. We have used the *in vivo* ornithine decarboxylase assay as a preliminary screen to establish whether compounds should be screened in the antipapilloma assay in the mouse.

Yuspa: It works just as well in culture as *in vivo* and it might tell you whether solubility or penetration is important in determining potency.

Breitman: Could you describe the 'super' retinoid? Where are the methyl groups required? What are the key elements of the structure?

Dawson: We have made progress in determining what structural features of the retinoid skeleton control biological activity, but additional synthetic and biological work must be done before retinoids with optimal therapeutic indices are identified. In addition, our studies indicate that the activity profiles of some retinoids differ in different systems.

In addition to this research area, we are interested in determining the location of the active site for retinoids. Although these studies could be undertaken with retinoic acid of high specific activity, this material is labile and difficult to use. Incorporation of a specific label into one of the aromatic retinoids would afford a very stable compound that could be easily manipulated in the laboratory and onto which a photo-affinity label could be readily attached. For example, (*E*)-4-[2-(5,6,7,8-tetrahydro-5,5,8,8-tetramethyl-2-naphthalenyl-1-propenyl]benzoic acid (TTNPB, **1**) is far more stable than retinoic acid, and, although there is the potential problem of isomerization about the double bond, radiolabelled material with a specific activity above 30 Ci/mmol could be prepared. Dr Sung Rhee at SRI has prepared **1** with a tritium label at the 6- and 7-positions of the tetrahydronaphthalene ring that has a specific activity of 24 Ci/mmol (unpublished work).

$\underline{1}$

Sherman: Dr A. Liebman and his colleagues at Roche (Malarek et al 1984) have found an improved way of making tritiated *all-trans*-retinoic acid at about that specific activity, and certainly in our hands it's quite stable. Even after storage at −20°C for more than three months we estimate that it is still about 90% *all-trans*-retinoic acid by high performance liquid chromatography.

Dawson: Some of the samples that have been sent to other workers have been shown to be polymeric on analysis.

Sporn: Where's the labelling?

Sherman: In the 11,12-position.

Sporn: Some years ago we had labelled retinoic acid with a specific activity of 6–7 Ci/mmol, but none of us ever wanted to use it because even when we had cleaned it up in the lab, it had deteriorated by the next day. For practical purposes, 1 Ci/mmol has been all that most investigators have wanted to work with.

Sherman: As I mentioned, the preparations that we have used recently have specific activities of 20–30 Ci/mmol, and they have been quite pure and quite stable. Dr A. Liebman and his colleagues (personal communication) believe that in the past sample instability was caused by small amounts of impurities in the retinoid mixture that acted catalytically.

Lotan: There has been some interest in whether or not new retinoids compete with retinoic acid for binding to the cellular binding protein. Have any of your compounds been tested in that sort of assay? I looked at 27 different analogues and found that all those with a free carboxylic acid group bound quite effectively to the protein, but any block of the carboxylic group decreased binding (Lotan et al 1980). Do your aromatic analogues still bind to the binding protein?

Dawson: Yes. Those aromatic retinoids with a free carboxylic acid terminus bind to cellular retinoic acid-binding protein (Sani et al 1984).

Koeffler: You examined the potency of your retinoids predominantly in one tissue type; have you looked to see if there is a parallel potency in another tissue type?

Breitman: Some of the compounds that are very active in the tracheal system are inactive in HL-60 (T.R. Breitman, unpublished work).

Dawson: And some of them have activity comparable to retinoic acid.

Sporn: Some of the retinoidal benzoic acid derivatives made by Fritz Frickel are more active than retinoic acid in F9 and some, but not the same substances, are more active than retinoic acid in HL-60. Every assay system will give you slightly different results.

Sherman: This gets us back to the question: if different cells respond in different ways to a given retinoid structure, are there different sites of action?

Sporn: I wouldn't think that the mechanisms are totally different, but the specific nature of the receptors may vary slightly depending on whether the cell is murine or human, haemopoietic or fibroblastic.

Sherman: What are the specific receptors?

Sporn: We don't know what the receptors are, but I hope that the chemical stability of this sort of molecule will allow us to introduce some sort of reactive atom to permit the affinity labelling of the true receptors.

Dawson: This research has been initiated. The photo-affinity label should be placed in a region of the molecule where it will not interfere with binding. Prof Koji Nakanishi has prepared 3-diazoacetoxy-9-*cis*-retinal and found that it forms a pigment with bovine rhodopsin (Sen et al 1982). Sheves (1984) has reported that 4-azido-(15-^{3}H)retinol binds to cellular retinol-binding protein from liver. Therefore, placement of a photo-affinity label at the 3- or 4-position in retinoic acid analogues may be effective. However, placement of a diazoacetoxy label at the 3-position of the tetrahydronaphthalene ring of 1 would probably interfere with binding and biological activity.

Frickel: I would like to continue the discussion on conformationally restricted retinoids by looking at the geometrical shape of some molecules. One can look at the biological activity of classical retinoids and of retinoidal benzoic acid derivatives in terms of reversal of keratinization in the tracheal organ culture assay (Sporn & Roberts 1984). The ED_{50} value of (1Z)-4-[2-methyl-4-(2,6,6-trimethyl-1-cyclohexen-1-yl)-1,3-butadienyl]benzoic acid is greater than 10^{-9} M, so this retinoid is virtually devoid of biological activity. This observation supports the hypothesis that a certain geometrical conformation is required to get high activity. This geometrical conformation can be assumed by *all-trans*-retinoic acid, possibly by 7,8-didehydro-retinoic acid and by the retinoidal benzoic acid derivative (E)-4-[2-(5,6,7,8-tetrahydro-3,5,5,8,8-pentamethyl-2-naphthalenyl)-1-propenyl]benzoic acid, which is at least as potent as retinoic acid (Table 1). We have seen a number of examples in which a

TABLE 1 (*Frickel*) Reversal of keratinization in retinoid-deficient hamster trachea (Sporn & Roberts 1984, M.B. Sporn, personal communication)

Retinoid	ED_{50} (M)
	3×10^{-11}
	5×10^{-10}
	1×10^{-11}
	$>10^{-9}$
	$>10^{-9}$

change in only the spatial arrangement of the conformer can dramatically increase or decrease the biological activity. For example, compound 2, which has a *trans* configuration, is one of the most active retinoids, but the corresponding *cis* isomer 3 is devoid of any biological activity in the papilloma system (Loeliger et al 1980).

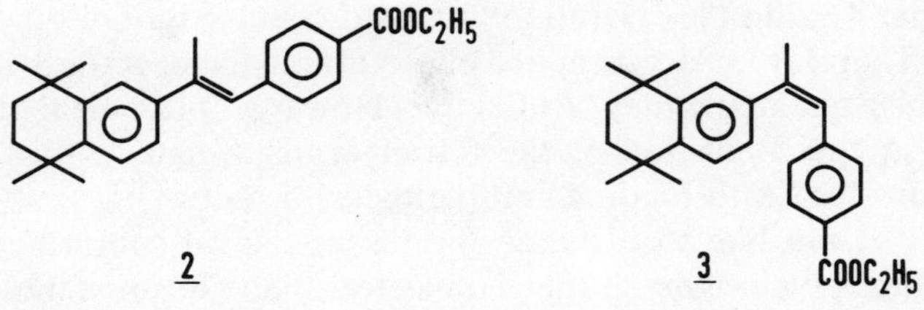

We have tried to interpret these observations by looking at the geometrical shape of the retinoids by X-ray crystallography. We looked at retinoidal benzoic acid derivatives and at *all-trans*-retinoic acid, and I think it is very impressive how these molecules fit (Fig. 1). The spatial arrangements shown in

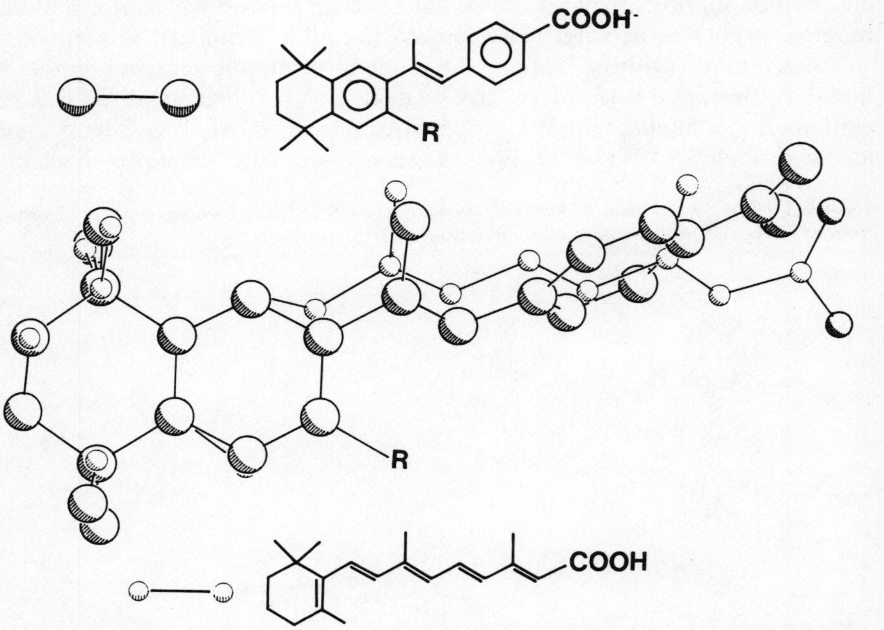

FIG. 1. (*Frickel*) Jointly projected structures of (*all-E*)-retinoic acid and a retinoidal benzoic acid derivative (R = H).

Fig. 1 are based on the X-ray data from the crystalline substances; this is important to note because some aspects of the conformations, for example the torsion angle between the two phenyl groups in the benzoic acid derivatives, may be slightly different on biological receptors. The observation that the biological activity of members of the retinoidal benzoic acid family is very sensitive to a minor change in the substitution pattern strongly supports the idea of a specific retinoid receptor. If we change the position of the carboxylic function in TTNPB from *para* to *meta*, the ability to induce differentiation in the HL-60 system (Table 2) drops dramatically (Strickland et al 1983); comparison of the X-ray structure of the *meta* compound with that of *all-trans*-retinoic acid (Fig. 2) provides a satisfactory explanation for this since even conformational changes of the two molecules cannot bring the carboxyl groups into spatially adjacent positions. Unless a metabolic activation of the corresponding *para* compound or deactivation of the *meta* substituted compound is

TABLE 2 (*Frickel*) **Induction of differentiation in human HL-60 promyelocytic leukaemia cells (Strickland et al 1983)**

Retinoid	ED_{50} (M)
(structure) —COOH	1×10^{-7}
(structure) —COOH	3×10^{-7}
(structure) —COOH	$>1 \times 10^{-6}$

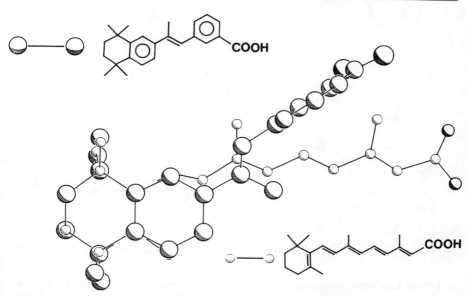

FIG. 2. (*Frickel*) Jointly projected structures of (*all-E*)-retinoic acid and *m*-carboxy analogue of TTNPB.

involved here, we are observing a marked structural specificity. A further example: the retinoidal cinnamic acid derivative **4**

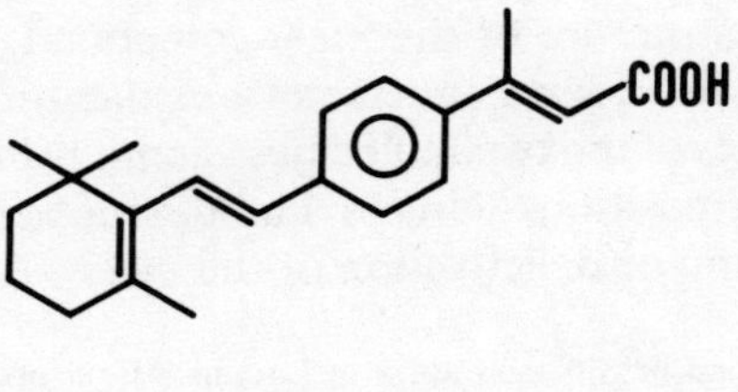

4

has no activity in the tracheal organ culture assay, even at 10^{-9} M, and X-ray analysis shows that, for this compound also, there is no fit between the carboxylic acid function and the carboxylic function of *all-trans*-retinoic acid (Fig. 3). We can now understand much better why this particular retinoid is without activity.

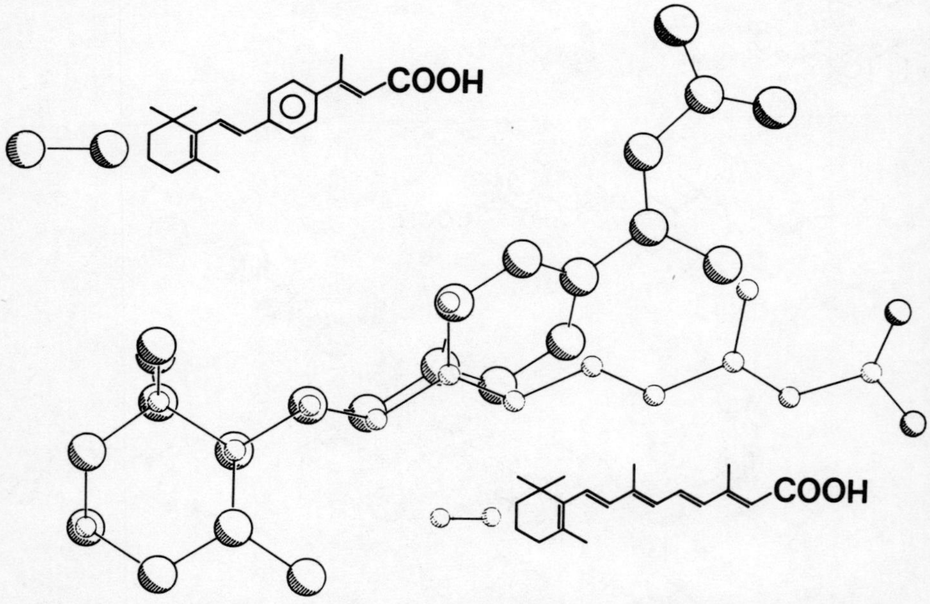

FIG. 3. (*Frickel* Jointly projected structures of (*all-E*)-retinoic acid and cinnamic acid derivative **4**.

I think that this comparative analysis of X-ray data on new retinoids supports the assumption that the activity of retinoids is transmitted from a highly organized state, in which the retinoid may be complexed to an as yet unknown receptor, a receptor that transmits a biological signal that is triggered by the retinoid.

Lotan: What is the X-ray structure of 13-*cis*-retinoic acid? In what position is the carboxylic acid group?

Frickel: There are two crystalline states of *all-trans*-retinoic acid as shown in Fig. 4. Both can be isolated. The triclinic structure is the more stable conformer; if one heats the monoclinic conformer to 80°C it turns into the triclinic structure. We did the X-ray analysis of 13-*cis*-retinoic acid and were rather surprised to see the monoclinic conformer. We do not know whether that is an unstable conformer or whether it is the most stable form of 13-*cis*-retinoic acid.

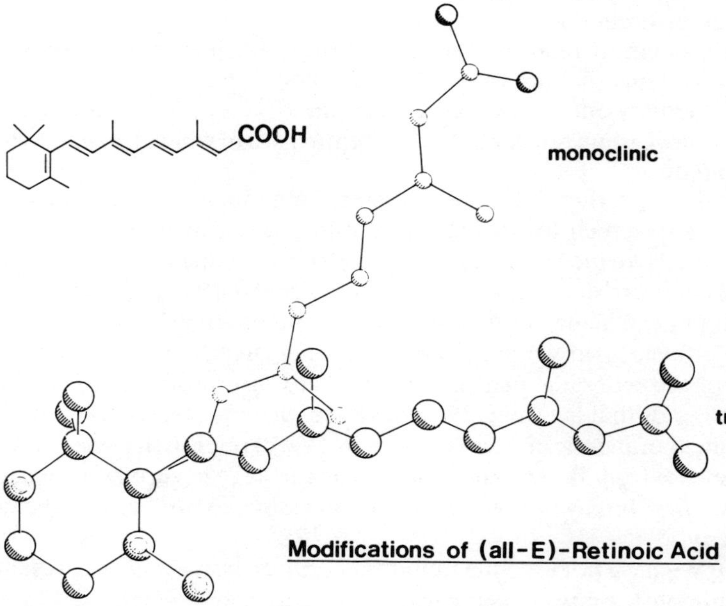

FIG. 4. (*Frickel*) Structures of the metastable monoclinic and the triclinic modifications of (*all-E*)-retinoic acid.

Hicks: Your assays for activity are defined by the biological system you are using; some retinoids are active in one assay and less active in another assay. Is a compound that is toxic in one particular *in vitro* assay toxic in all species, or are your toxicity results also defined entirely by the biological system in which you are working? If they are, what relevance has this in terms of the general design of molecules?

Sporn: We know that toxicity is species specific; for example, aflatoxin is extremely toxic to rats and relatively non-toxic to mice.

Hicks: So what are you using as your yardstick of toxicity, and why?

Frickel: There is no doubt that we are designing compounds for use in humans, but we have to run toxicity assays in animals and unfortunately we are limited to a fairly small number of assays.

Hicks: Can you use human cells *in vitro* instead of animal cells *in vitro*?

Frickel: It's very difficult to measure any *in vitro* toxicity that is of predictive value for *in vivo* toxicity.

Turton: In general, do the results from your *in vitro* tests for activity compare well with those from *in vivo* tests?

Dawson: There is generally a good correlation, but we have one retinoid that is inactive in the tracheal organ culture assay and the papilloma assay but is active in the ornithine decarboxylase assay.

Sporn: Excellent correlations can certainly be obtained between *in vitro* tests and *in vivo* systems such as the papilloma test system.

Hicks: In activity but not in toxicity. It is important to make sure that your toxicity tests have some relevance to the eventual use to which you intend to put the compounds.

Sporn: For any retinoid that has an interesting biological activity, the next thing to do is to screen for toxicity *in vivo* in rats and mice.

Peck: I would like more attention to be paid to screening for toxicity and with emphasis on bone toxicity and teratogenicity. From a dermatologist's perspective, we don't need a new retinoid that is more potent than isotretinoin for the treatment of acne, and we need a retinoid only slightly more effective than etretinate or the retinoidal benzoic acid derivative for psoriasis. However, we *do* need retinoids that are safer. What toxicities do you use, Dr Dawson, when you calculate therapeutic indices? Lotan et al (1980) refer to the assay system of sulphate release from the matrix of rabbit ear cartilage *in vitro* as a measure of retinoid potency. It may be possible to use this assay as a predictive indicator of bone toxicity.

Dawson: We have not used the rabbit ear cartilage lysis assay. At present, we are determining toxicity by comparing LD_{50} doses and weight loss in mice, using protocols similar to those of other National Cancer Institute investigators so that experimental results can be correlated. We also X-ray the animals at the time of death or at the end of the experiment to determine the extent of bone decalcification. We have found that this method is superior to the scoring system of Bollag (1979) that assesses long-bone breakage only.

Sporn: Compound **2** has been reported to have beneficial effects in psoriatic arthropathy.

Peck: The disease is difficult to study clinically because of its variable nature. Fritsch et al (1984a,b) felt that the compound (Ro 13-6298) was superior to etretinate for treatment of psoriatic arthropathy, but this was in a clinical study where the patients were also on non-steroidal anti-inflammatory agents. One goal of the study was to get the patients off these drugs and onto retinoid alone and this was accomplished. It would be best, however, to test a retinoid versus placebo in patients receiving no other therapy.

Sporn: The hydrocarbon **5** has been reported to have essentially no toxicity in the hypervitaminosis test in whole animals and is inactive in the papilloma test. However, although nowhere near as active as the carboxylic acid compounds, it does have significant activity in the tracheal organ culture assay. We presume that some sort of hydroxylation occurs in the terminal ring to introduce a polar function. It would be very interesting to make similar substances to see whether they are effective in breast cancer prevention. They should be quite lipophilic and non-toxic.

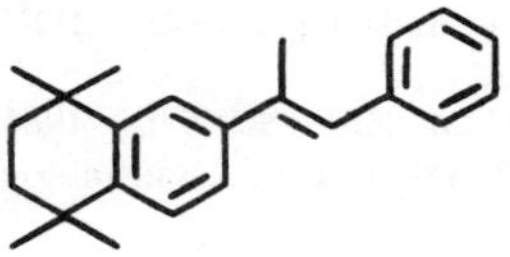

5

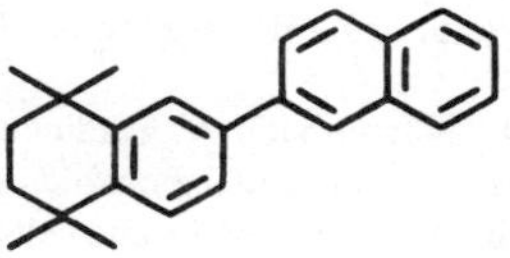

6

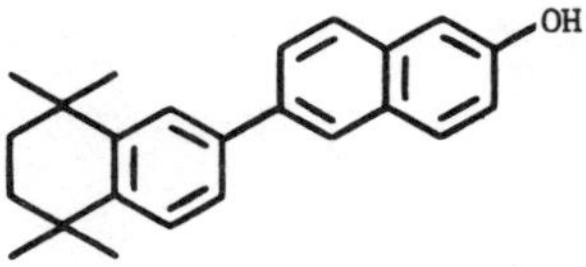

7

Dawson: We have prepared 2-(5,6,7,8-tetrahydro-5,5,8,8,-tetramethyl-2-naphthalenyl)naphthalene (**6**) and 6-(5,6,7,8-tetrahydro-5,5,8,8,-tetramethyl-2-naphthalenyl)-2-naphthalenol (**7**). Naphthalene **6** and naphthalenol **7** are inactive in the tracheal organ culture reversal of keratinization assay. These compounds have not been screened for toxicity.

REFERENCES

Bollag W 1979 Retinoids and cancer. Cancer Chemother Pharmacol 3:207-215

Fritsch P, Rauschmeier W, Neuhofer J 1984a Response of psoriatic arthropathy to arotinoid (Ro 13-6298): a pilot study. In: Cunliffe WJ, Miller AJ (eds) Retinoid Therapy. MTP Press, Lancaster, p 329-333

Fritsch P, Rauschmeier W, Zussner C 1984b Arotinoid in the treatment of psoriatic arthropathy. Dermatologica 169:250 (abstr)

Loeliger P, Bollag W, Mayer H 1980 Arotinoids, a new class of highly active retinoids. Eur J Med Chem Chim Ther 15:9-15

Lotan R, Neumann G, Lotan D 1980 Relationships among retinoid structure, inhibition of growth, and cellular retinoic acid-binding protein in cultured S91 melanoma cells. Cancer Res 40:1097-1102

Malarek DH, Burger W, Perry CW, Liebman AA 1984 An improved synthesis of high specific activity tritium labeled retinoic acid and its precursors. 188th Am Chem Soc Natl Meeting, Abstract ORGN 203

Sani BP, Dawson MI, Hobbs PD, Chan RLS, Schiff LF 1984 Relationship between binding affinities to cellular retinoic acid-binding protein and biological potency of a new series of retinoids. Cancer Res 44:190-195

Sen R, Carriker JD, Balogh-Nair V, Nakanishi K 1982 Synthesis and binding studies of a photoaffinity label for bovine rhodopsin. J Am Chem Soc 104:3214-3216

Sheves M, Makover A, Edelstein S 1984 Photo-affinity label for retinol-binding protein. Biochem Biophys Res Commun 122:577-582

Sporn MB, Roberts AB 1984 Biological methods for analysis and assay of retinoids—relationship between structure and activity. In: Sporn MB et al (eds) The retinoids. Academic Press, Orlando, vol 1:236-279

Strickland S, Breitman TR, Frickel F, Nürrenbach A, Hädicke E, Sporn MB 1983 Structure–activity relationship of a new series of retinoidal benzoic acid derivatives as measured by induction of differentiation of murine F9 teratocarcinoma cells and human HL-60 promyelocytic leukemia cells. Cancer Res 43:5268-5272

Inhibition by retinoids of neoplastic transformation *in vitro*: cellular and biochemical mechanisms

JOHN S. BERTRAM* and JOHN E. MARTNER†

Cancer Drug Center, Roswell Park Memorial Institute, Buffalo, New York 14263, USA

Abstract. We have demonstrated that non-toxic concentrations of retinoids can cause a dose-dependent inhibition of 3-methylcholanthrene-induced transformation of C3H/10T½ cells. On removal of the retinoid, transformed foci appear after a latent period of about four weeks at the same frequency as observed in controls treated with carcinogen only. Reasoning that this activity is compatible with the stabilization of the carcinogen-initiated state, we have succeeded in isolating from carcinogen-treated cultures a cell line which in the presence of retinyl acetate is similar to the parental 10T½ cells, but without retinyl acetate transforms at a high frequency after a latent period of about four weeks. Retinyl acetate treatment of this cell line (INIT/10T½) and the parental 10T½ cells induces an ultra-normal phenotype. In retinoid-deprived INIT/10T½ cells, the first sign of transformation (an increased thymidine-labelling index) occurs 16 days after retinyl acetate removal. We have detected by two-dimensional electrophoresis that concomitant with this there is an increase in phosphorylation of a protein of M_r 34 000 (34K) which may be associated with the cytoskeleton. This phosphoprotein has been found in all transformed lines examined. A second phosphoprotein, of about 38K, has also been detected in transformed cells. Retinyl acetate treatment of transformed cells alters the isoelectric point of this protein, a change compatible with decreased phosphorylation. Alkali-resistant phosphorylation, presumably on tyrosine, has been found on a second 34K protein of transformed cells. Retinyl acetate treatment specifically decreases this phosphorylation. Mechanisms for the altered tyrosine phosphorylation induced by retinyl acetate are as yet unresolved, but the decrease could be due to altered levels of substrate, kinase or phosphorylase. In view of the apparent role of tyrosine kinases as mediators of growth factors and as oncogene products, we consider the activity of retinoids as modulators of tyrosine phosphorylation to be of great potential significance.

1985 Retinoids, differentiation and disease. Pitman, London (Ciba Foundation Symposium 113) p 29–41

Present addresses: *Cancer Research Center of Hawaii, 1236 Lauhala Street, Honolulu, Hawaii 96813, and †McArdle Laboratory for Cancer Research, Madison, Wisconsin 53706, USA*

Vitamin A is required for vision, reproduction and the maintenance of normal epithelial differentiation in mammals. This last activity appears to correlate well with the ability of natural and synthetic analogues of vitamin A (retinoids) to prevent the development of experimentally induced carcinomas in laboratory animals (Sporn et al 1976). The prevention of experimental mammary cancer appears to be achieved by a delay in the onset of malignancy, i.e. by an increase in the latent period between caricnogen administration and tumour development (Moon et al 1979). This question has not been adequately addressed in other models.

Because of the many inherent limitations to carcinogenesis studies in animals, our group has aided in the development of cell culture systems in which to study the process of carcinogenesis in a quantitative and reproducible manner. In the studies described below we used the $C3H/10T\frac{1}{2}$ ($10T\frac{1}{2}$) mouse embryo fibroblast system (Reznikoff et al 1973). This line is widely used and has been the subject of a recent review (Yamaski et al 1984). Experimental procedures for using $10T\frac{1}{2}$ cells to study the activity of retinoids and to isolate carcinogen-initiated cells have been described by Merriman & Bertram (1979) and by Mordan et al (1982) respectively.

Retinoid inhibition of chemically induced neoplastic transformation

To induce neoplastic transformation in the $10T\frac{1}{2}$ system we generally expose sparsely seeded cultures to carcinogen for 24 h starting one day after seeding. Morphologically transformed foci do not appear in these treated cultures until about four weeks after seeding. An extra week is then allowed for the outgrowth of foci to macroscopic size before the fixation and staining of cultures. After cloning, most transformed foci injected into immunosuppressed syngeneic animals will form progressively growing sarcomas at the site of injection.

When retinyl acetate was applied to cultures treated seven days previously with a transforming concentration of 3-methylcholanthrene ($1\,\mu g/ml$), it caused a dose-dependent reduction in the number of transformed foci. Virtually 100% suppression of transformation was produced at a retinyl acetate concentration of $0.1\,\mu g/ml$. This concentration did not affect the plating efficiency or growth rate of either parental $10T\frac{1}{2}$ cells or their carcinogen-transformed derivatives, suggesting that the action of retinyl acetate was not due to toxicity. This was further confirmed in an experiment in which retinyl acetate was removed from carcinogen-treated cultures after four weeks of treatment. At the time of removal of retinyl acetate no transformed foci were present. However, after five weeks in the absence of retinyl acetate, transformed foci had appeared in equivalent numbers to those found in dishes treated with carcinogen only. We concluded that retinyl acetate reversibly inhibited the progression of initiated cells to neoplastically transformed cells (Merriman & Bertram 1979). The concept that retinoids reversibly inhibit the events leading to transformation *in vitro* fits well with *in vivo* data indicating that retinoid therapy only delays the onset of malignancy. In these experiments rats were first exposed to *N*-methyl-*N*-nitrosourea, a mammary carcinogen, and then to retinyl acetate to inhibit tumour formation. Cessation of retinoid treatment after 60 days led to a rapid increase in the number of tumour-bearing animals (Thompson et al 1979). In translating these results to a clinical setting, it is clear that a doubling of the tumour latent period, which is in the order of decades for exposure to bladder carcinogens and ionizing radiation (Doll & Peto 1981), would have a major impact on tumour incidence. However, in view of the reversible nature of retinoid effects on tumour inhibition, one cannot overlook the probable necessity for chronic administration of retinoids.

We have demonstrated (Bertram 1980) that the inhibitory activity of retinyl acetate towards neoplastic transformation in $10T\frac{1}{2}$ cultures is shared by other retinoids active as vitamin A substitutes in organ cultures of hamster trachea (Newton et al 1980). Two notable exceptions to this structure–activity correlation are the *all-trans-* and 13-*cis*-retinoic acids. At non-toxic doses these compounds increase the number of transformed foci developing in carcinogen-treated cultures. This lack of inhibitory activity of the retinoic acids in the $10T\frac{1}{2}$ cell system is perplexing since in many *in vivo* systems they are among the most potent retinoids. Boutwell's group has shown a similar phenomenon: although the induction of skin carcinomas in the mouse can be inhibited by *all-trans*-retinoic acid applied during a classical two-stage initiation/promotion regimen, the retinoid enhances transformation when applied during skin tumour induction by multiple exposures to the initiating agent 7,12-dimethylbenz[*a*]anthracene (Verma et al 1980). Similar enhancement has also been observed for UV-induced skin carcinogenesis (Forbes et al 1979). It is not known whether retinol or compounds capable of being converted to retinol (e.g. retinyl acetate) also enhance carcinogenesis when tested in similar *in vivo* protocols. If retinol produces different actions to retinoic acid, as occurs *in vitro*, the generally accepted belief that retinoic acid is the form of retinoids active in chemoprevention deserves re-examination.

Biochemical effects of retinoids on carcinogen-initiated cells

A major problem facing investigators seeking to identify biochemical events accompanying the conversion of initiated cells to neoplastically transformed cells is the low frequency induction of the initiating event in carcinogen-exposed populations, and the inability to separate initiated cells for study. Similarly, biochemical studies with retinoids are hampered by the unavailability of initiated cells, which appear to be the target cell population. A way around this impasse was suggested by the conclusion that retinyl acetate stabilizes the initiated state (Merriman & Bertram 1979). We thought that if this conclusion was correct it should be possible to isolate initiated cells from carcinogen-exposed cultures utilizing retinyl acetate to prevent transformation during cloning, testing and expansion of the isolated clone. This concept was tested and led to the isolation of a cloned line with the properties predicted of initiated cells. For a description of the isolation procedure and detailed characteristics of this line see Mordan et al (1982).

The most important characteristic of this line, which was designated INIT/10T$\frac{1}{2}$, is that in the absence of retinyl acetate most cells undergo morphological transformation after a latent period of three to four weeks. These transformed cells rapidly form tumours in nude mice, whereas the INIT/10T$\frac{1}{2}$ clone only slowly forms tumours and can have its malignant potential suppressed by dietary administration of retinoid. Because transformation occurs in the majority of cells over a narrow time window, we could use these cells to attempt to identify the biochemical changes that correlate in time with the onset of neoplastic transformation, and to determine the influence of retinoids on their target cells. As shown in Fig. 1, growth curves of INIT/10T$\frac{1}{2}$ cells are indistinguishable from those of similarly treated 10T$\frac{1}{2}$ cells until about the 21st day after seeding, when proliferation of INIT/10T$\frac{1}{2}$ cells resumes in cultures deprived of retinyl acetate. In cultures maintained with retinyl acetate, INIT/10T$\frac{1}{2}$ cells remain in a growth-inhibited state as a confluent monolayer.

Protein phosphorylation

There is currently intense interest in protein phosphorylation as a modulator of biological responses as diverse as those controlled by cyclic AMP (Greengard 1978, Kuo & Greengard 1969), and as a component of the response to mitogenic stimuli such as platelet-derived growth factor (PDGF) (Stiles 1983). This, together with the identification of the receptor for the tumour promoter 12-O-tetradecanoylphorbol-13-acetate (TPA) as a Ca^{2+}-dependent protein kinase (Kikkawa et al 1983) and the identification of several oncogene products as protein kinases (Bishop 1983), has led us to concentrate our initial studies on the detection of alterations in protein phosphorylation during neoplastic transformation.

Cultures of INIT/10T$\frac{1}{2}$ cells were seeded in the presence or in the absence of retinyl acetate, and at increasing time intervals cultures were placed in low phosphate medium containing $50\,\mu Ci/ml$ $^{32}PO_4$ for 3 h. Soluble proteins were then extracted from the cultures, and residual proteins run on two-dimensional gels (O'Farrell et al 1977). Exposure to low phosphate medium did not influence the growth rate of cells or prevent neoplastic transformation.

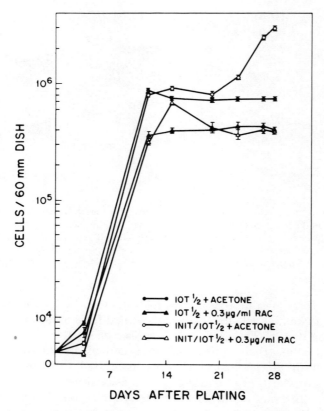

FIG. 1. Growth curves of parental $10T\frac{1}{2}$ cells and of $INIT/10T\frac{1}{2}$ cells in the presence or absence of retinyl acetate (RAC, $0.3\,\mu g/ml$). Cultures were seeded in basal Eagle's medium with 5% fetal calf serum at identical seeding densities of 5×10^3 cells/60 mm Petri dish. Acetone 0.5% was included as a solvent control.

Fig. 2 shows identical sections of autoradiographs of two-dimensional gels containing residual proteins extracted from $INIT/10T\frac{1}{2}$ cells cultured over a 21-day time course in the presence (lower row) or absence (upper row) of retinyl acetate. For the first 14 days of culture this region of the gels was the same for retinoid-treated and control cells; however, by day 16 a novel phosphoprotein of M_r 34000 (34K; $pp34_1$), and the more extensive phosphorylation of a protein of approximately 38K (pp38) were detected in cells in retinoid-free medium. These phosphorylation patterns were still evident on day 21. When we attempted to correlate these changes in phosphorylation with changes in cellular behaviour we found that, although no changes in

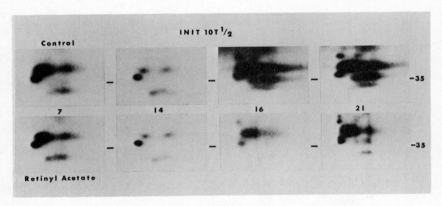

FIG. 2. Portions of autoradiographs of two-dimensional gel electropherograms. Initiated cells were cultured in the presence (bottom) or absence (top) of retinyl acetate. Sample cultures were taken 7, 14, 16 and 21 days after seeding and were labelled with $^{32}PO_4$ for 3 h. Cytoskeletal preparations were then solubilized and subjected to two-dimensional gel electrophoresis. The pI of proteins ranges from 5 to 6.5 from left to right and M_r ranges from 30 K to 42 K from bottom to top.

growth rate (Fig. 1) or morphology were evident on day 14, autoradiography of tritiated thymidine-labelled cultures revealed focal areas of high labelling beginning on day 15. These regions comprised about 10% of the area of the culture dishes and their appearance preceded by five to seven days the development of morphological changes associated with neoplastic transformation. For clarity, the entire two-dimensional gel autoradiographs are not shown; only minor alterations were detected in phosphorylation patterns in other regions of the gels.

Fig. 3 shows autoradiographs of phosphate-labelled residual proteins obtained from control and retinyl acetate-treated transformed cells derived from the transformation of INIT/10T$\frac{1}{2}$ cells. The phosphoprotein pp34$_1$, indicated by the triangle, has been detected in all transformed lines examined including those transformed by chemicals or by transfection with the human E J bladder oncogene or the Moloney viral oncogene. Although retinyl acetate treatment of already transformed cells does not apparently affect the auto-radiographic intensity or isoelectric point of pp34$_1$, the second protein, pp38, which becomes more heavily phosphorylated on transformation, was observed to alter its isoelectric point with retinoid treatment. This anodic shift is compatible with decreased phosphorylation but this has not been proved. We have demonstrated that transformed cells cannot be made to revert to the non-transformed phenotype by retinoids Merriman & Bertram 1979); however, drug treatment does cause some degree of cell flattening. On this basis

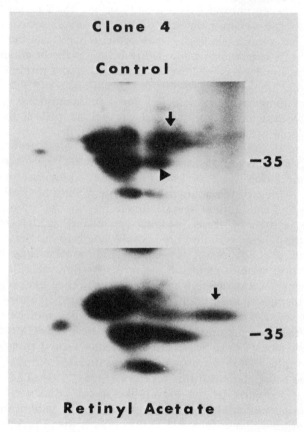

FIG. 3. Portions of autoradiographs of two-dimensional gel electropherograms. Established transformed cells were treated with 0.1 μg/ml retinyl acetate or acetone control. After four days, cells were labelled with $^{32}PO_4$ for 3 h. Cytoskeletal preparations were then solubilized and subjected to two-dimensional gel electrophoresis. The pI of proteins ranges from 5 to 6.5 from left to right and M_r ranges from 30 K to 42 K from bottom to top. Arrow pp38; triangle $pp34_1$.

one could speculate that phosphorylation of $pp34_1$ is associated with the irreversible attainment of neoplasia, whereas pp38 is perhaps involved in some reversible aspect of the phenotype such as cell shape.

Extensive studies of the nature of the phosphorylated residues have not yet been successfully completed because of the low abundance of the proteins of interest. Preliminary studies with one-dimensional sodium dodecyl sulphate–acrylamide gels have revealed the presence of a 34K protein which contains phosphate residues resistant to alkaline digestion in the absence but not the presence of retinyl acetate. Alkaline hydrolysis has been widely used to distinguish between phosphorylation on tyrosine, which is resistant to hydrolysis, and phosphorylation on serine and threonine, which is sensitive to hydrolysis. No similar alkali-resistant protein band has been detected in initiated cells before transformation. From the results of alkaline hydrolysis of two-dimensional gels, we have concluded that the protein called $pp34_1$, which becomes phosphorylated on transformation, is not the 34K phosphotyrosine-containing protein detected on one-dimensional gels. We are consequently designating this second protein as $pp34_2$. Its location on two-dimensional gels is currently being investigated.

The demonstration that an early event in the loss of growth control that accompanies neoplastic transformation of $INIT/10T\frac{1}{2}$ cells is an alteration in the phosphorylation state of two or possibly three proteins is perhaps appropriate in view of the many roles of protein phosphorylation referred to above. The fact that retinyl acetate apparently modifies the state of phosphorylation of two of these proteins (pp38 and $pp34_2$) strengthens the possibility that these proteins are closely involved in the transformation process, and are not mere passengers. The mechanisms by which the phosphorylation state is changed are presently obscure. Clearly the spontaneous events accompanying transformation could be due to protein kinase activation, to phosphatase inhibition or to enhanced synthesis or decreased degradation of the protein substrate. PDGF-like molecules have been detected in the supernatant from transformed $10T\frac{1}{2}$ cells (Bowden-Pope et al 1984). These factors may be secreted normally during transformation and could bind to the PDGF membrane receptor and activate the protein kinase known to be associated with this receptor (Stiles 1983). This explanation is only valid if secretion of the PDGF-like factor is an early event during transformation, matching the time course of the alterations in protein phosphorylation we have detected. Unfortunately the sequence of expression is currently unknown.

The regulation of protein phosphorylation by retinyl acetate would integrate well with its ability to antagonize many of the actions of the tumour promoter TPA. In cell culture TPA inhibits the stabilizing action of retinyl acetate on $INIT/10T\frac{1}{2}$ cells (Mordan et al 1982) and makes non-transformed cells mimic the transformed phenotype (Weinstein et al 1979), whereas retinyl acetate induces an opposite effect, that of accentuation of the non-transformed phenotype (Bertram 1983). The TPA receptor has recently been identified as a Ca^{2+}-dependent protein kinase (Kikkawa et al 1983) which when occupied enhances protein phosphorylation. As demonstrated above, retinyl acetate decreases phosphorylation of at least three protein species. It remains to be determined if retinoids and TPA exert opposing influences on the phosphorylation state of the same proteins. Attempts to identify the $pp34_1$ or $pp34_2$ as the pp34 substrate for pp60src (Erikson & Erikson 1980) have so far yielded equivocal results because the binding of antibody to the protein substrate lacks specificity.

In conclusion, investigations *in vitro* into the mechanisms of chemoprevention of cancer by retinoids are yielding clues to the biochemical basis for activity, and are providing useful tools to investigate the process of neoplastic transformation. In view of the growing evidence for the involvement of protein phosphorylation in the control of cell proliferation, we consider the activity of retinoids as modulators of phosphorylation of great potential significance.

Acknowledgement

This research was supported by USPHS grant CA 25489 from the US National Cancer Institute.

REFERENCES

Bertram JS 1980 Structure–activity relationships among various retinoids and their ability to inhibit neoplastic transformation and to increase cell adhesion in the C3H/10T½ CL8 cell line. Cancer Res 40:3141-3146

Bertram JS 1983 Inhibition of neoplastic transformation *in vitro* by retinoids. Cancer Surv 2:243-262

Bishop JM 1983 Cellular oncogenes and retroviruses. Ann Rev Biochem 52:301-354

Bowden-Pope DF, Vogel A, Ross R 1984 Production of platelet-derived growth factor-like molecules and reduced expression of platelet-derived growth factor receptors accompany transformation by a wide spectrum of agents. Proc Natl Acad Sci USA 81:2396-2400

Doll R, Peto R 1981 The causes of cancer: quantitative estimates of avoidable risks of cancer in the United States today. J Natl Cancer Inst 66:1191-1308

Erikson E, Erikson RL 1980 Identification of a cellular protein substrate phosphorylated by the avian sarcoma virus-transforming gene product. Cell 21:829-836

Forbes PD, Urbach F, Davies RE 1979 Enhancement of experimental photocarcinogenesis by topical retinoic acid. Cancer Lett 7:85-90

Greengard P 1978 Phosphorylated proteins as physiological effectors. Science (Wash DC) 149:146-152

Kikkawa U, Takai Y, Tanaka Y, Miyake R, Nishizuka Y 1983 Protein kinase C as a possible receptor protein of tumor-promoting phorbol esters. J Biol Chem 258:11442-11445

Kuo JF, Greengard P 1969 Widespread occurrence of an adenosine 3′,5′-monophosphate-dependent protein kinase in various tissues and phyla of the animal kingdom. Proc Natl Acad Sci USA 64:1349-1355

Merriman RL, Bertram JS 1979 Reversible inhibition by retinoids of 3-methylcholanthrene-induced neoplastic transformation in C3H/10T½ clone 8 cells. Cancer Res 39:1661-1666

Moon RC, Thompson HJ, Becci PJ et al 1979 N-(4-hydroxyphenyl)retinamide, a new retinoid for prevention of breast cancer in the rat. Cancer Res 39:1339-1346

Mordan LJ, Bergin LM, Budnick JL, Meegan RL, Bertram JS 1982 Isolation of methyl-cholanthrene-'initiated' C3H/10T$\frac{1}{2}$ cells by inhibiting neoplastic progression with retinyl acetate. Carcinogenesis (Lond) 3:279-285

Newton DL, Henderson WR, Sporn MB 1980 Structure–activity relationship of retinoids in hamster tracheal organ cultures. Cancer Res 40:3413-3425

O'Farrell PZ, Goodman HM, O'Farrell PH 1977 High resolution two-dimensional electro-phoresis of basic as well as acidic proteins. Cell 12:1133-1142

Reznikoff CA, Bertram JS, Brankow DW, Heidelberger C 1973 Quantitative and qualitative studies of chemical transformation of cloned C3H mouse embryo cells sensitive to post-confluence inhibition of cell division. Cancer Res 33:3239-3249

Sporn MB, Dunlop NM, Newton DL, Smith JM 1976 Prevention of chemical carcinogenesis by vitamin A and its synthetic derivatives (retinoids). Fed Proc 35:1332-1338

Stiles CD 1983 The molecular biology of platelet-derived growth factor. Cell 33:653-655

Thompson HT, Becci PJ, Brown CC, Moon RC 1979 Effect of the duration of retinyl acetate feeding on inhibition of 1-methyl-1-nitrosourea-induced mammary carcinogenesis in the rat. Cancer Res 39:3977-3980

Verma AK, Conrad EA, Boutwell RK 1980 Induction of mouse epidermal ornithine decarboxy-lase activity and skin tumours by 7,12-dimethylbenz[a]anthracene: modulation by retinoic acid and 7,8-benzoflavone. Carcinogenesis (Lond) 1:607-611

Weinstein IB, Lee L, Fisher PB, Muson A, Yamasaki H 1979 Action of phorbol esters in cell culture: mimicry of transformation, altered differentiation and effects on cell membranes. J Supramol Struct 12:195-208

Yamaski Y, Bertram JS, Landolph J et al 1984 Workshop on cell transformation. IARC (Int Agency Res Cancer)' Monogr Eval Carcinog Risk Chem Hum, in press

DISCUSSION

*Hicks:*There are a lot of similarities between your *in vitro* system and some of the *in vivo* animal experiments. You can maintain the initiated state *in vitro* as long as you have got the retinoid there, but once you take it away you get transformation. Similarly, you can use retinoids to stop the growth of trans-formed foci in animals *in vivo*, but once you take the retinoids away the foci will grow. After a latent period of about 10 weeks the foci spontaneously escape the retinoid control and, although the animals are maintained on the retinoid, tumours start to grow. Do your cells also eventually escape retinoid control if you keep them long enough?

Bertram: I can't satisfactorily answer that question because there are prob-lems *in vitro* in keeping cultures beyond three months; one starts getting sloughing of the cell monolayer. *In vivo* we have only gone up to 10 weeks. In our original studies *in vitro* we found that with weekly treatment with retinyl acetate we got some escape after 10 weeks (Merriman & Bertram 1979). In subsequent studies we found that retinyl acetate had a half-life of two to three days in culture medium and that if we gave retinyl acetate two or three times weekly we maintained almost 100% inhibition for 10 weeks.

Hicks: It would certainly be worthwhile doing this in animals for longer than 10 weeks.

Koeffler: What cloning efficiency for type 3 foci do you get in a given population of initiated cells?

Bertram: Of the surviving plated cells about 70% go on to become neoplastically transformed. Of that 70%, about 60% are type 3 and the remaining 40% are type 2, so the foci are not of a single morphological type. As in chemical transformation experiments, we get a mixture.

Sherman: How does this compare with a culture where you don't select initiated cells?

Bertram: The number of cells that are transformed is quite different; after carcinogen treatment only about 1% of the cells become transformed. However, of that 1%, 60% form type 3 foci and 40% form type 2 foci.

Sherman: So your INIT/10T½ cultures are enriched 60-fold or 70-fold with cells that will ultimately be transformed?

Bertram: Yes. I would like to add some information about the effects of oncogenes in 10T½ cells. The transforming oncogenes were first described in the 10T½ system by Weinberg's group (Shih et al 1979) and subsequently H-*ras* was found in at least one transformed cell. We have transfected the EJ *ras* gene into normal 10T½ cells, and this gives cells with a characteristic type 3 morphology that we can't seem to inhibit with retinoids. However, the phosphoprotein patterns are identical to those we see in chemically transformed cells, and when we add retinoid to these EJ-transformed cells we get the same shift in pI for pp38 and the same loss of the putative phosphotyrosine residues as we do in cells transformed with 3-methylcholanthrene. When we transform normal 10T½ cells with v-*mos* we get transformants with a different morphology from that seen with chemically transformed or *ras*-transformed cells, but we get the same pattern of phosphoproteins. However, when we add retinoids we get no change in phosphorylation. It seems therefore that the effect of retinoids is dependent upon oncogenes that are activated by chemicals during *de novo* transformation; activation of *ras* may well be important in this respect since the effects of retinoids on phosphorylation are seen when we transfect *ras* into 10T½ cells. However, when we put an atypical oncogene into the cells, retinoids don't have that same effect. This may relate to the different retinoid effects that we see in different cell types.

Yuspa: Have you done transfection studies in your initiated cell line to see if the cells remain responsive to retinoid after an oncogene is inserted?

Bertram: No, but we have tried to determine which oncogene is activated in initiated cells.

Sherman: You said that retinoids had no effect on the morphology of *ras*-transformed cells, but do they have any effect on the morphology of *mos*-transformed cells?

Bertram: No, but in some of the highly transformed cells (type 3) it is very difficult to show morphological changes in response to retinoids. There may be effects that are too subtle to detect.

Sherman: It is very interesting that although you see the difference in the phosphoprotein profile in one case and not the other, you don't see an effect on morphology in either case. Does this say something about the role of these phosphoproteins in transformation?

Bertram: I'm not sure, but it may say that the retinoids are acting before expression of the activated oncogene. Activation only takes place during the 'quantum leap' in transformation of an initiated cell to a neoplastically transformed cell; retinoids may act at an earlier stage.

Lotan: In what cell fraction do you find the phosphoproteins; are they associated with the cytoskeleton or with the cytoplasm? Have you checked to make sure that retinyl acetate is not merely shifting the compartmentalization of these proteins?

Bertram: We've looked at that by conventional ultracentrifugation analysis and we don't detect either of the 34K proteins in the cell cytoplasm. They always seem to be associated with the cytoskeleton. If we isolate the nucleus by alternative procedures, we don't find these proteins in the nuclear fraction either. So they are associated either with the cytoskeleton or with the organelles that remain attached to the cytoskeleton after conventional non-ionic detergent treatment. We are now trying to get enough of the proteins to raise antibodies to them so that we can ask more specific questions about localization.

Lotan: Are these proteins secreted by the cells?

Bertram: We haven't looked for secreted protein.

Yuspa: Have you tried treating retinoid-treated cells with 12-*O*-tetradecanoylphorbol-13-acetate and then looking for the same proteins?

Bertram: No.

Sporn: Have you any idea of the mechanism by which retinoids affect phosphorylation?

Bertram: It would be simple to say they are inhibiting a protein kinase, which would fit in with much of the data on growth factors and oncogenes. Alternatively, they could be stimulating a phosphatase; we know that in many systems phosphatases are the major regulators of protein phosphorylation. However, since we were not able to measure the amount of protein present because there was so little, we can't be sure that there is not a change in the protein substrate.

Lotan: Stimulation of alkaline phosphatase by retinoids has been reported, but the concentrations required are $\geq 10^{-5}$ M (Reese & Politano 1981).

Sporn: Those are very unconvincing reports, but it would be interesting if retinoids were shown to control phosphatase activity in some selective way at physiological concentrations.

Lotan: The concentrations used by Reese & Politano (1981) may have been toxic, but if you scale everything down 100-fold, the changes in the level of the enzyme at low retinoid concentrations might be sufficient to do the job. It's possible that there could be small changes in enzyme activity that one would not even be able to detect by alkaline phosphatase or other assays.

Sporn: We did some studies with retinoids and alkaline phosphatases in 10T½ cells and came up with nothing.

Lotan: Have you any idea why *all-trans*-retinoic acid is not active in your *in vitro* system?

Bertram: No, it's an enigma. We've looked at retinal, retinol, esters, amides and the new benzoic acid derivatives which are very active in the 10T½ system and will inhibit transformation at 10^{-9} or 10^{-10} M. These derivatives are not very toxic: we don't kill cells even at 10^{-5} M. They have crystal structures virtually identical to that of *all-trans*-retinoic acid, and yet with *all-trans*-retinoic acid and 13-*cis*-retinoic acid we get a high level of toxicity and no inhibition of transformation. In fact we see an enhancement of transformation. If we decrease the dose so we get no toxicity, we get no effect on transformation either. I don't understand it. We can find no binding proteins for retinoic acid but no binding proteins for retinol either, so a specific loss of binding protein cannot explain the lack of activity of retinoic acid. It may be that for some reason the 10T½ cell lacks the ability to metabolize retinoic acid to its active form, whatever that may be.

Sporn: Is retinoic acid methyl ester active?

Bertram: Yes, as are the amides, and *N*-(4-hydroxyphenyl)retinamide is more active than retinol.

Sporn: I think that the simplest suggestion is that the cells do not take up the retinoic acid because of some peculiar membrane effect.

REFERENCES

Reese DH, Politano VA 1981 Evidence for the retinoid control of urothelial alkaline phosphatase. Biochem Biophys Res Commun 102:322-327

Merriman RL, Bertram JS 1979 Reversible inhibition by retinoids of 3-methylcholanthrene-induced neoplastic transformation in C3H/10T½ clone 8 cells. Cancer Res 39:1661-1666

Shih C, Shilo B, Goldfarb MP, Dannenberg A, Weinberg RA 1979 Passage of phenotypes of chemically transformed cells via transfection of DNA and chromatin. Proc Natl Acad Sci USA 76:5714-5718

Role of retinoids in differentiation and growth of embryonal carcinoma cells

MICHAEL I. SHERMAN, MARY LOU GUBLER, URIEL BARKAI, MARY I. HARPER, GEORGE COPPOLA and JEFFREY YUAN*

Department of Cell Biology, Roche Institute of Molecular Biology, Roche Research Center, Nutley, New Jersey 07110, USA

Abstract. To study how retinoids promote differentiation and inhibit proliferation of embryonal carcinoma (EC) cells, we have followed their intracellular fate. Retinoic acid (RA) is effectively metabolized to more polar compounds by many EC lines. Unlike RA, retinol is slowly metabolized. Our inability to detect conversion of retinol to RA might indicate that the two retinoids elicit their effects on EC cells in different ways. Retinol added to cultures quickly appears in the nuclear fraction; the proportion associated with nuclei after detergent extraction is initially very low but increases with time. Retinol and RA might be translocated to nuclei by their respective binding proteins [cellular retinol-binding protein (CRBP) and cellular retinoic acid-binding protein (CRABP)]: isolated EC nuclei have specific, independent binding sites for both holoproteins but not their ligands. CRABP cannot be detected in the nucleoplasm of untreated EC cells, but activity is measurable after cells are exposed to RA. Interestingly, incubation with retinol promotes movement of both CRBP and CRABP into the nucleoplasmic fraction. Finally, we have demonstrated that brief exposure to RA dramatically reduces the cloning efficiency of EC cells. Since some cells are unaffected even by lengthy exposures to RA whereas the growth of their progeny is inhibited, we suggest that EC cells can become epigenetically refractory to RA.

1985 Retinoids, differentiation and disease. Pitman, London (Ciba Foundation Symposium 113) p 42–60

Embryonal carcinoma (EC) cells are the undifferentiated stem cells of teratocarcinomas. There is abundant evidence that EC cells are similar to pluripotent cells of the early embryo (see Jetten 1985 for a recent review). Because murine EC lines are readily available, these cells are useful as models for studying the events involved in differentiation. EC cells can

---

* *Present address:* Department of Molecular Biophysics and Biochemistry, Yale University, New Haven, Connecticut 06511, USA

differentiate in response to several types of physical and chemical stimuli (see Jetten 1985). Among these, retinoic acid (RA) is one of the few naturally occurring inducers and is also one of the most potent (Strickland & Mahdavi 1978, Jetten et al 1979). This laboratory has provided evidence that the cellular RA-binding protein (CRABP) mediates retinoid-induced differentiation in EC cells: (1) with few exceptions, there is a strong qualitative correlation between the ability of a retinoid to promote differentiation of EC cells and to compete for sites on CRABP (Jetten & Jetten 1979, Sherman et al 1983a); (2) several EC mutants have been generated that fail to differentiate in response to RA and have also lost most or all CRABP activity (Schindler et al 1981, McCue et al 1983, Wang & Gudas 1984); and (3) the reversal of the differentiation-defective phenotype in these mutant cells by hybridization or treatment with sodium butyrate is accompanied by the reappearance of CRABP activity (McCue et al 1984a,b).

Retinoids influence growth as well as differentiation of EC cells. The two effects are presumably different since the growth of some differentiation-defective EC mutants is nevertheless inhibited by RA (Schindler et al 1981). The combined effect of retinoids in suppressing growth and promoting differentiation in EC cells, as well as in promyelocytic leukaemia (see Amatruda & Koeffler 1985) and melanoma (see Lotan 1985) cells, makes natural retinoids and appropriate synthetic analogues (e.g. Jetten & Jetten 1979, Sherman et al 1983a) potentially useful as antineoplastic agents. To position ourselves more effectively in our attempts to elucidate the modes of action of retinoids in differentiation and growth, we have undertaken to learn more about the disposition of these molecules within the cell. A survey of our recent investigations follows.

Metabolism of retinoids

We began our investigations into retinoid metabolism as a result of our observation that retinol is capable of inducing the differentiation of several EC lines (Eglitis & Sherman 1983, Sherman et al 1983b). Since retinol is one of the few retinoids that do so but fail to compete for sites on the CRABP, and since it is markedly less potent than RA, we wished to test the possibility that retinol acts via conversion to RA. Accordingly, several EC lines were incubated for various periods of time with *all-trans*-[³H]retinol and the extracted labelled retinoids were analysed by high performance liquid chromatography (HPLC). A typical result is shown in Fig. 1 for cell line OC15 S1 (McBurney 1976), one of the most responsive lines to retinol (Eglitis & Sherman 1983). When [³H]retinol is incubated for 48 h in culture medium at 37 °C, some conversion to more polar derivatives (eluting at

1.5–3.5 min) is observed (Fig. 1A). There is also a shoulder on the retinol peak which coelutes with 13-*cis*-retinol after derivatization with acetic anhydride (M.L. Gubler & M.I. Sherman, unpublished work). At our limits of detection, we do not observe any labelled material coeluting with *all-trans*-RA or 13-*cis*-RA markers. After incubation of [³H]retinol for 48 h with OC15 S1 cells, the ³H profile from the culture medium is very similar to that from control medium (cf. Figs. 1A & C). The profile of cell-associated label (Fig. 1B) differs from that of the media: there appears to be an increase, or enrichment, of *cis* isomer(s) of retinol but a relative diminution of the very

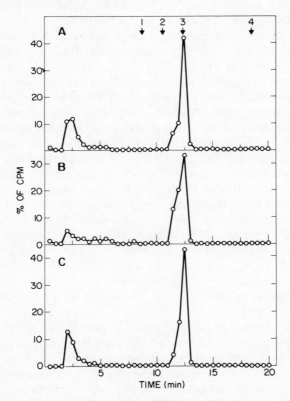

FIG. 1. Metabolism of *all-trans*-[³H]retinol by OC15 S1 cells. (A) Culture medium (no cells) was incubated with [³H]retinol (32 Ci/mmol; 2 μCi/ml) for 48 h at 37 °C. Labelled material was extracted and subjected to reverse-phase HPLC. The profile illustrates the time of elution of labelled material from the Zorbax ODS (Du Pont) column with acetonitrile/water (80:20, v/v). OC15 S1 cells (B) and their culture medium (C) were extracted, and labelled retinoids were analysed after 48 h incubation with [³H]retinol in the same way as (A). The arrows in (A) indicate the positions of elution of 13-*cis*-RA (1), *all-trans*-RA (2), *all-trans*-retinol (3) and *all-trans*-retinal (4) markers.

polar material; in the cell profile there is also a small but unique peak eluting at 8 min and two other peaks at 4.5 and 5.5 min, which were probably present, albeit at lower levels, in control media. As with the control medium, however, there is no indication of labelled material coeluting with the RA markers. Similar profiles have been obtained with several other EC lines (M.L. Gubler & M.I. Sherman, unpublished work). The identities of the labelled species that are more polar than the retinols are unclear. However, they are presumably alcohols, not acids, because their position of elution is altered by derivatization with acetic anhydride, which esterifies alcohols, but not with diazomethane, which esterifies acids (M.L. Gubler & M.I. Sherman, unpublished work).

Our results suggest that EC cells are incapable of converting significant amounts of retinol to RA. Similar conclusions have been reached recently by Williams & Napoli (1984). The differences between the cell and media HPLC profiles could well reflect differential penetrability of retinol and its derivatives into OC15 S1 cells: for example, the consistent observation of a substantially lower percentage of the very polar derivatives in cells than in media (Fig. 1; M.L. Gubler & M.I. Sherman, unpublished work) could signal that these derivatives are taken up inefficiently by EC cells; conversely, the higher ratio of 13-*cis*-retinol:*all-trans*-retinol in cells compared to media might reflect a greater tendency of the former retinoid to associate with cellular material. However, alternative interpretations (e.g. that EC cells actively isomerize *all-trans*-retinol) cannot be eliminated.

Our failure to detect conversion of retinol to RA by EC cells would suggest that such metabolism is not responsible for the differentiation-inducing ability of retinol in these cells. However, we cannot eliminate the possibility that there is a small amount (less than 1%) of enzymic or non-enzymic conversion of retinol to RA in EC cultures, since such low levels of RA could have escaped detection in our studies. Although we have found that the relative potency of retinol in inducing EC cell differentiation appears to be substantially greater than 1% of RA with some lines (Eglitis & Sherman 1983), we cannot unequivocally eliminate the possibility that our retinol-treated cultures contain low levels of RA which is responsible for the induction of differentiation. Our results do, however, suggest that differential metabolism of retinol does not explain the variability of response of different EC lines to retinol. At present, it would appear prudent to remain open-minded as to the possible mechanism by which retinol induces EC cell differentiation.

Unlike retinol, RA is extensively metabolized by many EC cells, including OC15 S1 (Fig. 2). In the absence of cells, most *all-trans*-[^{3}H]RA added to culture medium remains intact or isomerizes to 13-*cis*-RA (Fig. 2A). By contrast, neither of these retinoids is readily detectable in medium in which cells have been cultured for 18 h (Fig. 2C), and only small amounts of

all-trans-RA and 13-*cis*-RA are present in cell-associated labelled material
Fig. 2B). Other studies (Williams & Napoli 1984, M.L. Gubler & M.I.
Sherman, unpublished work) have confirmed that these results are, indeed,

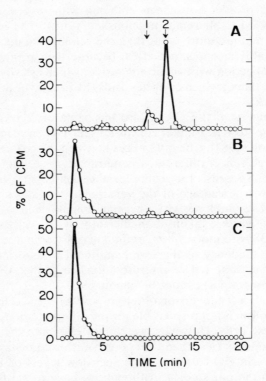

FIG. 2. Metabolism of *all-trans*-[³H]retinoic acid by OC15S1 cells. (A) Culture medium (no cells) was incubated with [³H]RA (32 Ci/mmol; 2 μCi/ml) for 18h at 37°C. Labelled material was analysed by HPLC as described in the legend to Fig. 1. (B, C) Profiles obtained from extracts of cells and their culture medium respectively, incubated with [³H]RA at the same time. The arrows in (A) indicate the positions of elution of 13-*cis*-RA (1) and *all-trans*-RA (2) markers.

due to the cellular metabolism of the RA. Several polar metabolites, most or all of them acids, are formed. In further analyses (M.L. Gubler & M.I. Sherman, unpublished work 1984), we have learned that the RA-metaboliz-ing system is constitutively high in some EC lines, but at lower basal levels in others. In some, but not all, of the latter cells the system can be induced to give very high levels of metabolism. It is perhaps notable that, of the EC lines we have surveyed (M.L. Gubler & M.I. Sherman, unpublished work), the one which metabolizes RA least efficiently is PCC4(RA)$^-$2, a differentiation-defective mutant line which has little or no CRABP activity (McCue et al 1983).

The significance of RA metabolism *vis-à-vis* its influence on differentiation and growth is unclear. Past efforts to identify RA metabolites that are biologically more potent than RA have been unsuccessful. Studies by Roberts and co-workers (see Roberts 1981) have demonstrated that RA is actively metabolized by some tissues *in vivo* and that this metabolic activity is inducible. The suggestion that the enzymes responsible are representatives of the cytochrome P-450 class of mixed-function oxidases (Roberts et al 1979) lends support to the view that the physiological purpose of this metabolism of RA is deactivation or detoxification (Roberts et al 1979, Roberts 1981). We do not yet know about the biological activity of RA metabolites generated by EC cells. If these metabolites are found to be inactive or much less active than RA itself, then the rapid metabolism of RA by some EC lines which nevertheless differentiate readily in response to the retinoid would suggest that the presence of RA is required primarily to initiate (and not necessarily to maintain) the differentiated state. Our finding that the arotinoid acid Ro 13-7410 is hardly metabolized by EC cells (M.L. Gubler & M.I. Sherman, unpublished work 1984), whereas it is a potent inducer of their differentiation (Sherman et al 1983a, Strickland et al 1983), indicates that retinoid metabolism is not an essential part of the induction process of EC cell differentiation.

Although the metabolism of RA by EC cells might not be required for the regulation of growth and differentiation by RA, such metabolism could nevertheless have considerable impact upon the potency of various retinoids, and this in turn might have substantial clinical relevance. We have, therefore, begun a study of the ability of various retinoids to induce the metabolic system and to act as substrates (M.L. Gubler & M.I. Sherman, unpublished work). Investigations with the arotinoid Ro 13-7410 have revealed that the two capabilities are not always shared (M.L. Gubler & M.I. Sherman, unpublished work 1984): as mentioned, this RA analogue is a poor substrate for the RA-metabolizing system, yet it effectively induces the enzyme(s) involved. The fact that Ro 13-7410 is only poorly metabolized might explain why this retinoid is notably more potent than RA at inducing differentiation at low concentrations (Sherman et al 1983a, Strickland et al 1983). Finally, we have obtained data that suggest that some metabolites of retinol and RA are prevented from entering cells efficiently and/or are more readily secreted than the parent compounds (M.L. Gubler & M.I. Sherman, unpublished work). It can be noted, for example, that the small amounts of *all-trans*-RA and 13-*cis*-RA remaining in OC15 S1 cultures after 18 h are almost entirely cell-associated (Fig. 2). Therefore, at a physiological level, the ability of a retinoid to reach, and remain associated with, its desired target cells might depend to a large degree upon its susceptibility to metabolism.

Intracellular movement of retinoids

Within several hours of exposure of EC cells to either [3H]retinol or [3H]RA, label is distributed among all subcellular fractions (Sherman et al 1983b). Similar observations have been made with other cell types (see Sherman 1985). Since retinol does not appear to be extensively metabolized to RA by EC cells, we were interested to learn whether retinol is translocated to the nucleus in these cells, as has been proposed for RA (Jetten & Jetten 1979). Our earlier studies suggested that after 16 h incubation, a large proportion of administered [3H]retinol or [3H]RA co-fractionated with nuclei (Sherman et al 1983b). Since non-specific association with nuclei might have occurred during disruption and fractionation of the cells, we have begun studies to determine whether retinol actually penetrates the outer nuclear membrane of EC cells. We have approached this issue by incubating intact F9 EC cells with [3H]retinol for various times, disrupting the cells and then preparing their nuclei with and without exposure to Triton X-100. This detergent treatment has been reported to remove the outer nuclear membrane and a large proportion of the nuclear phospholipid (Liau et al 1981). Table 1 illustrates that without detergent treatment the percentage of cell-associated [3H]retinol that co-fractionates with nuclei is independent of incubation time. The ratio of c.p.m. associated with detergent-treated nuclei to c.p.m. associated with untreated nuclei was very low after a 1 h incubation period but this ratio increased notably by 16 h. One interpretation of our results is that [3H]retinol

TABLE 1 Association of [3H]retinol with nuclei after incubation with F9 embryonal carcinoma cells

| Duration of incubation (h) | % of 3H associated with | | |
	Untreated nuclei (A)	Triton X-100-treated nuclei (B)	(B/A) × 100
1	10.0	0.6	6
5	12.7	1.0	8
16	10.0	2.1	21
24	11.7	1.5	13

Cells were incubated with [3H]retinol (2 μCi/ml) for the indicated time periods. Cells were washed, collected and disrupted. Nuclei were obtained with and without detergent treatment (0.5% Triton X-100) by sedimentation through sucrose. Values in columns A and B refer to the percentages of label in washed cells that were associated with the nuclear fraction.

associates rapidly with nuclear membranes either in the intact cell or
following disruption of the cells, whereas intranuclear localization occurs
gradually, over a period of several hours.

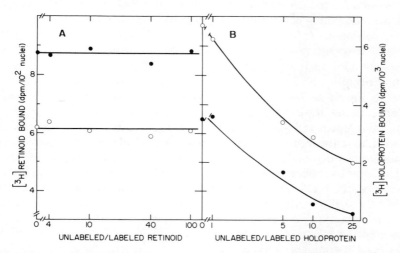

FIG. 3. Binding of retinoids and their holo-binding proteins to isolated nuclei from Nulli-SCC1
cells. Nuclei were obtained from confluent cultures of Nulli-SCC1 cells by the non-detergent
procedure described by Liau et al (1981). In (A), 10^6 nuclei were incubated at 25 °C for 2 h with
[³H]RA (●) [23.7 pmol (2.0×10^4 Bq)] or with [³H]retinol (○) [21.4 pmol (1.8×10^4 Bq)] with or
without excess amounts of unlabelled RA or retinol, respectively. The reaction was stopped by
placing the samples on ice. Nuclei were then collected on glass fibre filters, washed extensively
and subjected to liquid scintillation spectrometry. The experiment in (B) was carried out in the
same way except that the [³H]retinoids added were bound to their respective binding proteins.
The number of becquerels added per assay tube (containing 10^6 nuclei) was 2.8×10^3 for the
[³H]RA–CRABP complex and 1.3×10^3 for the [³H]retinol–CRBP complex. The retinoid-
binding proteins had been partially purified (about 500-fold) such that neither preparation bound
any of the heterologous ligand and each preparation contained only a single peak of bound
retinoid on hydroxylapatite chromatography.

McCormick et al (1984) have recently reported that the CRABP holoprotein binds specifically to isolated F9 EC nuclei. We have confirmed this report
and extended it to the CRBP (cellular retinol-binding protein) holoprotein.
Fig. 3A demonstrates that free [³H]retinol and [³H]RA bind in an unsaturable
manner to isolated nuclei from Nulli-SCC1 EC cells. When the labelled
retinoids are added as [³H]RA–CRABP or [³H]retinol–CRBP holoprotein
complexes, many fewer molecules are bound; however, under these circumstances unlabelled holoprotein can reduce the extent of binding of labelled
retinoids (Fig. 3B), which indicates that a finite number of binding sites exists.
In other experiments we have observed that neither free RA nor the reti-

nol–CRBP complex competes for sites with [³H]RA–CRABP; similarly, the nuclear binding of the [³H]retinol–CRBP complex is not reduced by the addition of free retinol or RA–CRABP (U. Barkai, unpublished work).

In summary, the interaction of retinoids and their holoprotein complexes with isolated Nulli–SCC1 nuclei appears to be similar to that described for F9 EC cell nuclei (McCormick et al 1984) and nuclei from other cell types, including liver and testis (e.g. Liau et al 1981, Cope et al 1984): both RA and retinol interact specifically with nuclei only when they are bound to their binding proteins, and neither free ligand nor heterologous holoprotein can compete for these specific nuclear sites. Liau et al (1981) have provided evidence that CRBP only delivers the retinol to its nuclear site, i.e. the apoprotein leaves the nucleus once interaction has occurred. This has led to the idea that the binding proteins assume the role of shuttles to transfer retinoids from cytoplasm to nucleus (McCormick et al 1984). However, this model has not yet been tested directly with either binding protein in EC cells.

Wiggert et al (1977) and Jetten & Jetten (1979) made nucleoplasmic preparations from retinoblastoma and EC cells, respectively, that had been exposed in culture to [³H]RA, and they observed labelled material which sedimented in a sucrose density gradient at 2 S, as would be expected for the RA–CRABP complex. Wiggert et al (1977) failed to demonstrate a similar effect with [³H]retinol, even though the retinoblastoma cells possessed CRBP. We have begun to assess this presumptive binding-protein-mediated translocation of retinoid from cytoplasm to nucleus in a somewhat different way: Nulli-SCC1 cells maintained in medium containing delipidated serum (which lacks detectable amounts of retinoids) are incubated for short periods of time (2–5 h) with retinol or RA. Nucleoplasmic extracts are then prepared, exposed to light (to generate apoprotein from any retinoid holoprotein that might be present) and challenged with [³H]RA or [³H]retinol. The results obtained indicate that in the absence of retinoids nucleoplasmic extracts contain very low levels of CRBP activity, whereas CRABP is undetectable by our assay. After incubation of cells with RA, however, substantial levels of CRABP, but not CRBP, activity are observed in the nucleoplasm. Retinol-treated cells show an equivalent movement of CRBP into the nucleoplasmic fraction, and interestingly, translocation of a somewhat smaller amount of CRABP also appears to occur (U. Barkai, unpublished work). This might reflect the conversion (enzymic or non-enzymic) in EC cultures of retinol to RA at levels which we cannot detect by HPLC analysis (see previous section).

Taken together with previous reports, our experiments strengthen the view that the specific association of retinoid holoproteins with isolated nuclei is of physiological significance. The nature of the retinoid-binding sites is unclear: McCormick et al (1984) have claimed that labelled RA is released from EC nuclei by DNase treatment, but the data of Cope et al (1984) with rat testis nuclei suggest that the retinoids interact predominantly with nuclear proteins. Although evidence for the nuclear localization of retinoids and their binding proteins mounts steadily, the biological significance of this phenomenon remains unclear. It has been proposed that retinoid-binding proteins deliver their ligands to specific DNA sites and modulate gene expression in a direct manner, analogous to steroid receptor proteins (see Chytil & Ong 1979). This is an attractive hypothesis and perhaps the simplest model to explain how retinoids influence growth and differentiation in several cell types. However, as is often the case with simple models, this one does not adequately account for all the available information. For example, if the hypothesis is correct, it is difficult to understand why the CRBP and CRABP holoproteins do not compete for the same nuclear binding sites, since the different retinoids that they carry are often effective in eliciting the same effects on cellular behaviour (albeit usually with different potencies). If it is argued that, at least in the case of EC cells, retinol-induced differentiation *in vitro* is actually caused by very low levels of RA, generated by the metabolism of retinol (Williams & Napoli 1984), then it is unclear why retinol and its binding protein should translocate to, and bind specifically to, the nucleus, as does the RA–CRABP complex. Indeed, one might expect instead that we would have failed to see movement of retinol to the nucleus, as did Wiggert et al (1977) in retinoblastoma cells. Finally, despite the compelling evidence mentioned at the beginning of this article in favour of a role for CRABP in RA-induced differentiation, there is the paradox that some cell types whose differentiation and growth are modulated by retinoids appear not to have detectable amounts of the appropriate binding proteins, at least at the levels of sensitivity of the assays employed (Lotan et al 1980, Douer & Koeffler 1982, Libby & Bertram 1982). Since retinoids are pleiotropic (see Lotan 1980), the possibility exists that these compounds can elicit similar effects on different cells in different ways (see Sherman 1985).

Effects of retinoic acid on clonal growth of embryonal carcinoma cells

If retinoid-binding proteins mediate the promotion of differentiation and the inhibition of growth of EC cells by retinoids, then one might expect that EC cells would begin to respond to retinoids soon after they were exposed to them, since there is evidence that the holoproteins are translocated to the nucleus within a few hours of retinoid treatment. Furthermore, if retinoids directly alter patterns of gene expression, it is conceivable that after triggering these changes their presence would no longer be required. Because we do not yet have appropriate markers for monitoring the biological action of retinoids within the first few hours of treatment, we are attempting to evaluate early retinoid effects by cloning studies. Since differentiated derivatives of EC cells generally fail to grow at clonal density, we expected that after a pulse exposure to RA only those cells that had not become irreversibly triggered to differentiate would continue to grow. In a preliminary experiment, we cultured Nulli-SCC1 cells in the absence of RA or in its presence (at 10^{-6} M) for either two days or nine days. Cells were then washed and seeded at clonal density either in liquid culture or in semi-solid methylcellulose medium (Stoker 1968). We observed a dramatic reduction in cloning efficiency when cells were treated with RA, even for two days (Table 2). Although the overall cloning efficiency was substantially lower in semi-solid medium than in liquid medium, the relative cloning efficiencies of RA-treated (compared to untreated) Nulli-SCC1 cells were very similar under the two conditions. It is notable that, even after nine days of RA treatment, small numbers of clones were present under both cloning conditions. When several of the latter clones were selected for expansion, they were found to possess a typical EC morphology (M. Harper & G. Coppola, unpublished work 1984).

We have begun to examine the effect of treating Nulli-SCC1 cells with RA for shorter periods of time. As Experiment 3 (Table 2) illustrates, a 6 h exposure to RA caused a marked reduction in cloning efficiency, although the extent of the effect was substantially less than that after treatment for two days (Table 2). On the other hand, EC cells exposed to RA during the wash period (less than 1 min) showed little or no diminution in cloning efficiency (M. Harper & G. Coppola, unpublished work 1984). In general, our results are in agreement with those obtained somewhat differently with PC13 EC cells by Rayner & Graham (1982).

Although these initial results pose more questions than they answer, they raise some interesting points. First, the data confirm our suspicion that a relatively short pulse with RA is adequate to elicit a change in the growth properties of Nulli-SCC1 cells. This would be expected if RA causes changes in gene expression in a direct manner which leads to an irreversible cascade of events culminating in a differentiated phenotype. However, other interpretations are possible. For example, RA associated with Nulli-SCC1 cells during the 6 h incubation period might fail to leave the cells during the washing procedure, perhaps because of strong interaction with CRABP. The reduction in cloning efficiency of the treated cells could then reflect either promotion of differentiation or adverse effects upon proliferation. Indeed, we have observed in other studies that EC cells incubated with [^{3}H]RA for 5 h still retain some labelled material (including small amounts of RA) 19 h later (M.L. Gubler & M.I. Sherman, unpublished work 1984). We are currently extending these experiments to differentiation-defective cells in order to determine whether reduction of cloning efficiency is mediated by effects upon growth or differentiation.

TABLE 2 Cloning efficiency of Nulli-SCC1 cells after exposure to retinoic acid[a]

| Duration of exposure | Relative cloning efficiency[b] | | |
	Experiment 1	Experiment 2	Experiment 3
0	100	100	100
6 h			58
2 days	15	13	16
9 days	8	5	

[a]Nulli-SCC1 cells in liquid culture were exposed to 10^{-6} M-RA for the indicated time periods, washed, removed from the dishes with trypsin-EDTA and counted. Cells were plated either in liquid culture at 10^3 cells/100 mm dish (Experiments 1 and 3) or in a semi-solid methylcellulose medium at 10^5 cells/100 mm dish (Experiment 2). The cells in Experiments 1 and 2 were derived from the same cultures; a different culture was used for Experiment 3.
[b]Cloning efficiency is the percentage of cells forming macroscopic clusters after 7–10 days. Absolute cloning efficiencies for control cultures (no exposure to RA) were 42% (Experiment 1), 9% (Experiment 2) and 85% (Experiment 3).

Another conclusion to be drawn from the data in Table 2 is that the response to RA is heterogeneous. The reduction in cloning efficiency after exposure to RA for 6 h is only about half that observed after 48 h treatments. Since the level of uptake of RA by EC cells reaches a plateau within a few hours (Schindler et al 1981, McCue et al 1983), it is unlikely that the amount of cell-associated RA differs in the 6 h and 48 h samples. It is possible that the cells themselves are heterogeneous, i.e. some cells in the culture might be refractory to RA action. Alternatively, there might be temporal differences in receptivity to RA; Yen & Albright (1984), for example, have suggested that HL-60 cells show cell-cycle dependence in their susceptibility to the differentiation-inducing effects of RA. We have attempted to obtain evidence for the presence of unresponsive EC cells by examining cells from some of the clones that were generated despite RA treatment for two or even nine days. Preliminary results show clearly that the subcloning efficiency of these initially unresponsive cells is markedly reduced in the presence of RA (M. Harper & G. Coppola, unpublished work 1984). We conclude that these clones, at least, were not originally refractory to RA because of genetic alterations.

Because the cell-cycle time of Nulli-SCC1 cells in 10^{-6} M-RA is 20 h (McCue et al 1983), one could reasonably attribute the incomplete response of cells to 6 h (compared with 48 h) RA treatment to differential potencies of RA at different phases of the cell cycle. However, this explanation alone, or some other simple stochastic model, seems inadequate to explain the persistence of some clones after nine days of pretreatment with RA, since an average Nulli-SCC1 cell would have passed through 10 cell cycles during this interval. We obtained similar results with PCC4.aza1R EC cells in earlier studies (Sherman et al 1981); we compared tumour incidence in mice injected with untreated or RA-treated PCC4.aza1R cells and found that, in mice given cells that had been exposed for 14 or even 28 days to the retinoid, tumour incidence was reduced, but not eliminated. The tumours that did persist had the histological properties of teratocarcinomas (Sherman et al 1981). We believe that the most likely explanation of our data is that a small proportion of EC cells in a culture can in some way be epigenetically blocked from responding to RA, but that this block can subsequently be reversed. Similarly, our ability to reverse the differentiation-defective phenotype and elicit the reappearance of CRABP in PCC4(RA)$^-$1 cells by treatment with sodium butyrate (McCue et al 1984b) could be explained by re-expression of an epigenetically silent CRABP gene (the homologue having presumably been genetically altered by chemical mutagenesis).

Whatever the mode of action of RA in suppressing clonal growth of EC cells in our studies (i.e. promotion of differentiation, suppression of proliferation or both), this type of retinoid action could have practical clinical application in oncology. However, one must be concerned about the small proportion of cells that escapes the suppressive effect of RA: these cells could either rapidly repopulate a tumour in the presence of retinoid or remain quiescent during therapy with the ability to proliferate rapidly once treatment had ceased. EC cells can be induced to differentiate by exposure to any one of several chemically unrelated agents (see Jetten 1985). It will be useful to test whether those cells that persist in forming clones even after lengthy treatment with RA are simultaneously refractory to one or more of these other inducers.

REFERENCES

Amatruda TT, Koeffler HP 1985 Retinoids and cells of the hematopoietic system. In: Sherman MI (ed) Retinoids and cell differentiation. CRC Press, Boca Raton, in press

Chytil F, Ong DE 1979 Cellular retinol- and retinoic acid-binding proteins in vitamin A action. Fed Proc 38:2510-2513

Cope FO, Knox KL, Hall RC 1984 Retinoid binding to nuclei and microsomes of rat testes interstitial cells: I—mediation of retinoid binding by cellular retinoid-binding proteins. Nutr Res 4:289-304

Douer D, Koeffler HP 1982 Retinoic acid inhibition of the clonal growth of human myeloid leukemia cells. J. Clin Invest 69:277-283

Eglitis ME, Sherman MI 1983 Murine embryonal carcinoma cells differentiate *in vitro* in response to retinol. Exp Cell Res 146:289-296

Jetten AM 1985 Induction of differentiation of embryonal carcinoma cells by retinoids. In: Sherman MI (ed) Retinoids and cell differentiation. CRC Press, Boca Raton, in press

Jetten AM, Jetten MER 1979 Possible role of retinoic acid binding protein in retinoid stimulation of embryonal carcinoma cell differentiation. Nature (Lond) 278:180-182

Jetten AM, Jetten MER, Sherman MI 1979 Stimulation of differentiation of several murine embryonal carcinoma lines by retinoic acid. Exp Cell Res 124:381-391

Liau G, Ong DE, Chytil F 1981 Interaction of the retinol/cellular retinol-binding protein complex with isolated nuclei and nuclear components. J Cell Biol 91:63-68

Libby PR, Bertram JS 1982 Lack of intracellular retinoid-binding proteins in a retinol-sensitive cell line. Carcinogenesis (Lond) 3:481-484

Lotan R 1980 Effects of vitamin A and its analogs (retinoids) on normal and neoplastic cells. Biochim Biophys Acta 605:33-91

Lotan R 1985 Retinoids and melanoma cells. In: Sherman MI (ed) Retinoids and cell differentiation. CRC Press, Boca Raton, in press

Lotan R, Ong DE, Chytil F 1980 Comparison of the level of cellular retinoid-binding proteins and susceptibility to retinoid-induced growth inhibition of various neoplastic cells. J Natl Cancer Inst 64:1259-1262

McBurney MW 1976 Clonal lines of teratocarcinoma cells in vitro: differentiation and cytogenetic characteristics. J Cell Physiol 89:441-455

McCormick AM, Pauley S, Winston JH 1984 F9 embryonal carcinoma cells contain specific nuclear retinoic acid acceptor sites. Fed Proc 43:788 (abstr)

McCue PA, Matthaei KI, Taketo M, Sherman MI 1983 Differentiation-defective mutants of mouse embryonal carcinoma cells: response to hexamethylenebisacetamide and retinoic acid. Dev Biol 96:416-426

McCue PA, Gubler ML, Maffei L, Sherman MI 1984a Complementation analyses of differentiation-defective embryonal carcinoma cells. Dev Biol 103:399-408

McCue PA, Gubler ML, Sherman MI, Cohen BN 1984b Sodium butyrate induces histone hyperacetylation and differentiation of murine embryonal carcinoma cells. J Cell Biol 98:602-608

Rayner MJ, Graham CF 1982 Clonal analysis of the change in growth phenotype during embryonal carcinoma cell differentiation. J Cell Sci 58:331-344

Roberts AB 1981 Microsomal oxidation of retinoic acid in hamster liver, intestine, and testis. Ann NY Acad Sci 359:45-53

Roberts AB, Frolik CA, Nichols MD, Sporn MB 1979 Retinoid-dependent induction of the in vivo and in vitro metabolism of retinoic acid in tissues of the vitamin A-deficient hamster. J Biol Chem 254:6303-6309

Schindler J, Matthaei KI, Sherman MI 1981 Isolation and characterization of mouse mutant embryonal carcinoma cells which fail to differentiate in response to retinoic acid. Proc Natl Acad Sci USA 78:1077-1080

Sherman MI 1985 How do retinoids promote differentiation? In: Sherman MI (ed) Retinoids and cell differentiation. CRC Press, Boca Raton, in press

Sherman MI, Matthaei KI, Schindler J 1981 Studies on the mechanism of induction of embryonal carcinoma cell differentiation by retinoic acid. Ann NY Acad Sci 359:192-199

Sherman MI, Paternoster ML, Taketo M 1983a Effects of arotinoids upon murine embryonal carcinoma cells. Cancer Res 43:4283-4290

Sherman MI, Paternoster ML, Eglitis MA, McCue PA 1983b Studies on the mechanism by which chemical inducers promote differentiation of embryonal carcinoma cells. Cold Spring Harbor Conf Cell Proliferation 10:83-95

Stoker M 1968 Abortive transformation by polyoma virus. Nature (Lond) 218:234-238

Strickland S, Mahdavi V 1978 The induction of differentiation in teratocarcinoma stem cells by retinoic acid. Cell 15:393-403

Strickland S, Breitman TR, Frickel F, Nürrenbach A, Hädicke E, Sporn MB 1983 Structure–activity relationships of a new series of retinoidal benzoic acid derivatives as measured by induction of differentiation of murine F9 teratocarcinoma cells and human HL-60 promyelocytic leukemia cells. Cancer Res 43:5268-5272

Wang S-Y, Gudas LJ 1984 Selection and characterization of F9 teratocarcinoma stem cell mutants with altered response to retinoic acid. J Biol Chem 259:5899-5906

Wiggert B, Russell P, Lewis M, Chader G 1977 Differential binding to soluble nuclear receptors and effects on cell viability of retinol and retinoic acid in cultured retinoblastoma cells. Biochem Biophys Res Commun 79:218-225

Williams JB, Napoli JL 1984 Metabolism of retinol and retinoic acid during differentiation of F9 embryonal carcinoma cells. Fed Proc 43:788 (abstr)

Yen A, Albright KL 1984 Evidence for cell cycle phase-specific initiation of a program of HL-60 cell myeloid differentiation mediated by inducer uptake Cancer Res 44:2511-2515

DISCUSSION

Strickland: Have you done any calculations, based on the number of nuclei you have and the extent of displaceable retinoid binding, to work out how many sites there are?

Sherman: One cannot do a really good calculation without a Scatchard analysis and for this one needs many more points than we have. Others have estimated that there are about 100 000 to 300 000 binding sites per nucleus for the retinoid-binding proteins; there are more for the retinol-binding protein (CRBP) than for the retinoic acid-binding protein (CRABP) (reviewed by Sherman 1985).

Moon: Mehta, in my laboratory, has found about 120 000 per nucleus in the mammary gland of pregnant rats and in nuclei of N-methyl-N-nitrosourea-induced mammary tumours (Mehta et al 1982).

Sporn: There has been some concern that there are too many binding sites to be true steroid-like receptors, and that their affinity is too low to make much physiological sense. Another problem is that nobody has yet shown that the combination of ligand, whether it be retinol or retinoic acid, with binding protein has any selective effect on chromatin, so there is a lot of scepticism about the function of these binding proteins.

Sherman: I don't know whether the affinity of the holoprotein for its nuclear binding sites is as strong as that of a steroid receptor for its DNA sites. However, with steroid receptors you can find large numbers of binding sites in total genomic DNA, and some of these irrelevant binding sites (i.e. those not regulated by the steroid receptor holoprotein) have affinities which differ by less than two logs from the small number of relevant binding sites (those regulated by steroid receptors) in the genome (see for example Mulvihill et al 1982). We do not yet know whether the nuclear binding sites for the retinoid-binding proteins are on DNA; we certainly have not identified specific DNA sequences whose transcription is regulated by retinoid-binding proteins, as has been achieved for steroid receptor proteins (Mulvihill et al 1982, Payvar et al 1983). However, the observed large number of nuclear sites for retinoid-binding proteins should not be taken as evidence against the existence of small numbers of physiologically significant DNA binding sites.

Moon: Mehta et al (1982) have found that in the mammary system retinoic acid associates with nuclear binding sites with a high affinity ($K_d = 2 \times 10^{-9}$ M), and that the retinoic acid–CRABP complex is necessary for the inhibition of RNA polymerase (Moon et al 1983).

Yuspa: What is the time course for the appearance of CRBP in the nucleus after you expose cells to retinoid?

Sherman: We have not done a detailed kinetic analysis, but after two hours we can already see evidence of movement of the binding protein into the nucleoplasmic compartment (U. Barkai, unpublished work). When one does very crude uptake studies on whole embryonal carcinoma cells, steady-state levels are reached in a very short time; within two hours, the cells appear to have taken up almost as much retinol as they will take up over a 72-hour period, in terms of counts associated with the cells after they have been trapped on a filter (J. Yuan & M.L. Gubler, unpublished work).

Breitman: You said that butyrate could affect CRABP levels. Does this mean that in the presence of butyrate you can get induction of metabolizing enzymes with much lower concentrations of retinoic acid than are normally required?

Sherman: We have not looked at the effects of different doses of retinoic acid in the presence and absence of butyrate, so I can't answer that. We have found that butyrate has differentiation-stimulating effects on two embryonal carcinoma cell lines (McCue et al 1984), but Levine et al (1984) have found that F9 cells do not respond in this way. The parental cell from which the binding protein-deficient mutants are obtained (PCC4.aza1R) is responsive, but how the butyrate acts we don't know.

Jetten: In fact Levine et al (1984) saw effects opposite to those you reported. They found that butyrate blocked the action of retinoic acid in F9 cells.

Lotan: In the cells that you treated with retinoids for long periods of time and that remained unresponsive, could you demonstrate changes in the levels of binding proteins? Could their expression be switched on and off in some physiological way as it can be by butyrate?

Sherman: We have been unable to look specifically at either unresponsive or responsive cells when we are measuring binding protein activity, because we don't have access to a binding-protein antibody. We could not expect to get definitive information by measuring total levels of retinoid-binding activity in mixed cultures.

Lotan: I thought that you could isolate the unresponsive clones and then expand them.

Sherman: Yes, but we have not tested binding protein activity in those expanded clones.

Lotan: You said that retinoic acid is metabolized by many embryonal carcinoma lines. If we treat S91 melanoma cells for 24 hours with retinoic acid labelled with either ^{14}C or tritium, and analyse them by high performance liquid chromatography (HPLC) after methanol extraction by the method of DeLuca, we find some isomerization of the retinoic acid by 24 hours but hardly any metabolism to anything that would move differently on the HPLC system (A. Clifford et al, unpublished work). I just want to make the point that different cells may metabolize retinoids differently.

Sherman: Yes, that is certainly true for embryonal carcinoma cells. In some cell lines metabolizing enzymes are constitutive whereas in others they can be induced, and at least one of the cell lines is very poor at metabolizing retinoic acid (M.L. Gubler & M.I. Sherman, unpublished work).

Bertram: Your observations on retinoid metabolism are very interesting in view of my comments about the lack of activity of *all-trans*-retinoic acid in the 10T½ system (see p 31). Similar rapid metabolism of *all-trans*-retinoic acid could explain the apparent lack of activity in 10T½ cells. Have you looked at the dose–response relationship for retinoic acid? Is there a concentration below which retinoic acid will not induce the metabolizing enzymes?

Sherman: We have found that retinoic acid at 10^{-8} M induces the metabolizing enzymes in F9 cells (M.L. Gubler & M.I. Sherman, unpublished work). We have not looked at lower concentrations.

Breitman: How active is retinal in inducing differentiation of your cells? Strickland & Mahdavi (1978) found that retinal is about 1000-fold less active than retinoic acid in inducing differentiation of F9 cells. Have you looked at the possible conversion of retinal to retinoic acid in your system?

Sherman: We have not studied retinal with our cells.

Breitman: The first step in the conversion of retinol to retinoic acid may be very slow and the second very fast.

Sporn: Are there separate dehydrogenases or just one that does the whole conversion?

Pitt: Two.

Strickland: Wouldn't it be very hard to see metabolism of retinol or retinal to retinoic acid in your system?

Sherman: It could be difficult if the retinoic acid, once formed, were to be rapidly metabolized. We know the positions of elution from the HPLC column of the metabolites of retinoic acid. We've looked in those positions after treating our cells with retinol but we don't find any label there. The problem is that there are several different metabolites from retinoic acid, so if there was 1% conversion of retinol to retinoic acid, that 1% could be split among these different possible products and we would not be able to see it by our methods. We have not yet studied retinal metabolism by embryonal carcinoma cells.

Moon: Is there any binding with retinal?

Jetten: There is no competition of retinal for the binding site on the retinoic acid-binding protein.

Pitt: Retinal can bind to lots of things in cells; it is taken up by many proteins and phospholipids, forming Schiff bases with their amino groups.

Strickland: Does retinol or retinal induce the metabolizing systems?

Sherman: We have not tested retinal. Retinol induces metabolism but it is 100 times less effective than retinoic acid (M.L. Gubler & M.I. Sherman, unpublished work). This fits well with the *in vivo* studies by Roberts et al (1979) on induction of retinoic acid metabolism in the hamster by retinyl acetate vs. retinoic acid.

Hartmann: DeLuca et al (1984) have reported that one of the most active compounds derived from retinoic acid metabolism in embryonal carcinoma F9 cells is a retinoyl-β-glucuronide. Could this be relevant to your system?

Sherman: Williams & Napoli (1984) have claimed that there is very little glucuronylation of retinoic acid in embryonal carcinoma cells. We have not looked at this directly but we have a good idea where the glucuronide would run in our HPLC profile and we don't have any label there.

Frickel: Have you tried dosing your cells with the alcohol analogue of (E)-4-[2-(5,6,7,8-tetrahydro-5,5,8,8-tetramethyl-2-naphthalenyl)-1-propenyl]-benzoic acid? You could compare the acid with the alcohol as you did retinoic acid with retinol.

Sherman: These experiments are in progress.

Sporn: Is the retinoidal benzoic acid derivative more effective than retinoic acid in inducing the metabolizing enzymes?

Sherman: We have not done a dose–response curve. We know that micromolar concentrations effectively induce the enzymes (M.L. Gubler & M.I. Sherman, unpublished work), but those levels are quite high for a retinoidal benzoic acid derivative.

REFERENCES

DeLuca HF, Silva D, Miller D, Jarnagin A 1984 Peripheral metabolism of retinoic acid. Dermatologica 169:219 (abstr)

Levine RA, Campisi J, Wang S-Y, Gudas LJ 1984 Butyrate inhibits the retinoic acid-induced differentiation of F9 teratocarcinoma stem cells. Dev Biol 105:443-450

McCue PA, Gubler ML, Sherman MI, Cohen BN 1984 Sodium butyrate induces histone hyperacetylation and differentiation of murine embryonal carcinoma cells. J Cell Biol 98:602-608

Mehta RG, Cerny WL, Moon RC 1982 Nuclear interactions of retinoic acid-binding protein in chemically induced mammary adenocarcinoma. Biochem J 208:731-736

Moon RC, Mehta RG, McCormick DL 1983 Suppression of mammary cancer by retinoids. In: Crispen RG (ed) Cancer: etiology and prevention. Elsevier Biomedical, Amsterdam, p 275-287

Mulvihill E, LePennec J-P, Chambon P 1982 Chicken oviduct progesterone receptor: location of specific regions of high affinity binding in cloned DNA fragments of hormone-responsive genes. Cell 28:621-632

Payvar F, DeFranco D, Firestone GL et al 1983 Sequence-specific binding of glucocorticoid receptor to MTV DNA at sites within and upstream of the transcribed region. Cell 35:381-392

Roberts AB, Frolik CA, Nichols MD, Sporn MB 1979 Retinoid-dependent induction of the in vivo and in vitro metabolism of retinoic acid in tisues of the vitamin A-deficient hamster. J Biol Chem 254:6303-6309

Sherman MI 1985 How do retinoids promote differentiation? In: Sherman MI (ed) Retinoids and cell differentiation. CRC Press, Boca Raton, in press

Strickland S, Mahdavi V 1978 The induction of differentiation in teratocarcinoma stem cells by retinoic acid. Cell 15:393-403

Williams JB, Napoli JL 1984 Metabolism of retinol and retinoic acid during differentiation of F9 embryonal carcinoma cells. Fed Proc 43:788 (abstr)

Regulation of differentiation of tracheal epithelial cells by retinoids

ANTON M. JETTEN and HENK SMITS

Cell Biology Group, Laboratory of Pulmonary Function and Toxicology, National Institute of Environmental Health Sciences, PO Box 12233, Research Triangle Park, North Carolina 27709, USA

Abstract. An *in vitro* culture system of rabbit tracheal epithelial cells has been developed to study the regulation of differentiation of the respiratory epithelium on the molecular level. At high density in the absence of retinoids these cells become squamous, stratify, and ultimately form cross-linked envelopes. Several factors influence this terminal differentiation: high Ca^{2+} concentrations and serum factors promote, whereas retinoids and medium conditioned by fibroblasts inhibit this process. Terminal squamous cell differentiation is accompanied by several biochemical changes: the synthesis of proteoglycans is dramatically reduced and the expression of keratin intermediate filaments is altered. Besides the eight major keratins expressed in undifferentiated cells, terminally differentiated cells also express a 48 kDa keratin. The expression of this keratin correlates well with squamous cell differentiation and appears to be under the control of retinoic acid. The level at which these biochemical changes are regulated has yet to be established. Specific retinol- and retinoic acid-binding proteins have been identified in these cells; the correlation between binding and biological activity of retinoids in this system is in agreement with a role for these binding proteins in mediating the action of these agents.

1985 Retinoids, differentiation and disease. Pitman, London (Ciba Foundation Symposium 113) p 61–76

The basic structure of the lining of the trachea *in situ* is a pseudostratified, columnar epithelium, which has been shown ultrastructurally to contain a diverse population of cell types (Breeze & Wheeldon 1977). The ciliated cell is one of the main cell types and is probably a non-proliferative, terminally differentiated cell. The non-ciliated population consists of a variety of secretory cells (goblet, Clara and/or serous cells, depending on species) and a non-secretory cell, the basal cell. Although the exact pathway(s) of differentiation are not yet fully defined, it is believed that the differentiated cells develop either directly from basal cells or indirectly via mucous cells (Lane & Gordon 1979, Chopra 1982, McDowell et al 1984). Under various conditions, such as mechanical or toxic injury, chronic respiratory infection or vitamin A deficiency, the basal and/or mucous cells express a different differentiation potential in the form of squamous metaplasia and keratinization (Wilhelm 1954, Wong & Buck 1971, Sporn et al 1975, Clark & Marchok 1979, McDowell et al 1984).

Recently, Wu & Smith (1982) have developed conditions to culture rabbit tracheal epithelial cells *in vitro* in serum-free medium. To understand the mechanism by which differentiation of these cells is regulated we have examined the factors that are important in the induction of differentiation of these cells. Since retinoids have been shown to be important regulators of differentiation of the respiratory epithelium *in vivo*, we have been concentrating on the action of these compounds on the differentiation of these tracheal epithelial cells in culture. The study of the regulation of differentiation on a molecular level requires biochemical markers. Therefore we have to determine: (1) the biochemical changes that accompany differentiation, (2) whether biochemical parameters can function as markers of differentiation and (3) the level at which the expression of these markers is controlled. Ultimately, an answer to these questions should be very helpful in elucidating the mechanism of action of retinoids.

Materials and methods

Isolation and culture of rabbit tracheal epithelial cells

Tracheal epithelial cells were isolated as described previously (Wu & Smith 1982). In our studies tracheae from 8–10-week-old male New Zealand white rabbits were used. Epithelial cells were obtained after protease digestion of the tracheal epithelium overnight at 4 °C. An average of about 2.5×10^6 cells per trachea was obtained. Cells were resuspended in cold Ham's F12 medium containing 10 µg/ml insulin, 5 µg/ml transferrin, 25 ng/ml epidermal growth factor and hypothalamus extract. Cells were plated in 35 mm or 60 mm dishes which were pretreated for 2–3 h with 1 ml or 2 ml phosphate-buffered saline containing 10 µg/ml fibronectin, 10 µg/ml bovine serum albumin and 30 µg/ml Vitrogen (Flow Laboratories, McLean, VA, USA). Cell counts were determined with a Coulter Counter (Coulter Electronics Inc, Hialeah, FL, USA).

Assay of cross-linked envelopes

For the determination of spontaneous formation of cross-linked envelopes, medium and cells were collected and centrifuged and the pellet was resuspended in 20 mM-dithiothreitol and 2% sodium dodecyl sulphate (SDS). Detergent-resistant envelopes were counted as described by Rice & Green (1979). To determine the competence of the cells to form cross-linked envelopes, cells were first treated for 2 h with 50 μg/ml calcium ionophore X537A (Hoffmann-La Roche, Nutley, NJ, USA).

Preparation of keratin-enriched fractions

Cells were labelled with [^{35}S]methionine (25 μCi/ml; 1097 mCi/mmol) for 18 h. Keratins were isolated essentially according to a procedure described by Franke et al (1978) with some minor modifications. After labelling, cells were washed twice with phosphate-buffered saline (PBS) and collected in PBS by scraping with a rubber policeman. After centrifugation at 14 000*g* for 1 min, cells were homogenized in 1 ml of 20 mM-Tris-HCl pH 7.6 containing 140 mM-NaCl, 5 mM-MgCl$_2$ and 0.6 M-KCl. After 30 min on ice the insoluble material was pelleted and extracted once more in the same buffer supplemented with 1% Triton X-100. After 30 min on ice the insoluble material, which was enriched in keratins, was pelleted and lyophilized.

Analysis of keratins

Keratins were analysed by two-dimensional electrophoresis with non-equilibrium pH-gradient electrophoresis (NEPHGE) in the first dimension and an 8% polyacrylamide–SDS gel in the second dimension as described by O'Farrell et al (1977).

Binding protein assay

For the assay of retinol- and retinoic acid-binding proteins, cytosolic fractions from rabbit tracheal epithelial cells were prepared. The presence of the binding proteins was determined by a rapid gel-filtration assay (Matthaei et al 1983). Cytosolic fractions (76 μg protein) were incubated with 50 nM-[^{3}H]retinoic acid (32 Ci/mmol) or 50 nM-[^{3}H]retinol (52 Ci/mmol).

Results and discussion

The cell population obtained after Pronase digestion of the epithelium of the rabbit trachea is heterogeneous and contains about 50% ciliated cells, 20%

basal cells and 10% goblet cells. About 10% of the cells stain positively with nitroblue tetrazolium, indicating the presence of Clara cells, and 10% are unidentifiable. We have established primary cultures of these epithelial cells according to a modified procedure described by Wu & Smith (1982). After a

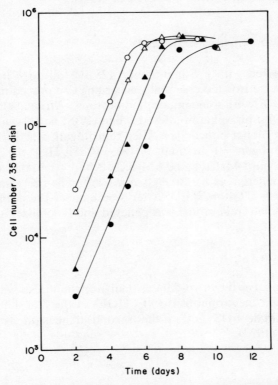

FIG. 1. Growth curves of rabbit tracheal epithelial cells. Cells were plated at different densities on day 0, and at different times cell number was determined. ○, 200 000 cells plated; △, 100 000 cells plated; ▲, 33 000 cells plated; ●, 10 000 cells plated.

few days in culture the epithelial cell population appears homogeneous when morphological observations are made by phase-contrast and electron microscopy. Ciliated cells apparently do not attach to the dish and are absent at the start of the culture. The cultured cells have no other differentiative characteristics; they do not contain secretory granules, do not produce mucous glycoproteins and have a large ratio of nucleus to cytoplasm. The growth curves of tracheal epithelial cells plated at different densities are shown in Fig. 1. During the exponential growth phase cells proliferate rapidly with a doubling time of about 19 h. Cells appear as rounded, single cells and are very migratory. During this period cells can be easily subcultured with an attachment efficiency of 86–90% and a cloning efficiency of 40–60%. When the cells reach high density, they cease proliferating and undergo a change in morphology. At this stage cell–cell contacts are established and cells become less migratory. Also, it rapidly becomes more difficult to subculture the cells, as shown by a decline in colony-forming efficiency. The final density appears independent of the number of cells initially plated. When cells are maintained at high density for longer times, they gradually become squamous, form cross-linked envelopes and slough off into the medium. This process of terminal differentiation is very similar to terminal differentiation of keratino-cytes from the skin, conjunctivum and cornea in culture (Green & Watt 1982, Phillips & Rice 1983).

What is the origin of these rapidly proliferating undifferentiated cells? We used centrifugal elutriation to fractionate the cell population obtained after proteolytic digestion of the tracheal epithelium, and we were able to obtain a 90% pure basal-cell preparation (Hook & Jetten, unpublished work 1984). The morphology and growth characteristics of these cells in tissue culture appeared very similar to the characteristics of the rapidly proliferating undifferentiated cells. Cells from the other elutriated fractions either did not attach or did not grow under the conditions used. This result strongly suggests that the proliferative epithelial cells are derived from basal cells.

Control of proliferation and terminal differentiation

Several factors affect the proliferation and terminal differentiation of rabbit tracheal epithelial cells in culture. Proliferation of the cells is dependent on the presence of calcium. Optimal growth occurs at $0.2–0.6\,mM\text{-}Ca^{2+}$. At higher concentrations a gradual decrease in growth rate is observed; cells appear flatter and cell–cell contacts are established. When cells reach high density, increasing calcium concentrations promote cornification (Fig. 2A). Fetal bovine serum also reduces cell proliferation, increases squamous appearance and promotes the formation of cross-linked envelopes (Fig. 2B). Similar results have been described for human bronchial epithelial cells in culture (Lechner et al 1984). We have tested several other factors such as hydrocortisone and the phorbol ester 12-O-tetradecanoylphorbol-13-acetate, but these appeared to have little effect on terminal differentiation. When medium conditioned by BALB/c 3T3 fibroblasts is used in a 1:1 combination with Ham's F12 medium, cells reach a fourfold higher density and the induction of the squamous morphology and cross-linked envelopes is delayed. This result indicates that certain mesenchymal factors promote growth and affect differentiation of these tracheal cells.

Vitamin A and its analogues have been shown to control differentiation of tracheal epithelial cells *in vivo* and in organ culture (Wong & Buck 1971, Sporn et al 1975, McDowell et al 1984). When rabbit tracheal epithelial cells are grown in the presence of retinoic acid, a small increase in cell growth occurs at 10^{-10} M. Concentrations higher than 10^{-7} M gradually decrease cell proliferation. Retinoic acid inhibits the formation of squamous cells and

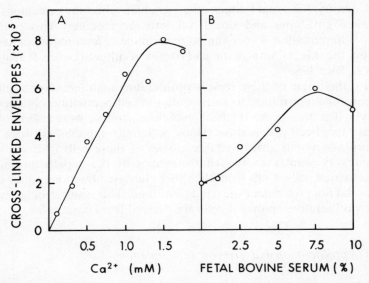

FIG. 2. Stimulation of the formation of cross-linked envelopes in rabbit tracheal epithelial cells by Ca^{2+} and fetal bovine serum. Cells were grown to subconfluence and treated with various concentrations of Ca^{2+} (A) and fetal bovine serum (B); eight days later the spontaneous formation of cross-linked envelopes was determined.

cross-linked envelopes. This inhibition occurs at concentrations of retinoic acid as low as 10^{-9} M and is optimal at about 10^{-7} M (Fig. 3). It has been shown that the formation of cross-linked envelopes depends on the synthesis of specific protein components of the cross-linked envelope and on the enzyme transglutaminase, which upon activation by calcium ions cross-links these proteins. Retinoids could prevent the formation of cross-linked envelopes by inhibiting the synthesis of either these structural proteins of the envelope or the enzyme transglutaminase, or by inhibiting the activation of this enzyme. The competence of cells to form cross-linked envelopes can be determined by treating cells with the Ca^{2+} ionophore X537A. In the presence of X537A, cells at all stages of the growth curve, even those that have been treated with retinoic acid, are induced to form cross-linked envelopes (Fig.

3). This indicates that the proteins of the cross-linked envelope and the enzyme transglutaminase are synthesized. Consistent with this result is the observation that tracheal epithelial cells contain transglutaminase activity. However, the differentiated cells contain about 20-fold higher levels of transglutaminase activity than the undifferentiated and retinoic acid-treated cells. The high levels of transglutaminase in the differentiated cells are

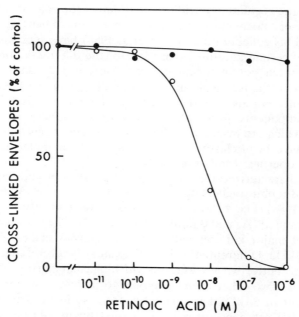

FIG. 3. Action of retinoic acid on the formation of cross-linked envelopes of rabbit tracheal epithelial cells. Cells were plated at 50 000 cells/60 mm dish and at near confluence were treated with various concentrations of retinoic acid. ○, Spontaneous formation of cross-linked envelopes after day 10; ●, formation of cross-linked envelopes after treatment with calcium ionophore X537A.

probably related to the formation of cross-linked envelopes, whereas the low levels of transglutaminase in retinoic acid-treated cells may be one of the causes of the inhibition of spontaneous cross-linked envelope formation by retinoids. However, these low levels are apparently high enough to induce cross-linked envelopes in the presence of the Ca^{2+} ionophore X537A.

The effects of retinoids on the differentiation of tracheal cells may be influenced by the substrate on which the cells are grown. On collagen gels, tracheal cells can be induced by retinoids to differentiate normally into mucin-secreting cells. Electron micrographs show that secretory granules are present and we have analysed the secretory products biochemically to confirm that mucin is produced. Without retinoids, cells grown on collagen gels become squamous and form cornified envelopes.

Biochemical markers

To examine the regulation of differentiation and the role of retinoids at the molecular level, biochemical markers are necessary. Fuchs & Green (1981) and Sun and his collaborators (Weiss et al 1984) have successfully used the expression of certain keratins as markers for the differentiation of epithelial cells of the cornea, conjunctivum and skin. Moreover, Fuchs & Green (1981) have shown that the terminal differentiation of human keratinocytes and the expression of keratins are regulated by vitamin A. Keratins form the structural components of the intermediate filaments and are a diverse group of proteins which are insoluble at high salt concentrations and in detergent, and which have been classified on the basis of their molecular weight and isoelectric properties (Moll et al 1982).

We have characterized the expression of keratins in undifferentiated and differentiated rabbit tracheal epithelial cells (H. Smiths & A. Jetten, unpublished work 1984). Fig. 4 shows the two-dimensional electrophoresis patterns of undifferentiated (A) and differentiated (B) cells. The undifferentiated cells express eight major keratins with molecular masses of 58, 56, 54, 52, 50, 48, 43 and 40 kDa. Differentiation is accompanied by several changes in keratin pattern: consistently, two new keratins with molecular masses of 48 and 54 kDa are expressed. Two other changes appear to be less consistent: an increase in the amount of 56 kDa keratin and a decrease in the amount of 52 kDa keratin. High Ca^{2+} concentrations and fetal bovine serum induce similar changes in keratin expression. Cells treated with retinoic acid exhibit a keratin pattern similar to undifferentiated cells except that a 54 kDa keratin is present (Fig. 4C). These studies (Table 1) show that the expression of the 48 kDa keratin correlates well with the squamous cell differentiation of rabbit tracheal epithelial cells in culture and suggest that this keratin might function as a marker of squamous cell differentiation. Preliminary observations have shown that the number of desmosomal junctions increases during differentiation. Morphological studies have indicated that these structures serve as nexi

FIG. 4. Expression of keratins by undifferentiated cells (A, culture day 5), differentiated cells (B, culture day 10) and cells treated with 3×10^{-7} M-retinoic acid (C, culture day 10; retinoic acid treatment from day 6). Keratins were isolated and analysed by two-dimensional electrophoresis as described in 'Materials and methods'. Note the different amounts of keratins 6 (56 kDa), 8 (52 kDa), 13 (54 kDa) and 16 (48 kDa) in A, B and C; a, actin.

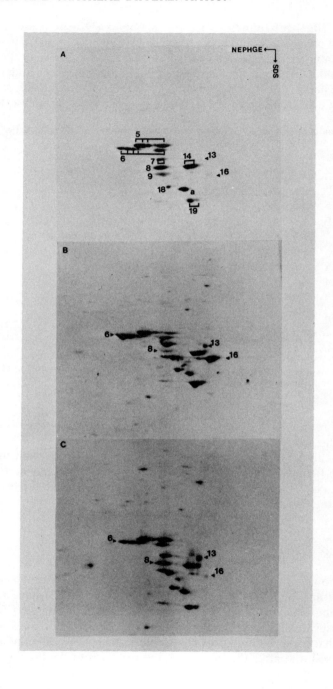

(attachment sites) for numerous keratin filaments (Drochmans et al 1978). Whether the change in keratins is related to this increase in desmosomes has yet to be established. Our results also show that the expression of the 48 kDa keratin is under the control of retinoic acid. Experiments are under way to determine whether this control occurs at the level of the mRNA synthesis. Studies by Fuchs & Green (1981) in human epidermal keratinocytes have

TABLE 1 Action of retinoic acid on keratin expression in rabbit tracheal epithelial cells

| Keratin | | Cells | | |
kDa	No.	Undifferentiated (culture day 5)	Differentiated (culture day 10)	Treated with retinoic acid[a] (culture day 10)
58	5	+	+	+
56	6	+	↑	+
54	7	+	+	+
52	8	+	↓	+
48	9	+	+	+
54	13	−	+	+
50	14	+	+	+
48	16	−	+	−
43	18	+	+	+
40	19	+	+	+
45	(actin)	+	+	+

Keratins were isolated and analysed by two-dimensional electrophoresis as described in 'Materials and methods'; +, present; −, absent; ↑, increased compared to undifferentiated cells; ↓, decreased compared to undifferentiated cells (see also Fig. 4).
[a]Cells were treated with retinoic acid (3×10^{-7} M) from day 6 on.

shown that retinol directly or indirectly affects the expression of a 67 kDa keratin at the level of the mRNA synthesis.

Another change that occurs during differentiation is a fivefold to 10-fold reduction in the synthesis of proteoglycans. This alteration may be related to the cellular change in morphology. We are examining whether other components of the extracellular matrix are affected.

Mechanism of action of retinoids

Little is known about the exact mechanism by which retinoids influence biochemical and biological functions. Chytil & Ong (1983) have identified specific retinol- and retinoic acid-binding proteins (designated CRBP and CRABP respectively) in cytosolic preparations of various tissues and have suggested that these binding proteins mediate the actions of retinoids. They have proposed that after the binding of the retinoids to the binding protein, the complex is translocated to the nucleus where it interacts with specific sites on the chromatin and induces alterations in gene transcription. This hypothesis offers an attractive mechanism to explain changes in gene expression induced by retinoids during differentiation.

The tracheal epithelial cell system seems a suitable and relevant model for the study of the mechanism of action of retinoids. We asked the following

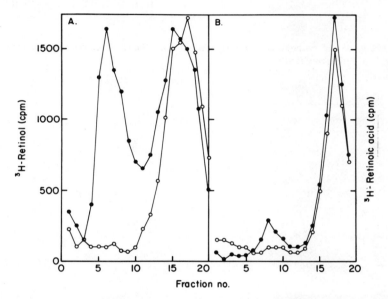

FIG. 5. Presence of retinol- and retinoic acid-binding proteins (CRBP and CRABP) in rabbit tracheal epithelial cells. Cells were grown in 150 cm^2 dishes and collected at subconfluence. Cells were homogenized and CRBP and CRABP determined by rapid gel-filtration assay, as described by Matthaei et al (1983). (A) Cytosolic fractions were incubated with [^{3}H]retinol in the presence (○) or absence (●) of unlabelled retinol. (B) Cytosolic fractions were incubated with [^{3}H]retinoic acid in the presence (○) or absence (●) of unlabelled retinoic acid.

questions. Are CRABP and CRBP present in these cells? What is the structure–function relationship of retinoids? Are retinoids directly or indirectly altering gene expression, and are the binding proteins involved? Some of these questions we have begun to address. We have prepared cytosolic fractions of tracheal epithelial cells and determined the levels of CRABP and CRBP. Fig. 5 shows the elution profiles of cytosolic fractions incubated with radiolabelled retinoic acid or retinol. The levels of retinol- and retinoic acid-binding proteins were 2.21 and 0.29 pmol/mg protein respectively, giving a CRBP:CRABP ratio of 7.6:1. These levels of binding proteins do not correlate well with the activities of retinol and retinoic acid since retinoic acid is about 50 times more active than retinol.

To investigate further the possibility that the observed binding proteins are involved in the action of retinoids on rabbit tracheal epithelial cells, the ability of various retinoids to inhibit keratinization was determined and compared with the capacity of the retinoids to bind to the binding proteins (Table 2). The retinoids that are active in inhibiting the formation of

TABLE 2 Action of retinoids on the formation of cross-linked envelopes and comparison with binding to CRBP and CRABP

Retinoid[a]	Cross-linked envelopes[b] (% of control)	Binding[c]	
		CRBP	CRABP
(structure) CH_2OH	9	+	−
(structure) CO_2H	<1	−	+
(structure) CO_2H	<1	−	+
(structure) CO_2H	61	−	−
CH_3O (structure) CO_2H	<1	−	+
(structure) CO_2H	91	−	−
(structure) CO_2H	11	−	+
(structure) CO_2H	<1	−	+
(structure)	95	−	−

[a] Retinoids were tested at 10^{-7} M.
[b] Formation by tracheal epithelial cells *in vitro*.
[c] Adapted from Jetten & Jetten (1979).
CRBP, retinol-binding protein; CRABP, retinoic acid-binding protein.

cross-linked envelopes *in vitro* also bind to either CRBP or CRABP (Jetten & Jetten 1979, Jetten 1981). The same retinoids also reverse keratinization in hamster tracheal organ culture (Newton et al 1982). The relationship between structure and function is in agreement with a role for these binding proteins in mediating the action of retinoids in these cells.

Summary

In summary, we have established an *in vitro* system that enables us to study at the molecular level the differentiation of tracheal epithelial cells into squamous cornifying cells. Mesenchymal factors, substratum and retinoids appear to be important factors in the regulation of this differentiation. Changes in intermediate filaments and proteoglycan synthesis accompany squamous cell differentiation. We now have to determine the level and mechanism by which this process is controlled.

REFERENCES

Breeze RG, Wheeldon EB 1977 The cells of the pulmonary airways. Am Rev Resp Dis 116:705-777

Chopra DP 1982 Squamous metaplasia in organ culture of vitamin A-deficient hamster trachea: cytokinetic and ultrastructural alterations. J Natl Cancer Inst 69:895-905

Chytil F, Ong DE 1983 Cellular retinol- and retinoic acid-binding proteins. Adv Nutr Res 5:13-29

Clark JN, Marchok AC 1979 The effect of vitamin A on cellular differentiation and mucous glycoprotein synthesis in long term rat tracheal organ cultures. Differentiation 14:175-183

Drochmans P, Freudenstein C, Wanson JC et al 1978 Structure and biochemical composition of desmosomes and tonofilaments isolated from calf muzzle epidermis. J Cell Biol 79:427-448

Franke WW, Weber K, Osborn M, Schmid E, Freudenstein C 1978 Antibody to prekeratin: decoration of tonofilament-like arrays in various cells of epithelial character. Exp Cell Res 116:429-445

Fuchs E, Green H 1981 Regulation of terminal differentiation of cultured human keratinocytes by vitamin A. Cell 125:617-625

Green H, Watt FM 1982 Regulation by vitamin A of envelope cross-linking in cultured human keratinocytes derived from different human epithelia. Mol Cell Biol 2:1115-1117

Jetten AM, Jetten MER 1979 Possible role of retinoic acid binding protein in retinoid stimulation of embryonal carcinoma cell differentiation. Nature (Lond) 278:180-182

Jetten AM 1981 Action of retinoids and phorbol esters on cell growth and the binding of epidermal growth factor. Ann NY Acad Sci 359:200-217

Lane B, Gordon RE 1979 Regeneration of vitamin A-deficient rat tracheal epithelium after mild mechanical injury. Differentiation 14:87-93

Lechner JF, Haugen A, McLendon I, Shamsuddin A 1984 Induction of squamous differentiation of normal human bronchial epithelial cells by small amounts of serum. Differentiation 25:229-237

Matthaei KI, McCue PA, Sherman MI 1983 Retinoid binding protein activities in murine embryonal carcinoma cells and their differentiated derivatives. Cancer Res 43:2862-2867

McDowell EM, Keenan KP, Huang M 1984 Effects of vitamin A-deprivation on hamster tracheal epithelium. Virchows Arch B Cell Pathol 45:197-219

Moll R, Franke WW, Schiller DL, Geiger B, Krepler R 1982 The catalog of human cytokeratins: patterns of expression in normal epithelia, tumors and cultured cells. Cell 31:11-24

Newton DL, Henderson WR, Sporn MB 1982 Structure–activity relationships of retinoids in hamster tracheal organ culture. Cancer Res 40:3413-3425

O'Farrell RQ, Goodman GHM, O'Farrell PH 1977 High resolution two dimensional electrophoresis of basic as well as acidic proteins. Cell 12:1133-1142

Phillips MA, Rice RH 1983 Convergent differentiation in cultured rat cells from nonkeratinized epithelia: keratinocyte character and intrinsic differences. J Cell Biol 97:686-691

Rice RH, Green H 1979 Presence in human epidermal cells of a soluble protein precursor of the cross-linked envelope: activation of the cross-linking by calcium ions. Cell 18:681-694

Sporn MB, Clamon GH, Dunlop NM, Newton DL, Smith JM, Saffiotti U 1975 Activity of vitamin A analogues in cell cultures of mouse epidermis and organ cultures of hamster trachea. Nature (Lond) 253:47-50

Weiss RA, Eichner R, Sun T-T 1984 Monoclonal antibody analysis of keratin expression in epidermal diseases: a 48- and 56-kdalton keratin as molecular markers for hyperproliferative keratinocytes. J Cell Biol 98:1397-1406

Wilhelm DL 1954 Regeneration of the tracheal epithelium in the vitamin A-deficient rat. J Pathol Bacteriol 67:361-365

Wong Y, Buck RC 1971 An electron microscopic study of metaplasia of the rat tracheal epithelium in vitamin A deficiency. Lab Invest 24:55-66

Wu R, Smith D 1982 Continuous multiplication of rabbit tracheal epithelial cells in a defined, hormone-supplemented medium. In Vitro (Rockville) 18:800-812

DISCUSSION

Hicks: Your description of the effects of retinoids on tracheal cells grown on collagen gels reminds me of the very old system described by Rous and his colleagues. If you paint a rabbit's ear with 3-methylcholanthrene, papillomas develop and, as they grow, they become keratinized. If you paint the tumours with what used to be called vitamin A they become mucus secreting, but if you stop the vitamin A treatment the keratin reappears; there appears to be a nice flip-flop system between keratin and mucin. You have now reproduced exactly the same thing in the tracheal cultures, but in a more refined system.

Peck: What do you think the collagen is doing to allow mucin formation?

Jetten: I don't know, but it's not an unprecedented observation. Haeuptle et al (1983) found that if they plated mammary cells onto collagen gel the organization of the cells was affected in some way that allowed differentiation. When the cells were plated onto plastic no differentiation occurred. The molecular mechanism of this induction is not clear.

Peck: I'm not aware of any cell culture system in which you would see mucin formation in epithelial cells without collagen gel.

Jetten: No, only in malignant cells. There are some adenocarcinoma cells that will produce mucin while growing on plastic (A.M. Jetten, unpublished work).

Sporn: Can you increase ciliation in your system?

Jetten: In the rabbit system we get very limited induction into ciliated cells but in hamster tracheal cell cultures under the same conditions many ciliated cells are formed. The problem is that we don't have a biochemical marker for these cells to monitor differentiation. With the phase-contrast microscope we can see cilia beating, but for quantitative measurements we need markers. We are trying to obtain an antibody against dynein, to quantitate the differentiation into ciliated cells.

Moon: In your system you get proliferation of basal cells that are undifferentiated, but do the cells that do differentiate also proliferate?

Jetten: No. When the cells differentiate into squamous cells they stop proliferating.

Moon: Can you stimulate the proliferation of differentiated cells in any way?

Jetten: Not as far as we know. The differentiation is not reversible. However, if we add 3T3-conditioned medium to proliferating cells they grow to a much higher density and are maintained in the undifferentiated state for a much longer time. We thought it possible that some of the transforming growth factors (TGFs) could be responsible for this inhibition of differentiation, so we are looking at the effects of TGF-α and TGF-β on cellular differentiation in our system.

Sporn: In a human bronchial epithelial system, TGF-β strongly inhibits cell proliferation.

Jetten: That's what we have found too. If we add TGF-β to rabbit tracheal cells, cell growth is inhibited.

Lotan: You said that epidermal growth factor (EGF) was one of the constituents of your culture medium; do you see increased EGF binding to your cells?

Jetten: We have not looked at that because if we omit EGF, the cells become squamous. They are then a different cell type and it becomes difficult to interpret the results.

Lotan: What does the bovine hypothalamic extract that you include in your culture medium contain?

Jetten: We have repeated our experiments in the absence of the hypothalamic extract and we get very similar results except that the density that the cells reach is higher if we use the extract. I don't know what it contains.

Yuspa: I think you should be cautious about interpreting the effect of the Ca^{2+} ionophore. You are probably using high concentrations in accordance with the method of Rice & Green (1979). At these levels the ionophore is cytotoxic and kills the cells very rapidly. It dissolves membranes and redistributes intracellular contents; it is certainly not Ca^{2+} specific. In fact Rice & Green (1979) showed that envelope cross-linking can be enhanced as effectively with Triton as with the Ca^{2+} ionophore.

Jetten: We are cautious for another reason. Simon & Green (1984) have demonstrated that in almost every cell line the Ca^{2+} ionophore induces the formation of cross-linked envelope-like structures. We are looking into this to see what kind of proteins are involved; it may be that the proteins in the cross-linked envelopes that form spontaneously are different from those in the structures that form after Ca^{2+} ionophore treatment.

Yuspa: That is certainly a possibility. Another possibility is that the calcium ionophore and detergents alter the subcellular localization of transglutaminases so that they can get to the active site for cross-linking.

Sporn: Could you speculate about how retinoids may affect the mobilization of calcium? Is this related in any way to phosphatidylinositol?

Jetten: That is one way to mobilize calcium. We are now examining the effects of retinoids on the synthesis of phosphatidylinositol phosphates.

Sporn: Is the keratin 16 that you mentioned the same keratin that Howard Green has shown to be under retinoid control?

Jetten: No. The keratins that are affected in our system are different from keratins that are affected in epidermal keratinocytes. For example, we don't see the appearance of high molecular mass class keratins, like the 67 kDa keratin. Moreover, no changes in expression of the 48 kDa keratin have been observed during epidermal differentiation. Our results indicate that keratin expression in tracheal epithelial cells is regulated in a different way from that described for epidermal cells.

REFERENCES

Haeuptle M, Suard Y, Bogenmann E, Reggio H, Racine L, Kraehenbuhl J 1983 Effects of cell shape change on the function and differentiation of rabbit mammary cells in culture. J Cell Biol 96:1425-1434

Rice RH, Green H 1979 Presence in human epidermal cells of a soluble protein precursor of the cross-linked envelope: activation of the cross-linking by calcium ions. Cell 18:681-694

Simon M, Green H 1984 Participation of membrane-associated proteins in the formation of the cross-linked envelope of the keratinocyte. Cell 36:827-834

Inhibition of epidermal terminal differentiation and tumour promotion by retinoids

ULRIKE LICHTI and STUART H. YUSPA

Laboratory of Cellular Carcinogenesis and Tumor Promotion, National Cancer Institute, Bethesda, Maryland 20205, USA

Abstract. Retinoids are physiological regulators of growth and differentiation for a number of epithelial tissues. In several of these, retinoids also act as pharmacological anti-carcinogens. Retinoids are most effective as anticarcinogens in the post-initiation portion of carcinogenesis. In mouse skin, retinoids are inhibitors of phorbol ester-mediated tumour promotion and can cause regression of pre-existing benign tumours. Studies *in vivo* and *in vitro* have indicated that phorbol ester-mediated skin tumour promotion results from selective clonal expansion of initiated cells. We have proposed that the biological basis for selection resides in the induction of terminal differentiation in subpopulations of keratinocytes while other keratinocytes, including initiated cells, are stimulated to proliferate. Terminal differentiation is accelerated by phorbol esters through the induction of epidermal transglutaminase and consequent cornification. Retinoids inhibit terminal differentiation of keratinocytes. Retinoids also induce transglutaminase in epidermis, but they inhibit cornification. Recent results suggest a biochemical basis for this paradox. The phorbol ester-induced transglutaminase is primarily particulate but the retinoid-induced enzyme is cytosolic. The induced enzymes differ in kinetic parameters, thermal stability and in elution from ion-exchange columns. Induction of the retinoid enzyme is associated with suppression of the induction of transglutaminase by phorbol esters. The retinoid-induced epidermal transglutaminase could interfere with normal or promoter-induced differentiation by inappropriately cross-linking precursor proteins before their assembly at the cell periphery. This could explain one aspect of the inhibitory action of retinoids on tumour promotion.

1985 Retinoids, differentiation and disease. Pitman, London (Ciba Foundation Symposium 113) p 77–89

Vitamin A was first shown to reduce tumour development in several stratified squamous epithelia [see references cited in Yuspa (1983) for the following brief review]. More recently, synthetic retinoids have been used to inhibit cancer development in other lining epithelia (Yuspa 1983, Moon et al 1983). In these models retinoids appear to exert most of their inhibitory effects in the phase of tumour development after initiation or carcinogen exposure. Retinoids are pharmacologically effective as tumour inhibitors in those tissues in which they exert a physiological influence on growth and differentiation.

Of all the model systems currently used to study the inhibitory effects of retinoids on carcinogenesis, mouse skin is probably the most extensively characterized and most amenable to experimental manipulation. Tumour development can be divided operationally and mechanistically into at least two stages, initiation and promotion (Yuspa 1981, 1982). Evidence from several laboratories strongly suggests that the action of retinoids is confined to the inhibition of promotion by phorbol esters; retinoids do not influence the initiation phase of skin carcinogenesis, nor are they effective against the promoting action of agents other than phorbol esters (Yuspa 1983, Gensler & Bowden 1984). Furthermore, to be effective in blocking promotion, retinoids must be applied before or immediately after the phorbol ester; retinoids do not inhibit if applied several hours after the promoter. A similar time dependence applies to the strong inhibition of phorbol ester-induced ornithine decarboxylase (ODC) activity by retinoids *in vivo* and *in vitro*. Since retinoids do not inhibit the binding of the phorbol ester to its receptor, direct interference by retinoids with phorbol ester binding is unlikely. For a series of synthetic retinoids, there is a strong association between inhibition of ODC and inhibition of tumour promotion; however, retinoids do not influence the hyperplastic response to phorbol esters. In addition to effectively preventing tumour formation in skin, retinoids are also potent at causing regression of existing tumours. Since retinoids do not significantly alter the labelling index of cells in existing tumours, their primary effect may be to diminish in number the most differentiated cells, as suggested by a markedly reduced granular cell layer and stratum corneum in retinoid-treated tumours. Epithelial necrosis and stromal enlargement are also observed and the tumour glycoprotein pattern is altered (Levin et al 1983). Retinoids do not inhibit, and may in fact enhance, the induction of benign and malignant tumours by repeated application of polycyclic aromatic hydrocarbons. Furthermore, retinoic acid itself can promote tumours if applied repeatedly after carcinogen initiation at concentrations that would inhibit tumour promotion if given together with phorbol esters (Hennings et al 1982). Taken together, the biological data indicate a specific antagonistic effect of retinoids on phorbol ester-mediated tumour promotion.

Before we can suggest how retinoids might inhibit tumour promotion, we need a working hypothesis to explain the cellular basis for promotion. We have proposed that tumour promotion by phorbol esters is best explained as a selective clonal expansion of initiated cells (Yuspa et al 1981a,b, Yuspa 1984). That cell selection is involved in promotion is strongly suggested by the monoclonal origin of papillomas, the remodelling of the epidermis during prolonged promoter exposure, and the requirement for repeated promoting stimuli before the development of tumours. Our studies on differentiation and proliferation in cultures of epidermal basal cells indicate that phorbol esters produce a balanced and programmed heterogeneous response in epidermis (Yuspa et al 1982a). Phorbol esters are potent inducers of epidermal terminal differentiation in only a subset of basal keratinocytes (Yuspa et al 1982a, 1983a, Reiners & Slaga 1983); they stimulate proliferation in others. This heterogeneity allows for selection and expansion of the proliferating population with each promoter exposure but loss of the differentiating population, so that remodelling of the target tissue occurs. Initiated cells would undergo clonal expansion if they were among the subpopulation stimulated to proliferate while normal basal cells were induced to differentiate. Indeed our results suggest that initiated cells are uniformly resistant to the induction of terminal differentiation by phorbol esters (Yuspa 1984, Yuspa et al 1983c). Therefore any agent that interferes with the ability of phorbol esters to accelerate keratinocyte terminal differentiation might also interfere with tumour promotion.

Retinoids are known to modulate squamous differentiation in keratinocytes (Table 1). Under the influence of retinoids epidermal cells may be capable of expressing properties of secretory epithelia, as indicated by morphological criteria, a marked change in glycoprotein synthesis and new synthesis of keratins characteristic of secretory tissues (Yuspa & Harris 1974, Yuspa et al 1977, Adamo et al 1979, Elias et al 1983, Steinert & Yuspa 1978, Fuchs & Green 1981, Eckert & Green 1984). Most importantly, retinoids are capable of interfering with the terminal differentiation of keratinocytes induced by phorbol esters (Yuspa et al 1981c, 1983b). Current studies of the mechanism by which retinoids act in this system are the subject of this paper.

TABLE 1 Phenotypic modifications of keratinocytes by retinoids

(1) Morphological conversion to a secretory epithelium (Yuspa & Harris 1974, Yuspa et al 1977)
(2) Increased synthesis of galactose-, mannose- and glucosamine-containing glycopeptides (Adamo et al 1979, Elias et al 1983)
(3) Increased activity of soluble transglutaminase (Yuspa et al 1980)
(4) Marked reduction in cornified envelope formation (Yuspa et al 1982b, Yaar et al 1981, Green & Watt 1982)
(5) Decreased transcription of the gene for a 67 kDa keratin (squamous differentiation) (Fuchs & Green 1981)
(6) Increased transcription of the genes for 40 kDa and 52 kDa keratins (secretory differentiation) (Eckert & Green 1984)
(7) Modulation of synthesis and subcellular localization of pemphigus and pemphigoid antigens (Thivolet et al 1984)
(8) Increased desquamation of superficial cells (Milstone et al 1982)
(9) Inhibition of terminal differentiation induced by Ca^{2+} and phorbol esters (Yuspa et al 1981c, 1983a)

Materials and methods

Cell culture methods

Epidermal cells were prepared from BALB/c new-born mice and grown in medium containing $0.05 \, mM$-Ca^{2+} (Eagle's minimum essential medium plus 8% Chelex-treated fetal calf serum) as reported by Hennings et al (1980). Growth in this 'low calcium' medium selects for the basal cell population. To induce differentiation, the concentration of Ca^{2+} was increased to $> 0.1 \, mM$ by addition of appropriate amounts of $0.3 \, M$-$CaCl_2$. All media contained 1% antibiotic–antimycotic solution (Gibco, Grand Island, NY). At appropriate times after plating, cells were exposed to the test agents [12-O-tetradecanoyl-phorbol-13-acetate (TPA), retinoids or Ca^{2+}]. At the end of the exposure interval, medium was removed and the cells were washed three times and frozen.

Biochemical assays

Transglutaminase activity was measured as described by Yuspa et al (1980) or by a modified method with increased sensitivity (Lichti et al 1985). Trans-glutaminase extracts were obtained from cell lysates after sonication, brief incubation with Triton X-100 where indicated, and centrifugation at $122\,000\,g$ for 15–20 min to remove particulate material. Cell extracts were chromato-graphed on the anion exchanger Mono-Q (HR 5/5) in the Pharmacia FPLC system (Pharmacia, Piscataway, NJ). Protein was eluted with a linear gradient of NaCl from 0.0 to $0.5 \, M$ in $10 \, mM$-Tris at pH 7.5, 0.2% Triton, $5 \, mM$-dithiothreitol and $0.5 \, mM$-EDTA (Lichti et al 1985).

Results and discussion

The discovery that extracellular ionic calcium is a key regulator of epidermal growth and differentiation has facilitated experimental studies of the response of proliferating keratinocytes to pharmacological agents. In medium with low calcium concentrations (0.02–$0.09 \, mM$), only epidermal basal cells grow, whereas cells committed to terminal differentiation detach from the culture dish. In the basal cells there are few cornified envelopes and low trans-glutaminase activity (Yuspa et al 1980, 1983a). The Ca^{2+}-dependent epidermal transglutaminase is responsible for the cross-linking of the cell envelope found in the final stages of keratinocyte terminal differentiation and is localized in the suprabasal keratinocytes (Peterson & Wuepper 1984). Both transglutaminase activity and the number of cornified cells increase during terminal differentiation induced by increasing the concentration of Ca^{2+} in the medium to > 0.1 mM (Yuspa et al 1980). In fact many aspects of differentiation *in vivo* are mimicked by the calcium-induced changes (Hennings et al 1980). One of the most striking effects of tumour promoters on basal cells *in vitro* and *in vivo* is the induction of transglutaminase activity and a consequent increase in the number of cornified cells (Yuspa et al 1980, 1983a, Reiners & Slaga 1983). We consider this programmed response to be essential for tumour promotion in epidermis since it leads to a selective loss of normal basal cells. We reasoned that any interference in this differentiation pathway could inhibit tumour promotion.

Retinoid-induced transglutaminase

Retinoids seemed likely to be antipromoting agents because of their profound effects on epidermal differentiation. However, initial studies yielded the paradoxical finding that retinoic acid, like TPA, induced transglutaminase activity (Yuspa et al 1980, 1981c). With retinoic acid, the high enzyme activity was associated with a marked reduction in cornified envelopes (Yuspa et al 1982b, Yaar et al 1981, Green & Watt 1982). As predicted, simultaneous exposure to retinoic acid and TPA protected the epidermal cell population from the terminal differentiation and cell loss that occur in response to promoter alone (Yuspa et al 1983b). When a series of retinoids were evaluated for potency as inducers of transglutaminase, the ranking paralleled their reported antipromoting potencies. Several characteristics were shared by the transglutaminase activity induced by retinoids and that induced by TPA: both required new RNA and protein synthesis, the induction of both activities was blocked by protease inhibitors (Kawamura et al 1983), and both activities had similar K_m values for the substrate putrescine at neutral pH. Yet there were differences as well. The time course for induction was quite different for each treatment. The TPA-induced enzyme activity reached a maximum by 8–10 h and then diminished as cornified cells sloughed from the culture, whereas the retinoid-induced enzyme activity increased more slowly and attained very high levels, which were maintained as long as fresh retinoid was applied every 48 h. When applied simultaneously, retinoic acid and TPA were antagonistic: their inductive effects were less than additive (Yuspa et al 1983b). Under these conditions, both the rapid transglutaminase increase due to TPA and the slower increase due to retinoic acid were reduced.

Recent studies have clarified the enigma of transglutaminase induction by retinoids (Lichti et al 1985). Subcellular fractionation of lysates from cells induced by retinoic acid, Ca^{2+} or TPA showed that more than 80% of the retinoic acid-induced activity is in the soluble compartment, whereas most of the activity of the Ca^{2+}-induced or TPA-induced cells is particulate in the absence of detergents. Further analysis indicates that enzyme activity from all three inducers has a pH optimum of 9.0–9.5, but the retinoid-induced enzyme is thermolabile at that pH. At 37 °C there is rapid decay of only the retinoid-induced activity (half-life ≈ 3 min) whereas all three activities are linear at 25 °C for 20–30 min. Redetermination of enzyme kinetics at high rather than neutral pH showed the K_m for putrescine for the retinoid-induced enzyme to be consistently half that for the other two enzymes. When detergent extracts both from cultured cells exposed to the inducers and from epidermis obtained from untreated new-born mice were fractionated by anion-exchange chromatography, two major peaks of transglutaminase activity could be resolved. One peak elutes at 0.25 M-NaCl and is common to the mouse skin extract and extracts from TPA- and Ca^{2+}-induced cells. A second peak elutes at 0.4 M-NaCl and is characteristic of retinoic acid-induced cells. Mouse epidermis treated with retinoic acid *in vivo* yields two peaks of activity at 0.25 M-NaCl and 0.4 M-NaCl. These results are summarized in Table 2.

TABLE 2 **Properties of transglutaminases induced by calcium, 12-*O*-tetradecanoylphorbol-13-acetate and retinoic acid**

	Transglutaminase	
Property	*Ca^{2+}-induced or TPA-induced*	*Retinoic acid-induced*
Subcellular distribution	Particulate	Soluble
Heat stability at pH 9.0, 37 °C	Stable	Labile
K_m (putrescine) at pH 9.0, 25 °C	0.125 mM	0.065 mM
Elution from Mono-Q anion exchanger	0.25 M-NaCl	0.40 M-NaCl
Immunological cross-reaction with tissue transglutaminase	Negative	Positive

TPA, 12-*O*-tetradecanoylphorbol-13-acetate.

The retinoic acid-induced transglutaminase differs from the normal epidermal enzyme (i.e. that detected in mouse skin extract), but the TPA- and Ca^{2+}-induced enzymes are identical with the normal epidermal transglutaminase. This suggests that the retinoic acid-induced transglutaminase is either a different enzyme or a markedly altered form of the epidermal enzyme. It probably represents the 'tissue' transglutaminase that is present in high levels in liver (for a recent review see Folk 1983). Preliminary studies with an antibody to tissue transglutaminase (Murtaugh et al 1983) showed that the activity of the retinoid-induced enzyme, but not of the TPA- or calcium-induced enzymes, can be specifically immunoprecipitated (U. Lichti & S. Yuspa, unpublished work 1984). The tissue transglutaminase appears to be a ubiquitous enzyme since it is present or can be induced in many cell types, although immunological cross-reactivity with the liver enzyme has not always been tested. No specific function is known for this enzyme.

The antagonism between retinoic acid and calcium or TPA leads us to speculate that cells synthesize either one transglutaminase or the other but not both. In cells pretreated with retinoic acid, the induction of the epidermal enzyme by TPA is almost totally suppressed and the induction by calcium is markedly reduced. Conversely, when cells are first treated with TPA or Ca^{2+} and then with retinoic acid, the appearance of the retinoic acid-induced enzyme is prevented (U. Lichti & S. Yuspa, unpublished work 1984). It appears therefore that the uncommitted epidermal cell can respond to the retinoid by an increase in transglutaminase activity. It is probable that *in vivo* the basal cells of the epidermis are responsible for the synthesis of transglutaminase in response to retinoic acid.

Modulation of differentiation and promotion by retinoids

One mechanism by which retinoids could counteract the differentiation stimulus of phorbol esters is by inhibiting the synthesis of the epidermal transglutaminase. The assembly of the cornified envelope could also be prevented if retinoids inhibited the synthesis of cornified-envelope precursors, caused cross-linking of precursors before their proper assembly at the cell periphery (through the action of the soluble retinoid-induced transglutaminase), or prevented access of calcium to the epidermal transglutaminase. Whatever the precise mechanism, the inhibition of cornification and the consequent cell loss would eliminate a critical element in the tissue response to tumour promoters. In a tissue such as epidermis, the clonal expansion of a specific cell type requires the compensatory loss of other cells. When the cell loss results from programmed accelerated maturation rather than direct cytotoxicity, specific pharmacological inhibitors would be effective antipromoters. The opposing influences of phorbol esters and retinoids on the maturation programme of keratinocytes could account for the specific antagonistic action of retinoids on phorbol ester-mediated tumour promotion.

We do not suggest that the induction of a unique transglutaminase is the sole response to retinoids responsible for the modulation of differentiation. The suppression by retinoids of the synthesis of the 67 kDa keratin (Steinert & Yuspa 1978, Fuchs & Green 1981), a protein which is highly correlated with squamous differentiation, is also an indication of the major role retinoids play in keratinocyte differentiation, as is the transcriptional activation of 40 kDa and 52 kDa keratins (Fuchs & Green 1981, Eckert & Green 1984) commonly seen in secretory but not squamous epithelia. Retinoids produce other significant changes in gene expression and cellular metabolism (Table 1) that have dramatic phenotypic effects on keratinocytes. This extensive reprogramming of a committed cell type by a single agent can be best explained through an action on a second messenger system. The lessons learned in studies with phorbol esters, whose specific binding to and activation of protein kinase C lead to reprogramming in a wide variety of cell types (Blumberg et al 1984), may be applicable to the retinoid problem. It seems likely that the retinoids act similarly, by activating a second messenger through a specific binding site. The elucidation of such a mechanism of action would provide major insights into retinoid action and the control of cellular differentiation.

Acknowledgements

The authors are grateful to Theresa Ben for technical assistance and to Maxine Bellman for typographical work. The antiserum to tissue transglutaminase was kindly provided by Dr Peter Davies, Department of Pharmacology, University of Texas Medical School, Houston, Texas, USA.

REFERENCES

Adamo S, De Luca L, Silverman-Jones C, Yuspa SH 1979 Mode of action of retinol: involvement in glycosylation reactions in cultured mouse epidermal cells. J Biol Chem 254:3279-3287

Blumberg PM, Jaken S, Konig B et al 1984 Mechanism of action of the phorbol ester tumor promoters: specific receptors for lipophilic ligands. Biochem Pharmacol 33:933-940

Elias PM, Chung JC, Topete RO, Nemanic MK 1983 Membrane glycoconjugate visualization and biosynthesis in normal and retinoid-treated epidermis. J Invest Dermatol 81:81s-85s

Eckert RL, Green H 1984 Cloning of cDNAs specifying vitamin A-responsive human keratins. Proc Natl Acad Sci USA 81:4321-4325

Folk JE 1983 Mechanism and basis for specificity of transglutaminase-catalyzed ε(γ-glutamyl)lysine bond formation. Adv Enzymol Relat Areas Mol Biol 48:1-56

Fuchs E, Green H 1981 Regulation of terminal differentiation of cultured human keratinocytes by vitamin A. Cell 25:617-625

Gensler H, Bowden GT 1984 Influence of 13-cis-retinoic acid on mouse skin tumor initiation and promotion. Cancer Lett 22:71-75

Green H, Watt FM 1982 Regulation by vitamin A of envelope cross-linking in cultured keratinocytes derived from different human epithelia. Mol Cell Biol 2:1115-1117

Hennings H, Michael D, Cheng C, Steinert P, Holbrook K, Yuspa SH 1980 Calcium regulation of growth and differentiation of mouse epidermal cells in culture. Cell 19:245-254

Hennings H, Wenk M, Donahoe R 1982 Retinoic acid promotion of papilloma formation in mouse skin. Cancer Lett 16:1-5

Kawamura H, Strickland JE, Yuspa SH 1983 Inhibition of 12-O-tetradecanoylphorbol-13-acetate induction of epidermal transglutaminase activity by protease inhibitors. Cancer Res 43:4073-4077

Levin LV, Clark JN, Quill HR, Newberne PM, Wolf G 1983 Effect of retinoic acid on the synthesis of glycoproteins of mouse skin tumors during progression from promoted skin through papillomas to carcinomas. Cancer Res 43:1724-1732

Lichti U, Ben T, Yuspa SH 1985 Retinoic acid induced transglutaminase in mouse epidermal cells is distinct from epidermal transglutaminase. J Biol Chem 260:1422-1426

Milstone LM, McGuire J, LaVigne JF 1982 Retinoic acid causes premature desquamation of cells from confluent cultures of stratified squamous epithelia. J Invest Dermatol 79:253-260

Moon RC, McCormick DL, Mehta RG 1983 Inhibition of carcinogenesis by retinoids. Cancer Res 43:2469s-2475s

Murtaugh MP, Mehta K, Johnson J, Myers M, Juliano RL, Davies PJA 1983 Induction of tissue transglutaminase in mouse peritoneal macrophages. J Biol Chem 258:11074-11081

Peterson LL, Wuepper KD 1984 Epidermal and hair follicle transglutaminases and crosslinking in skin. Mol Cell Biochem 58:99-111

Reiners JJ, Slaga TJ 1983 Effects of tumor promoters on the rate and commitment to terminal differentiation of subpopulations of murine keratinocytes. Cell 32:247-255

Steinert P, Yuspa SH 1978 Biochemical evidence for keratinization by mouse epidermal cells in culture. Science (Wash DC) 200:1491-1493

Thivolet CH, Hintner HH, Stanley JR 1984 The effect of retinoic acid on the expression of pemphigus and pemphigoid antigens in cultured human keratinocytes. J Invest Dermatol 82:329-334

Yaar M, Stanley JR, Katz SI 1981 Retinoic acid delays terminal differentiation of keratinocytes in / suspension culture. J Invest Dermatol 76:363-366

Yuspa SH 1981, 1982 Chemical carcinogenesis related to the skin. Prog Dermatol 15:1-10, 16:1-10

Yuspa SH 1983 Retinoids and tumor promotion. In: Roe D (ed) Diet, nutrition and cancer: from basic research to policy implications. Alan R Liss, New York, p 95-109

Yuspa SH 1984 Molecular and cellular basis for tumor promotion in mouse skin. In: Fujiki et al (eds) Cellular interactions of environmental tumor promoters. Jap Sci Soc Press, Tokyo, p 315-326

Yuspa SH, Harris CC 1974 Altered differentiation of mouse epidermal cells treated with retinyl acetate in vitro. Exp Cell Res 86:95-105

Yuspa SH, Elgjo K, Morse MA, Wiebel FJ 1977 Retinyl acetate modulation of cell growth kinetics and carcinogen–cellular interaction in mouse epidermal cell cultures. Chem Biol Interact 16:251-264

Yuspa SH, Ben T, Hennings H, Lichti U 1980 Phorbol ester tumor promoters induce epidermal transglutaminase activity. Biochem Biophys Res Commun 97:700-708

Yuspa SH, Hennings H, Kulesz-Martin M, Lichti U 1981a The study of tumor promotion in a cell culture model for mouse skin, a tissue which exhibits multistage carcinogenesis in vivo. In: Hecker et al (eds) Cocarcinogenesis and biological effects of tumor promoters. Raven Press, New York, p 217-230

Yuspa SH, Hennings H, Lichti U 1981b Initiator and promoter induced specific changes in epidermal function and biological potential. J Supramol Struct Cell Biochem 17:245-257

Yuspa SH, Lichti U, Ben T, Hennings H 1981c Modulation of terminal differentiation and tumor promotion by retinoids in mouse epidermal cell cultures. Ann NY Acad Sci 359:260-274

Yuspa SH, Ben T, Hennings H, Lichti U 1982a Divergent responses in epidermal basal cells exposed to the tumor promoter 12-O-tetradecanoylphorbol-13-acetate. Cancer Res 42:2344-2349

Yuspa SH, Ben T, Steinert P 1982b Retinoic acid induces transglutaminase activity but inhibits cornification of cultured epidermal cells. J Biol Chem 257:9906-9908

Yuspa SH, Ben T, Hennings H 1983a The induction of epidermal transglutaminase and terminal differentiation by tumor promoters in cultured epidermal cells. Carcinogenesis (Lond) 4:1413-1418

Yuspa SH, Ben T, Lichti U 1983b The regulation of epidermal transglutaminase activity and terminal differentiation by retinoids and phorbol esters. Cancer Res 43:5707-5712

Yuspa SH, Kulesz-Martin M, Ben T, Hennings H 1983c Transformation of epidermal cells in culture. J Invest Dermatol 81:162s-168s

DISCUSSION

King: Do you know what happens to the particulate transglutaminase after retinoid treatment?

Yuspa: We don't know what happens to the protein itself; we can only measure activity.

King: So you can't distinguish between the cytosolic transglutaminase induced by retinoids and the enzyme that would normally be there?

Yuspa: No. We know only that a particular fraction does not have activity after retinoid treatment, but we cannot determine whether or not the phorbol ester-inducible enzyme is still there. It is possible that A23187, Triton and other agents stimulate envelope cross-linking by the redistribution of the transglutaminase. The soluble transglutaminase induced in keratinocytes, and perhaps also that induced in tracheal cells, could then cross-link a precursor at a membrane site that the enzymes would normally be restricted from reaching. Ulrike Lichti (unpublished work) has looked to see whether the retinoid-induced enzyme might be a modified normal transglutaminase rather than a different protein, but antibody experiments and kinetic studies suggest that it is a distinct enzyme. There is no evidence for phosphorylation or for sugar modifications of the transglutaminase induced by retinoids.

Lotan: We have looked at two human sarcoma cell lines; in one of them retinoids had no effect at all on transglutaminase activity but in the other there was a twofold increase in activity after treatment with retinoic acid (Lotan et al 1984). The reason we looked at this was that we found changes in the mobility of membrane proteins after retinoic acid treatment and we thought that an increase in transglutaminase activity might explain the restriction on the reorganization of membrane components.

Breitman: It has also been found that retinoic acid induces a cytosolic transglutaminase in HL-60 cells (Davies et al 1984).

Hartmann: You said that in your system the transglutaminase induced by 12-*O*-tetradecanoylphorbol-13-acetate (TPA) appeared very fast but that the retinoic acid-induced enzyme took about 30 hours to appear. Does this mean that the TPA-induced enzyme would produce cornification?

Yuspa: The time course of events in culture does not exactly follow the time course *in vivo*; within 24 hours of retinoic acid exposure *in vivo* the retinoid enzyme is induced. If you give retinoic acid and TPA together to cells in culture, the peak activity of the TPA enzyme is markedly reduced. So there is some antagonism that occurs very rapidly between the two inducers despite the fact that maximum induction takes longer for the retinoid enzyme. This is important to consider when one is trying to work out the mechanism of antipromotion because the antipromoting effects of retinoic acid *in vivo* show a very strong time dependence. Something must happen rather quickly after retinoid administration, and our observations on transglutaminase activity are consistent with this time requirement.

Bertram: Since retinoic acid is presumably involved in the normal differentiation of the keratinocyte, and since you say there is no soluble transglutaminase found in normal cells, can we deduce that the induction by retinoic acid of the soluble enzyme is a pharmacological effect? Or is it simply that your methods don't allow detection of the enzyme in normal cells?

Yuspa: Detection problems are certainly a possibility. However, retinoic acid may be involved in the control of differentiation in the skin in a negative way; it may be that only in the absence of retinoic acid can the normal epidermal transglutaminase be expressed. The permissive environment for the expression of this enzyme (which is not found in tissues other than skin) perhaps reflects a normal state of relative vitamin A deficiency in epidermis. When pharmacological doses of retinoids are administered the expression of the vitamin A-deficient phenotype would be altered.

Bertram: But if there is this balance between soluble and membrane-bound enzyme, one might expect to see soluble enzyme in basal cells because in these cells differentiation is inhibited.

Yuspa: It's possible, but it has been very difficult to look at that possibility *in vivo* because most transglutaminase activity is in the differentiating cells and they are the most numerous in the skin. One would be trying to assay a minor population of cells. In fact ion-exchange chromatography shows that there is a little peak of soluble enzyme activity in normal skin *in vivo*, so I suspect that in basal cells there is some expression of the soluble enzyme. We have received antibody to the soluble enzyme from Peter Davies (University of Texas), so we will be able to study this question by immunofluorescence. I hope we will find some basal cells *in vivo* that are expressing that soluble transglutaminase antigen.

Strickland: Is there any chance that the transglutaminase could be related to the metabolism of the retinoid?

Yuspa: Yes, it's possible that the retinoid could be a substrate for transglutaminase. The enzyme normally catalyses an acyl transfer reaction. In the presence of calcium it takes protein-bound glutamine and a primary amine like lysine and simply transfers a carboxyl group. The only reason for thinking that retinoic acid could be a substrate is that one of the metabolites of retinoic acid is retinyl taurine which could be a substrate for a transglutaminase reaction.

Sporn: But that's just a degradative product. It is formed in the end stages of the degradation of retinoic acid, after hydroxylation.

Yuspa: But transglutaminase could well be involved in that kind of degradation of retinoic acid. We know that retinoic acid is broken down very rapidly in keratinocytes in culture: within about 20 hours most of it is gone.

Sporn: How does TPA break DNA in keratinocytes?

Yuspa: We have no evidence that TPA breaks DNA. DNA breaks are a consequence of epidermal terminal differentiation; TPA does not break DNA directly, at least in keratinocytes, but it accelerates epidermal terminal differentiation. During the normal process of epidermal differentiation there is nuclear degradation, and DNA breakage may simply be caused by released nucleases. Retinoids do not directly prevent the breakage of DNA; they will not, for example, prevent the breakage of DNA by benzoyl peroxide. However, they prevent terminal differentiation and therefore you don't see the breakage of DNA that usually accompanies differentiation.

Strickland: Can you be sure that you see breakage of DNA and not unscheduled synthesis? Varshavsky (1981) has reported induction of unscheduled DNA synthesis by TPA. I don't think your assay would distinguish between the two possibilities.

Yuspa: The time course would be very unusual for the induction of unscheduled synthesis; that usually happens quite rapidly after something damages DNA directly. We don't see any evidence of DNA breakage for 24 hours after TPA exposure, until the cells are fully terminally differentiated, so I think unscheduled synthesis is unlikely.

Lotan: How do culture conditions influence differentiation? Anton Jetten has used Ham's F12 medium, which has a normal calcium concentration, for his rabbit tracheal epithelial cells and he gets almost exclusively basal-type cells (Jetten & Smits, this volume). You have used reduced calcium concentrations for keratinocytes and find that the cells differentiate. Could you comment on the possibility that calcium has different roles in the differentiation of these two cell types?

Yuspa: I can only guess. In human bronchial epithelium cultures differentiation is regulated by calcium: in defined medium the cells will undergo squamous differentiation to a limited extent if the calcium concentration is changed from low to high. In that model, there is a marked synergistic effect of calcium and serum—apparently there are permissive factors or differentiation factors in serum which work in concert with calcium to cause extensive terminal differentiation (Lechner et al 1984). Perhaps in epidermis calcium alone is sufficient; we can get the same response to calcium in serum-free medium as we get in serum-containing medium. However, in bronchial epithelium and trachea, factors in addition to calcium are needed to regulate differentiation. Have you ever cultured your cells in low calcium conditions in the presence of serum, Dr Jetten, to see if they will differentiate?

Jetten: Yes, we have used 0.1 mM-Ca^{2+} with calcium-poor serum and under these circumstances keratinization is still promoted. I think that certain serum factors other than Ca^{2+} influence differentiation in these cells.

REFERENCES

Davies PJA, Murtaugh MP, Moore WT, Lucas D 1984 Retinoic acid induced expression of tissue transglutaminase in human promyelocytic leukemia (HL-60) cells. J Biol Chem, in press

Jetten AM, Smits H 1985 Regulation of differentiation of tracheal epithelial cells by retinoids. In: Retinoids, differentiation and disease. Pitman, London (Ciba Found Symp 113) p 61-76

Lotan R, Meromsky L, Marikovsky Y 1984 Prevention by retinoic acid of anionic site redistribution on the surface of cultured human sarcoma cells. Biol Cell 51:147-156

Lechner JF, Haugen A, McClendon IA, Shamsuddin AM 1984 Induction of squamous differentiation of normal human bronchial epithelial cells by small amounts of serum. Differentiation 25:229-237

Varshavsky A 1981 On the possibility of metabolic control of replicon 'misfiring': relationship to emergence of malignant phenotypes in mammalian cell lineages. Proc Natl Acad Sci USA 78:3673-3677

General discussion I

The retinoid 'receptor'

Malkovský: Retinoids have been around for many years, but there is still no system in which we can even suggest a mechanism of action or identify the target structure. On the one hand this is disappointing, but on the other hand it is encouraging because the major breakthroughs are to come. Retinoids stimulate some systems but they inhibit others. Some people try to compare them with hormones or phorbol esters and suggest that there is a specific structure, a kind of receptor that mediates their actions. But is it conceivable that in fact there is no such structure, no receptor, and that retinoids simply bind, on the basis of their hydrophobicity or other physico-chemical properties, to various proteins? Cell function might or might not be influenced; this would depend on the function of the particular protein.

Sporn: I think that's somewhat unlikely because these agents work at very low concentrations. In the tracheal organ culture system, *all-trans*-retinoic acid is active at 10^{-11} M, and some of the retinoidal benzoic acid derivatives are active at 10^{-12} M. We don't know of any other substances that will produce the same effects as retinoids. Their chemical specificity is extremely high: Leonard Schiff has shown that although TTNPB [(E)-4-[2-(5,6,7,8-tetrahydro-5,5,8,8-tetramethyl-2-naphthalenyl)-1-propenyl]benzoic acid], which has a carboxyl group in the *para* position on the benzene ring, is highly active at 10^{-12} M, you lose all biological activity if you move the carboxyl group to the *meta* position (unpublished work). That does not fit in with the notion that retinoids do something wherever they go, but suggests that there is some specific 'receptor'. We just don't know what it is yet.

It is true that retinoids have a number of different effects: some things go up, some things go down, some grow, some stop growing. There are analogies with steroids, in particular with glucocorticoids, which have had a tremendous influence as practical agents in clinical medicine. I'm not saying that the receptor types or the mechanisms of action are the same, but steroids, like retinoids, have pleiotropic effects. They cause enzyme induction, may induce or block growth and have permissive interactions with all sorts of other growth-controlling substances. Although the discovery of retinoids and the elucidation of their chemical structures goes a long way back, we are a generation behind the people working on steroids in terms of sophisticated molecular biology.

There is a lot to be done to explain all the paradoxes of retinoids, and I think they will turn out to be just as complicated as the steroids.

Malkovský: So you believe that there is a retinoid receptor?

Sporn: Yes. There may not be just one; there may be a set of them.

Sherman: As far as we know there is no surface receptor for steroids, and it is now becoming clear that the steroid receptors may never get out of the nucleus (King & Greene 1984, Welshons et al 1984). So the pleiotropic effects of steroids might all be elicited by a nuclear action on various genes. Steroids might not need to interact with surface receptors or with growth factor receptors to have a profound influence on the growth properties of the cell. The situation may be no different for retinoids.

Sporn: One of the major questions is: how do retinoids, by themselves or with some sort of regulatory protein, interact with the genome? We must consider the regulation of the genome, enhancer elements and so on. Some of these questions are beginning to be answered for the steroids, so, without pushing the analogy between retinoids and steroids too far, I think that all of us who are interested in retinoid mechanisms should certainly keep an eye on what is happening in steroid research.

Lotan: The idea that retinoic acid might bind to various proteins by virtue of its hydrophobic properties is not supported by the observations of those looking for cytoplasmic binding proteins. They find only the 2 S cytoplasmic cellular retinoic acid-binding protein, which may or may not be functional; they do not find anything else. On the other hand, the 2 S binding protein cannot account for all the retinoic acid that is incorporated into cells. In a melanoma cell there are perhaps 50000 molecules of the retinoic acid-binding protein, but millions of molecules of retinoic acid can be taken up by a cell. There are obviously many molecules that are not bound to the binding protein, but what are they bound to? They are definitely in the membrane fraction but what are they doing there? Can all the activity of the retinoid be accounted for by the 40000–50000 molecules bound to the binding protein? Do the rest of the molecules just sit there inactive? These are open questions.

We have some data that indicate that retinoids may cause physico-chemical changes in the properties of the membrane, different from those that are seen with only very high doses. For example, the mobility of negatively charged molecules on the surface of sarcoma and melanoma cells is inhibited. This is not an immediate, detergent-like effect. The retinoids just prevent the clustering of the negatively charged molecules (Lotan et al 1984). This should be investigated, because several hormones work by binding to and clustering receptors. One could suggest that retinoids increase the sialylation of certain glycoproteins, as a result of which the mobility of some of the charged molecules in the plane of the membrane is decreased. Some of these may be hormone receptors, so retinoids could have an indirect effect on the response of cells to other factors.

Malkovský: My point is that perhaps the 'receptor' is not a unique structure. There may be a set of receptors; there could be 10 up to 200 or so.

Sporn: It seems reasonable to assume that there is a defined receptor or set of receptors. We don't know what the receptor is but the opportunity now exists to use the techniques of modern molecular biology to investigate this. Using a highly stable retinoidal benzoic acid derivative with an N_3 group or some other appropriately reactive group attached, we should be able to get affinity labelling of the true receptor. We should be able to fish that receptor out and clone it. Whether we will find it in a vitamin A-deficient quail embryo or in some other system in which retinoids are required for a biological response, I don't know.

Breitman: We should not forget that the receptor may be an enzyme.

Sporn: Yes, it could be an enzyme like C-kinase. There may be many parallels with the phorbol ester system. Now that the phorbol ester receptor has been identified as a C-kinase, we have a much better idea of the biochemical mechanism of action of phorbol esters. The original ligand used in the tumour promotion work was 12-*O*-tetradecanoylphorbol-13-acetate (TPA), but while this was used in binding studies little progress was made. Labelled TPA binds to everything, to all sorts of membrane structures. However, once the synthetic ligand phorbol dibutyrate was introduced, progress was much more rapid, and the phorbol ester receptor was identified as a protein kinase.

There must be some kind of 'receptor' for retinoids. I use the word 'receptor' in the most general way. It could be any sort of specific cellular target: it could be a membrane-bound receptor, a cytosolic protein or a nuclear structure. It does not have to be one molecule; it may be two or three. But the fact that organic chemists have made 300 active retinoids does not mean that there are 300 different types of receptor. There may be a very small number of alternative chemical configurations that these compounds can adopt to bind to whatever triggers the biological response. We don't know what the natural ligand for the receptor is. I don't think that many people believe that retinol is really the active substance; it is more likely to be retinoic acid or one of its metabolites. If we label one of these natural ligands we see binding to all sorts of things including the so-called 'binding proteins', but many of us don't believe in the binding protein story as it has evolved. I don't think that these proteins are the true receptors. They may be transport systems, shuttle systems or they may represent a mechanism by which the cell protects itself against accumulation of undesirably high levels of retinoic acid, which is a very toxic substance.

We are waiting for chemists to make us a suitable ligand for identifying the receptor. It may not be TTNPB but it will be something like it that can be made in radiolabelled form with a very high specific activity, perhaps 50–100 Ci/mmol. Once the appropriate molecule is made, we should then be able to get affinity labelling of the receptor, to sequence it and to find out what its real function is. The problem now is one of methodology: as long as we use retinol, retinoic acid and its metabolites we are going to make little progress. In spite of all the research on the cell biology of retinoids and the development of very elegant new systems, for example for culturing tracheal epithelial cells, we still don't have a means of getting at the mechanism of action of retinoids. We need a new ligand like the phorbol derivative, phorbol dibutyrate, which is the key to the great progress that has been made on phorbol mechanisms.

Yuspa: The most important contribution that has come from phorbol ester research has been an understanding of the vast importance of protein kinase C. For some years we knew of the many effects of phorbol esters but had absolutely no idea how they worked. When it became clear that these effects were mediated by activation of protein kinase C, the many activities of this kinase were displayed. I suspect the same thing will happen with retinoids.

Sherman: There are two reasons why the retinoid story might be different from the TPA story. First, we have two binding proteins that do have high affinities, one for retinoic acid and the other for retinol. Those affinities are respectable and comparable to the affinities of steroid receptor proteins for their ligands. Some of us would like to think that the retinoid-binding proteins work at the gene level although nobody has good evidence for that. However, the fact that we have not identified how these binding proteins work does not mean that their function is trivial or unrelated to the mechanism of action of retinoids. Second, I can understand your enthusiasm for having a probe that will bind specifically to a 'true' retinoid receptor, but when we add retinol or retinoic acid to cellular extracts, we get binding to the 2S proteins; if another binding protein exists that has a similar ligand affinity for retinoids, why have we failed to detect it? It is not as though we are looking into a morass of proteins that bind non-specifically to a ligand, as may have been the case with TPA. We already have specific binding proteins for the retinoids; if there were another one, the 'true' receptor, why has it not shown itself?

Jetten: If there is a receptor in the membrane, how would you detect it? Retinoids are very hydrophobic and you might only measure non-specific binding of retinoids to the membrane. It may therefore be that we are not able to detect specific binding in the membrane, so I think the existence of a cell surface receptor cannot be ruled out. In this respect we may have similar problems to those reported for TPA binding.

Sherman: If you look at every fraction of the cell as you fractionate it and you don't find specific binding, other than to something already recognized as the 2S retinoid-binding proteins, then perhaps the elusive receptor doesn't exist.

Sporn: But one can get a lot of binding of retinoic acid to cell membranes. Some of that may be low affinity but there may be a very small but biologically significant contribution from membrane proteins that are functional retinoid receptors, which cannot be identified for the same reasons that we could not identify phorbol receptors with TPA.

Yuspa: There is an important difference between phorbol esters and retinoids with regard to receptors. Every cell that responds to phorbol esters has the receptor; there are no exceptions. Not every cell that has a phorbol ester receptor responds to the ligand, but there may be reasons for that downstream from the receptor. However, in a number of cells that respond to retinoids, the retinoid-binding proteins do not exist or cannot be measured.

Sporn: Like the HL-60 system.

Sherman: Strickland et al (1983) have identified a retinoid analogue that has poor activity with HL-60 cells but has a 100-fold lower ED_{50} with F9 embryonal carcinoma cells; conversely, a closely related analogue has a 10-fold lower ED_{50} in HL-60 cells than in F9 cells. Marcia Dawson has cited the case of a retinoid that has activity in the ornithine decarboxylase assay but does not have activity in the keratinization assay in tracheal organ culture (see p 26). The fact that retinoids have different potencies in different test systems makes one wonder whether it is necessarily true that the *same* mechanisms must operate in different cell types. Retinoids may alter cell behaviour at several different levels. They may achieve similar results in two systems, but by different mechanisms, so the binding proteins need not exist in every retinoid-sensitive cell type.

Jetten: We have looked at the binding of retinoic acid in the HL-60 system, and have found that in these cells the lipid content is very high. The non-specific binding is also extremely high and it may not be possible to detect specific binding to receptors above this level.

Koeffler: The concentration of *all-trans*-retinoic acid needed to induce differentiation of HL-60 cells is very high: $10^{-6}-10^{-8}$ M, suggesting that the compound perhaps does not induce differentiation via a receptor. Most receptor–ligand interactions occur at $10^{-9}-10^{-11}$ M. For example, $1\alpha,25$-dihydroxycholecalciferol induces about 50% of HL-60 cells to differentiate at 3×10^{-9} M, and the K_d for the receptor–ligand is almost at the same level (Mangelsdorf et al 1984).

Strickland: To get differentiation of embryonal carcinoma cells in the presence of serum you do need high concentrations of retinoids, but only because most of the retinoid is bound to serum proteins. If you do the experiment in the absence of serum, the dose–response curve shifts by about two log units down to a lower concentration range (Rizzino & Crowley 1980), so at least in these cells the retinoids do work at about 10^{-10} M.

Sporn: And the ED_{50} of retinoic acid in the tracheal organ culture system is $10^{-10}-10^{-11}$ M.

Breitman: There is now much evidence from experiments *in vitro* that retinoic acid does not necessarily act alone. In the HL-60 system, a combination of a physiological concentration of retinoic acid (10 nM) with either prostaglandin E (at physiological concentrations) or gamma interferon gives full induction of differentiation (Breitman & Keene 1982, H. Hemmi & T.R. Breitman, unpublished work). A whole series of compounds can act in concert with retinoic acid, but retinoic acid must be there, although at a very low concentration. It is a mistake to think on the basis of experiments *in vitro* that retinoic acid acts alone *in vivo*. I believe that *in vivo* retinoic acid acts in combination with other mediators, perhaps even with vitamin D_3.

Lotan: Schroder et al (1980) have shown that pretreatment of cultured murine Swiss 3T3 cells with retinoic acid makes the cells competent to respond to other mitogens, so its effects are in line with what you have shown in HL-60.

Breitman: Strickland et al (1980) showed a similar effect in F9 cells. They primed the cells with retinoic acid and then treated them with a cyclic AMP-inducing agent in the absence of retinoic acid. There was a marked induction of differentiation. This induction was achieved in two stages; the obligatory order in sequential treatment was retinoic acid followed by cyclic AMP. Olsson and I have reported similar results with HL-60 (Olsson et al 1982).

Lotan: This kind of complementation raises questions about the retinoid-binding proteins. You could say that in HL-60 you do not need the binding protein because retinoic acid acts epigenetically to allow the cells to respond to something that works at the gene level. When we are considering the nature of retinoid 'receptors' it is worth remembering the physiology of vitamin A. It has not been established that there is a membrane receptor for retinol uptake, although such an entity has been proposed by Chen & Heller (1977). However, we know that retinoids can partition into the membrane, so one could easily imagine that retinoic acid and its analogues just bind non-specifically to hydrophobic pockets of some membrane component that either exerts epigenetic effects or reaches the nucleus by membrane flow. It would be interesting to see whether HL-60 cells have any binding sites for retinoid–protein complexes or even for the retinoid-binding proteins.

Bertram: We must not forget the physiology of retinol delivery to the cell. It is delivered attached to a binding protein that presumably recognizes receptors on the cell and then transfers the retinol. I don't know whether we can assume that the moment retinol is transferred it becomes distributed widely throughout the cell. It may be compartmentalized; it may be distributed very narrowly within the cell. Retinol may be converted to retinoic acid at discrete sites, and perhaps these discrete sites contain the receptors. One of the problems of our approach in looking for specific receptors is that we simply dump in free retinoid. That is not what the cell normally sees, so it is perhaps not surprising that we get masses of non-specific binding.

Sporn: That's all the more reason for getting a ligand that is active at tiny concentrations and has a very high specific activity.

Bertram: Perhaps we should deliver that ligand bound to the serum binding protein.

Hicks: Is this relevant to retinoic acid? You are talking about retinol, but retinol is less active than retinoic acid and retinoic acid is not transported in serum on a binding protein.

Sherman: You can artificially put retinoic acid on a plasma retinol-binding protein and then deliver it to the cell in an *in vitro* system.

Sporn: But it appears that, physiologically, the retinoic acid within a cell is generated *in situ* from retinol.

REFERENCES

Breitman TR, Keene BR 1982 Growth and differentiation of human promyelocytic cell line HL-60 in a defined medium. Cold Spring Harbor Conf Cell Proliferation 9:691-702

Chen CC, Heller J 1977 Uptake of retinol and retinoic acid from serum retinol-binding protein by retinal pigment epithelial cells. J Biol Chem 252:5216-5221

King WJ, Greene GL 1984 Monoclonal antibodies localize oestrogen receptor in the nuclei of target cells. Nature (Lond) 307:745-747

Lotan R, Meromsky L, Marikovsky Y 1984 Prevention by retinoic acid of anionic site redistribution on the surface of cultured human sarcoma cells. Biol Cell 51:147-156

Mangelsdorf DJ, Koeffler HP, Donaldson CA, Pike JW, Haussler MR 1984 1,25-Dihydroxyvitamin D_3-induced differentiation in a human promyelocytic leukemia cell line (HL-60): receptor-mediated maturation to macrophage-like cells. J Cell Biol 98:391-398

Olsson IL, Breitman TR, Gallo RC 1982 Priming of human myeloid leukemic cell lines HL-60 and U-937 with retinoic acid for differentiation effects of cyclic adenosine 3′:5′-monophosphate-inducing agents and a T-lymphocyte-derived differentiation factor. Cancer Res 42:3928-3933

Rizzino A, Crowley C 1980 Growth and differentiation of embryonal carcinoma cell line F9 in defined media. Proc Natl Acad Sci USA 77:457-461

Schroder EW, Chou IN, Black PH 1980 Effects of retinoic acid on plasminogen activator and mitogenic responses of cultured mouse cells. Cancer Res 40:3089-3094

Strickland S, Smith KK, Marotti KR 1980 Hormonal induction of differentiation in teratocarcinoma stem cells: generation of parietal endoderm by retinoic acid and dibutyryl cAMP. Cell 21:347-355

Strickland S, Breitman TR, Frickel F, Nürrenbach A, Hädicke E, Sporn MB 1983 Structure-activity relationships of a new series of retinoidal benzoic acid derivatives as measured by induction of differentiation of murine F9 teratocarcinoma cells and human HL-60 promyelocytic leukemia cells. Cancer Res 43:5268-5272

Welshons WV, Lieberman ME, Gorski J 1984 Nuclear localization of unoccupied oestrogen receptors. Nature (Lond) 307:747-749

Effects of arotinoid ethyl ester on epithelial differentiation and proliferation

D. TSAMBAOS*, R. STADLER, K. HILT*, B. ZIMMERMANN† and C.E. ORFANOS

Department of Dermatology, University Medical Center Steglitz, Berlin and †Institute of Toxicology and Embryonic Pharmacology, The Free University of Berlin, Federal Republic of Germany

Abstract. In recent years the search for new retinoids, safer and more potent than the available compounds, has led to the development of arotinoids. In preliminary clinical trials, arotinoid ethyl ester [(E)-4-[2-(5,6,7,8-tetrahydro-5,5,8,8-tetramethyl-2-naphthalenyl)-1-propenyl]benzoic acid ethyl ester; TTNPB ethyl ester] was found to be highly effective in the treatment of severe and etretinate-resistant dermatoses. Using adult hairless mice, embryonic mouse limb-bud cultures and keratinocyte cultures as experimental models, we have performed morphological, autoradiographic and biochemical studies on the effects of arotinoid ethyl ester on epithelial differentiation *in vivo* and *in vitro*. Arotinoid ethyl ester stimulates proliferation of both embryonic and adult mouse epidermis. However, it inhibits the differentiation of embryonic epidermis and enhances that of adult epidermis. Arotinoid ethyl ester induces a decrease in the cyclic AMP content of cholera toxin-stimulated keratinocytes but fails to alter cyclic AMP concentrations in keratinocytes not treated with cholera toxin.

1985 Retinoids, differentiation and disease. Pitman, London (Ciba Foundation Symposium 113) p 97–116.

The successful clinical application of oral isotretinoin and etretinate in the management of severe cutaneous disorders has introduced a new era in systemic dermatotherapy. In recent years the search for new retinoids, safer and more potent than the available compounds, has led to the development of arotinoids (Loeliger et al 1980). One of the most active retinoids of this group is the arotinoid ethyl ester [(E)-4-[2-(5,6,7,8-tetrahydro-5,5,8,8-tetramethyl-2-naphthalenyl)-1-propenyl]benzoic acid ethyl ester; TTNPB ethyl ester]; minute oral doses of this arotinoid have been found to exert favourable

---

* *Present address: Department of Dermatology, University of Essen, Hufeland Straße 55, D-4300 Essen 1, Federal Republic of Germany*

therapeutic effects on chemically induced skin papillomas and carcinomas of mice (Bollag 1981). In this paper we present the results of experimental studies on the effects of arotinoid ethyl ester on: (1) the differentiation and proliferation of embryonic and adult mouse epidermis and (2) the cyclic AMP-stimulated proliferation and intracellular cyclic AMP content of neonatal mouse keratinocytes (Stadler et al 1984, Tsambaos 1984).

Materials and methods

Embryonic mouse epidermis

To study the effects of arotinoid ethyl ester on the differentiation and proliferation of embryonic mouse epidermis we used limb-bud cultures of 11-day-old mouse embryos as an experimental model. In this highly repro-ducible system epidermogenesis *in vitro* proceeds in a well-organized fashion, simulating the corresponding processes *in vivo* (Schultz-Ehrenburg 1975). Forelimb buds were dissected from 11-day-old NMRI mouse embryos (day $0 =$ day of conception) and were grown for six days in an organ culture as described by Zimmermann (1978). The culture medium consisted of 75% Bigger's medium BLC 127 supplemented with 25% fetal calf serum, 0.85 mM-ascorbic acid, 1.3 mM-glutamic acid, 50 IU/ml penicillin and 50 IU/ml streptomycin (IU = international unit). The cultures were incubated at 37 °C in a water-saturated 5% CO_2-air atmosphere.

Arotinoid ethyl ester was dissolved in 0.1% dimethylsulphoxide (DMSO) and added to the medium to give final concentrations of 0.1, 1.0. 10.0 and 15.0 ng/ml. Corresponding amounts of 0.1% DMSO were added to the medium of the control cultures. Limb buds were exposed to arotinoid on days 1–3. Some limb buds were then processed for electron microscopy. Others were transferred to the arotinoid-free control medium and cultivated for three additional days and were then processed for electron microscopy and autoradiography ($[^3H]$thymidine labelling) by standard techniques.

Adult mouse epidermis

One hundred and ninety-five adult female hairless Ng/− mice (mean weight 33 g) divided into 13 groups of 15 animals were treated by an oesophageal tube with 0.1, 0.05, 0.01 and 0.004 mg/kg per day arotinoid ethyl ester (dissolved in arachis oil) for a period of one, two, three and four weeks. Sixty control animals divided into four equal groups received 0.5 ml/day arachis oil for one, two, three and four weeks respectively. All animals were kept at

21 °C and a relative humidity of 55% and were allowed free access to food and water. On completion of treatment, 10 mice from each arotinoid-treated and control group were killed. Biopsies from back skin were processed for light and electron microscopy (three animals) and autoradiography ([³H]thymidine labelling; seven animals) by standard techniques. The remaining five mice in each group were killed four weeks after the end of treatment. Skin biopsies were then processed for light and electron microscopy (two animals) and autoradiography (three animals).

Neonatal mouse keratinocytes

Cultures of neonatal mouse keratinocytes were prepared as described by Marcelo et al (1978). One day after plating, the following drugs were added to quadruplicate sets of Petri dishes: (1) 0.2% DMSO; (2) 0.5 mM-8-bromo cyclic AMP; (3) 50 μg cholera toxin plus 50 μM-1-methyl-3-isobutylxanthine (MIX); (4) 0.5 μg/ml arotinoid ethyl ester in 0.2% DMSO; (5) 0.5 mM-8-bromo cyclic AMP plus 0.5 μg/ml arotinoid ethyl ester in 0.2% DMSO; (6) 50 μg cholera toxin plus 50 μM-MIX plus 0.5 μg/ml arotinoid ethyl ester in 0.2% DMSO (3 h later). The growth medium was changed every two days. The concentration of cyclic AMP in cells exposed to drugs for six days was determined as described by Marcelo (1979). After six or 14 days' continuous exposure to drugs, keratinocytes were pulse-labelled with [³H]thymidine and proliferation was estimated as described by Marcelo et al (1978).

Embryonic mouse epidermis

Control cultures

After three days in the control medium, the embryonic epidermis (except for the apical epidermal ridge which is not considered here) consisted of three to five cell layers organized in three strata. The cuboidal cells of the stratum basale had round or oval nuclei and contained several mitochondria, small amounts of glycogen and a few cisternae of rough endoplasmic reticulum. Desmosomes and hemidesmosomes were rare, free tonofilaments could not be detected and most intercellular spaces were wide. The cells of the stratum intermedium were polygonal or slightly flattened with their long axes parallel to the surface of the epidermis; they revealed more immature desmosomes than the basal cells and were covered by extremely flattened, electron-dense peridermal cells, which contained spindle-shaped nuclei, several mitochondria, Golgi membranes and many membrane-bound vesicles.

After six days in the control medium the embryonic mouse epidermis revealed the morphological features of a normally keratinizing squamous epithelium and consisted of 8–12 cell layers organized in four clearly defined strata. The cells of the stratum basale were cuboidal or cylindrical and their nuclei were oriented with the long axis perpendicular to the continuous basal lamina, which revealed frequent hemidesmosomes. Several mitochondria, considerable amounts of glycogen and some profiles of rough endoplasmic reticulum were visible. Small numbers of desmosomes and free tonofilaments were seen; the intercellular spaces were wide. In the lower stratum spinosum the epithelial cells were polygonal, whereas in the stratum spinosum they were flattened. In contrast to the basal cells, the cells of the stratum spinosum revealed an advanced stage of differentiation and contained large amounts of free tonofilaments, keratinosomes and several keratohyaline granules. In the narrow intercellular spaces interdigitating cell processes could be observed, held together by desmosomes, which were more numerous than in the stratum basale. Occasionally gap junctions were seen. The cells of the stratum granulosum were markedly flattened with their long axes oriented parallel to the surface of the epidermis. They contained large numbers of free tonofilaments, keratinosomes and keratohyaline granules. The intercellular spaces were narrow and numerous desmosomes were seen connecting neighbouring cells. The stratum corneum consisted of four to six flattened corneocytes, which showed thickened plasma membranes and were mostly loosely packed. Occasionally, keratohyaline granules and vacuoles were present in some corneocytes; in some places intercellular desmosomal discs could be recognized. Most corneocytes contained uniformly dense filaments in an arrangement that closely resembled a normal keratin pattern.

Tritium-labelled nuclei were found in basal and occasionally in suprabasal cells. The labelling index of the embryonic mouse epidermis after six days *in vitro* was 5.6% ± 1.9% (x ± SD).

Arotinoid-treated cultures

0.1 ng/ml. In limb buds treated with this concentration over the first three days and processed for electron microscopy either immediately or after a further three-day cultivation in arotinoid-free control medium, the epidermis revealed no significant alterations compared to the controls. After six days *in vitro* (three-day exposure to 1.0 ng/ml arotinoid ethyl ester and three days in the control medium), the labelling index of the epidermis was 5.2% ± 2.4%.

1.0 ng/ml. In limb buds that were investigated immediately after three days' treatment with this concentration, the embryonic epidermis was thicker than in controls and consisted of five to nine cell layers. In some places the basal lamina revealed gaps and reduplications. In the lower epidermal layers there was a marked condensation of cells and an extensive formation of gap junctions. Hemidesmosomes and immature desmosomes were only rarely

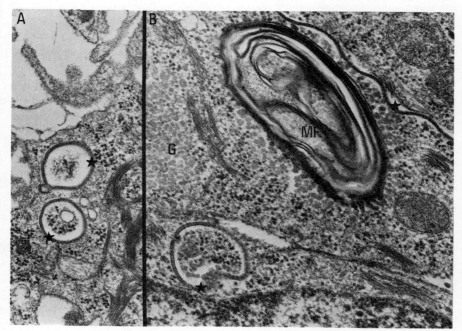

FIG. 1. Gap junctions (⋆) in mouse embryonic epidermis after three days' exposure to arotinoid ethyl ester (0.1 ng/ml). G, glycogen; ML, myelin figure. A, × 67 230; B, × 87 150.

seen, whereas in most cells increased numbers of Golgi membranes, long and straight microtubules and centrioles could be observed. In some places, fusion of plasma membranes was detected and coiled membrane figures were visible near the cell surface. In limb buds cultivated for three additional days in control medium after three days' exposure to this arotinoid concentration the thickening of the epidermis was a prominent feature. Lying on a basal lamina that showed gaps and reduplications, the densely packed epidermal cells exhibited very low amounts of free tonofilaments, desmosomes and hemidesmosomes but large numbers of gap junctions (Fig. 1). Only rarely could keratohyaline granules and keratinosomes be detected. In most specimens cornification was completely absent. The focal fusion of plasma membranes was prominent. Coiled membrane figures were often seen in the intercellular spaces or intracellularly. After six days *in vitro* (three days' exposure to 1.0 ng/ml arotinoid and three days in the control medium) the labelling index of the embryonic epidermis was 11.7% ± 3.1%.

10.0 ng/ml. In limb buds which were exposed to this arotinoid concentration for three days and were investigated either immediately or after a further three-day culture in the control medium, the thickened embryonic epidermis revealed no evidence of differentiation. Through the gaps in the basal lamina, cytoplasmic projections of mesenchymal cells and collagen fibrils were found in close apposition to the cells of the stratum basale. The epithelial cells closely resembled the underlying mesenchymal cells. Tonofilaments, desmosomes, keratinosomes and keratohyaline granules were completely absent. Cell contacts of the gap junction type were markedly increased. Pronounced fusion of the plasma membranes in some places made it very difficult to define the limits of the cellular surface of adjacent epithelial cells. Occurrence of coiled membrane figures was a prominent feature. Cornification was completely absent.

After six days *in vitro* (three days' exposure to 10.0 ng/ml arotinoid and three days in the control medium) the embryonic epidermis revealed a labelling index of 14.3% ± 4.5%.

15.0 ng/ml. In limb buds exposed to this arotinoid concentration for three days with or without a subsequent three-day cultivation in the control medium the embryonic epidermis showed massive necrotic changes which made the evaluation of arotinoid effects impossible.

Adult mouse epidermis

Control animals

During the four-week administration of arachis oil, the light and electron microscopic features of the epidermis revealed no alterations compared to the untreated animals and will not be described here. As shown in Fig. 2 the labelling index of the epidermis of the control animals also did not significantly change during the four weeks of treatment.

Arotinoid-treated animals

0.004 mg/kg per day. After one week of treatment with this dose the adult mouse epidermis revealed an enhanced desquamation and occasionally a focal thickening of the stratum granulosum. In particular the basal and suprabasal keratinocytes showed slightly enlarged nuclei with prominent nucleoli, increased numbers of ribosomes and proliferation of the endoplasmic reticulum. In most specimens an enhanced formation of gap junctions could be observed. In the subsequent weeks of arotinoid administration all these alterations became more pronounced. In addition, hypergranulosis and an increase in the numbers of tonofilaments and desmosomes were seen. Changes in the basal lamina and dermal vessels could not be detected. There was no evidence of an inflammatory infiltrate.

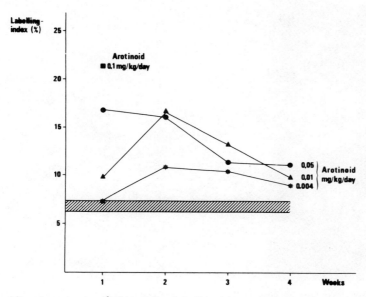

FIG. 2. Alterations in the [³H]thymidine labelling index (median values) of adult mouse epidermis after oral administration of arotinoid ethyl ester for four weeks. Hatched area: animals treated with arachis oil only.

0.01 mg/kg per day. After one week of treatment the epidermal alterations induced by this arotinoid dose closely resembled those seen in animals treated with 0.004 mg/kg per day. However, after the second week slight oedematous and vacuolar changes could be seen in the lower epidermis. No intracellular or intercellular deposits of PAS-positive material were found. In the upper stratum spinosum and in the stratum granulosum the numbers of keratinosomes were increased. After the third and fourth weeks of arotinoid administration the oedematous and vacuolar changes were clearly reduced or absent. The numbers of tonofilaments, desmosomes and gap junctions were increased; hypergranulosis and acanthosis were the most prominent

epidermal features. Apart from a slight dilatation of dermal vessels in the first
two weeks of treatment, no alterations were found in the dermis at this dose.

0.05 mg/kg per day. After the first weeks of arotinoid administration marked
acanthosis, desquamation and parakeratosis were seen in the mouse

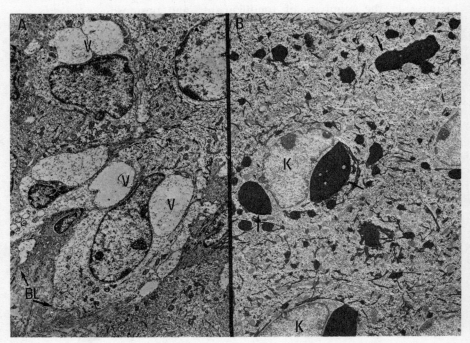

FIG. 3. (A) Oedematous and vacuolar (V) changes in adult mouse epidermis after oral
administration of arotinoid ethyl ester (0.05 mg/kg per day) for two weeks. BL, basal lamina;
× 5670. (B) Hypergranulosis in adult mouse epidermis after oral administration of arotinoid
ethyl ester (0.05 mg/kg per day) for four weeks. Arrows point to keratohyaline granules. K,
keratinocytes; × 6560.

epidermis. In the enlarged nuclei of keratinocytes the size and the numbers of
nucleoli were increased. Oedematous and vacuolar changes were observed in
the cytoplasm of most keratinocytes (Fig. 3). The number of gap junctions
was increased, whereas that of tonofilaments and desmosomes seemed to be
reduced; most tonofilaments were retracted at the periphery of cells. In the
widened intercellular spaces of the stratum basale and stratum spinosum, a
fine granular, PAS-negative material (diameter of granules: 8–18 nm) was
seen. Under the parakeratotic horny layer there was a marked increase in the
numbers of keratohyaline granules and keratinosomes. Most dermal capillaries were dilated and engorged with erythrocytes. Occasionally, several

lymphocytes and mononuclear cells invaded the epidermis through the gaps in the basal lamina. Though these alterations were more intense after the second week, in the subsequent two weeks of arotinoid administration the oedematous and vacuolar changes of keratinocytes became much less pronounced. Acanthosis, desquamation and particularly hypergranulosis were the most prominent features of the epidermis (Fig. 3). The numbers of tonofilaments, desmosomes and gap junctions were clearly increased, whereas those of keratinosomes showed a tendency to normalization. Vascular changes were only rarely seen; no inflammatory infiltrate could be detected.

0.1 mg/kg per day. Because severe side-effects occurred at 0.1 mg/kg within the first two days of application, arotinoid was administered to the animals for one week only. The epidermis of the treated mice revealed a marked acanthosis, desquamation and parakeratosis. The enlarged nuclei of keratinocytes showed an abnormal chromatin distribution and irregular configurations and contained prominent nucleoli. Massive oedematous and vacuolar changes were seen in the cytoplasm of most epidermal cells. The numbers of intact tonofilaments and desmosomes were greatly reduced. Particularly in the widened intercellular spaces of the stratum basale and stratum spinosum, but also intracellularly, varying amounts of a fine granular PAS-negative material intermingled with rests of cell organelles could be seen. The upper stratum spinosum and the markedly thickened stratum granulosum revealed an excessive increase in the numbers of keratinosomes and intercellular lamellar bodies and a thickening of their plasma membranes. Numerous lipid droplets and rests of nuclei or other cell components were found in the horny layer, which was sometimes completely absent. Most capillaries were markedly dilated; occasionally, small numbers of lymphocytes and mononuclear cells, together with some eosinophils, were found in the perivascular area and in some instances they invaded the epidermis.

The alterations in the epidermal [³H]thymidine labelling index (median values) during the four-week administration of arotinoid are shown in Fig. 2. All other arotinoid doses apart from the 0.004 mg/kg dose led to a rise in the labelling index which was statistically significant at 0.05 and 0.1 mg/kg per day. After two weeks of treatment the labelling index was increased in all arotinoid-treated groups; this increase was statistically significant at 0.01 and 0.05 mg/kg per day. In the subsequent two weeks of arotinoid administration the increase in the labelling index of mouse epidermis became less pronounced; it remained, however, statistically significant at 0.01 and 0.05 mg/kg per day arotinoid.

In skin biopsies obtained four weeks after the end of arotinoid administration no significant alterations in the morphology and labelling index could be found in the epidermis of the treated animals.

Neonatal mouse keratinocytes

Incubation of keratinocytes with cholera toxin and MIX led to an irreversible stimulation of adenylate cyclase activity and to a marked increase in the intracellular cyclic AMP levels (Table 1). Arotinoid ethyl ester (0.5 µg/ml)

TABLE 1 Effects of arotinoid ethyl ester, cholera toxin and 1-methyl-3-isobutylxanthine on intracellular cyclic AMP concentrations in keratinocytes from neonatal mice

	Cyclic AMP activity as % of DMSO control	
Treatment	*Per µg DNA*	*Per mg protein*
Arotinoid ethyl ester	94.7	125.0
Cholera toxin + MIX	576.9	501.3
Arotinoid ethyl ester + cholera toxin + MIX	290.4	305.6

Keratinocytes were exposed to drugs (0.5 µg arotinoid ethyl ester/ml; 50 µg cholera toxin; 50 µM-MIX) for six days *in vitro*. Dimethylsulphoxide (DMSO; 0.2%) was used as a control. MIX, 1-methyl-3-isobutylxanthine. For details see 'Materials and methods'.

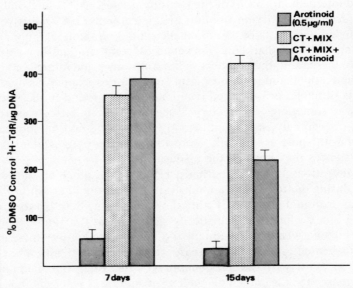

FIG. 4. Effects of arotinoid ethyl ester (0.5 µg/ml) on the *in vitro* proliferation of neonatal mouse keratinocytes stimulated with cholera toxin (CT) and 1-methyl-3-isobutylxanthine (MIX). Keratinocytes were exposed to the drugs for six or 14 days before being labelled with [³H]thymidine. Proliferation was measured as [³H]thymidine (TdR) incorporation into DNA and is expressed as a percentage of the value for keratinocytes treated with dimethylsulphoxide (DMSO) only. Bars show SD.

alone failed to alter the levels of intracellular cyclic AMP in seven-day keratinocyte cultures; nevertheless, it inhibited the increase in cyclic AMP induced by cholera toxin by about 50%. As shown in Figs. 4 & 5 arotinoid ethyl ester (0.5 μg/ml) significantly decreased the proliferation rate of neonatal mouse keratinocytes and inhibited the stimulation of keratinocyte proliferation produced by cholera toxin plus MIX and by 8-bromo cyclic AMP.

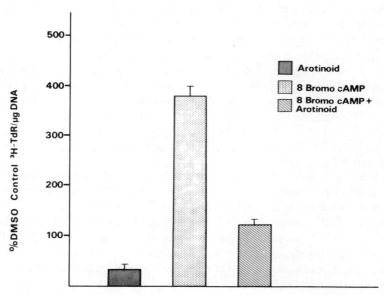

FIG. 5. Effect of arotinoid ethyl ester (0.5 μg/ml) on the 8-bromo cyclic AMP-stimulated proliferation of neonatal mouse keratinocytes. Cells were exposed to the drugs for six days *in vitro* before being labelled with [³H]thymidine. Proliferation was measured as [³H]thymidine (TdR) incorporation into DNA and is expressed as a percentage of the value for keratinocytes treated with dimethylsulphoxide (DMSO) only. Bars show SD.

Discussion and conclusions

Embryonic mouse epidermis

Our results show that the arotinoid ethyl ester induced profound changes in the differentiation and proliferation of mouse embryonic epidermis *in vitro*. A three-day treatment of the mouse limb buds with the arotinoid resulted in a dose-dependent inhibition of epidermal differentiation, which persisted after incubation of the limb buds in the control medium for three additional days.

Skin from the upper lip of 13.3-day-old mouse embryos treated in organ culture with excess vitamin A revealed an inhibition of keratinization and mucus production (mucous metaplasia) (Hardy 1967, Hardy & Bellows 1978). In our material we were unable to find any ultrastructural evidence for mucus production by or in the epithelial cells, even in limb buds treated with the highest concentration of arotinoid. It seems reasonable, therefore, to suggest that the embryonic mouse epidermis has distinct site-dependent variations in its susceptibility to the action of retinoids; these variations may be due to differences between various skin regions with respect to genetic information, density of retinoid receptors and influence of adjacent tissues. Since in our experiments only 11-day-old mouse embryos were cultivated, we cannot exclude the possibility that mucus production may also appear in the epidermis of the mouse limb bud if it is exposed to arotinoid ethyl ester at an earlier or later stage of embryonic development. Evidence in favour of this possibility comes from New (1963), who could demonstrate mucous metaplasia of rat tongue in 16-day-old embryos treated with vitamin A but not in 20-day-old embryos, which responded only with an inhibition of their epidermal differentiation.

Gaps in the basal lamina have been previously reported in organ cultures of embryonic mouse and chick skin treated with vitamin A and retinoic acid (Hardy 1974, Peck et al 1977, Hardy et al 1978). Since this phenomenon preceded all other vitamin A-induced changes in the embryonic epithelium, Hardy et al (1978) suggested that the basal lamina may be primarily affected by this compound. Its defects could be related to the mucous metaplasia of the epidermis, since they may have allowed 'inappropriate' dermal signals to reach the epidermal cells and modulate their differentiation. This hypothesis was supported by the results of another study (Hardy et al 1973), which revealed that areas of hair follicle walls with intact basal lamina maintained their normal structure in an organ culture of mouse skin in the presence of excess vitamin A, whereas areas with vitamin A-induced breakdown of basal lamina showed metaplastic changes of the epithelial cells. In the present study we could detect no differences in the expression of arotinoid effects on embryonic mouse epidermis between areas with basal lamina breakdown and those with intact basal lamina. On the other hand, inhibition of differentiation can be observed also in isolated epidermis after treatment with retinoids *in vitro* (unpublished work). It seems likely, therefore, that epithelial–mesenchymal interactions through the gaps in the basal lamina are not of primary importance for the retinoid action on embryonic epithelium. Moreover, since the basal lamina is thought to be produced by the basal epithelial cells (Banerjee et al 1977), its breakdown under retinoids may represent the consequence rather than the cause of alterations in the epithelial cells.

As shown in this study, fusion of adjacent plasma membranes, an increase in the numbers of gap junctions and coiled membrane figures are prominent features of the arotinoid-treated embryonic epidermis, probably reflecting changes in the structure and turnover of epithelial membranes. It is unclear whether these membrane changes are of primary importance for the mechanisms of arotinoid action or simply represent secondary events.

The number of autoradiographic studies of embryonic epidermis is surprisingly small. Wessels (1963) reported that in 11- and 12-day-old chick embryos epidermal cells of all layers were labelled with [³H]thymidine, whereas in 13-day-old embryos only the basal cells were labelled. In 18-week-old human embryos (Gerstein 1971) uptake of [³H]thymidine was found only in basal cells, which revealed a mean labelling index of 5.2%. Herken & Schultz-Ehrenburg (1981) investigated the cell kinetics of embryonic mouse epidermis in limb-bud cultures. They found that in the first three days *in vitro* epithelial cells of all layers were labelled with [³H]thymidine. On the fourth day *in vitro* a switching from horizontal to vertical proliferation occurred and two days later the epidermis of the embryonic mouse limb bud revealed the typical proliferation pattern of adult epidermis. To our knowledge, no autoradiographic investigations have been performed on the effects of retinoids on the proliferation of embryonic epidermis. In the present study it was shown that a three-day treatment of limb-bud cultures with arotinoid ethyl ester resulted in a dose-dependent increase of labelling index of the embryonic mouse epidermis which did not reverse within a subsequent three-day cultivation of the limb buds in the arotinoid-free control medium. A much longer incubation of the limb buds in the medium may have been necessary for normalization of the epidermal proliferation rate. This possibility is supported by our recent observation that the effects of arotinoid ethyl ester on epidermal proliferation in 11-day-old mouse embryos *in vitro* are completely reversible. On the other hand, at this or some earlier stage of embryonic development, a three-day treatment of mouse limb buds with arotinoid ethyl ester results in an inhibition of epidermal differentiation that seems to be irreversible. In conclusion, our results show that the arotinoid ethyl ester is capable of inducing a dose-dependent inhibition of differentiation and a dose-dependent stimulation of proliferation of the embryonic mouse epidermis *in vitro*. The mouse limb-bud organ culture seems to be a useful *in vitro* model for screening retinoid effects on functional and structural parameters of embryonic epithelium.

Adult mouse epidermis

We have shown that oral arotinoid ethyl ester induces significant alterations in the morphology and proliferation of the adult mouse epidermis; these effects are dose- and time-dependent, and reversible. These alterations could be classified into two developmental phases. During the first phase (first and second week of drug administration), the epidermis of the treated animals was characterized by acanthosis, desquamation, parakeratosis, increased numbers of keratohyaline granules and keratinosomes, oedema, vacuolization and accumulation of a PAS-negative fine granular material. The numbers of tonofilaments and desmosomes seemed to be reduced but this reduction may be related to the intense cytoplasmic disintegration. Under the lower arotinoid doses all these alterations were either much less pronounced or completely absent. Finally, a markedly enhanced formation of gap junctions could be observed. The epidermal labelling index revealed a dose-dependent increase and achieved its maximum values after one or two weeks of treatment. During the second phase (third and fourth week of drug administration), acanthosis, desquamation and parakeratosis were still present but became less pronounced. Hypergranulosis was the most prominent feature of the treated epidermis. The numbers of keratinosomes, the extent of oedematous and vacuolar changes and the amounts of PAS-negative fine granular material were clearly reduced compared to the first phase; on the other hand, the numbers of desmosomes, tonofilaments and gap junctions were increased. Also, the rise in the labelling index was less pronounced.

Labilization of lysosomal membranes is regarded as an important aspect of the biological action of vitamin A and retinoic acid (Lazarus et al 1975, Jarrett et al 1978). Increased lysosomal enzyme activities were found in normal human and animal epidermis after topical application of these compounds (Braun-Falco & Christophers 1969, Jarrett 1980) and in human epidermis after oral administration of etretinate (Marks et al 1984), suggesting that epidermal lysosomes may represent common targets for the action of etretinate and other synthetic retinoids. The oedematous and vacuolar changes of keratinocytes observed in the present study at the higher arotinoid doses could therefore be due to an arotinoid-induced release of lysosomal proteases and hydrolases. Also, the PAS-negative granular material may be a product of intracellular cytotoxic processes mediated by lysosomal enzymes. Even the marked desquamation of mouse epidermis observed under arotinoid may be related to an enhanced lysosomal activity. Indeed, the numbers of keratinosomes, which are regarded as specialized lysosomes and participate in the keratinization processes (Gonzalez et al 1976), revealed a marked increase in the first phase of the experiment, possibly resulting in a reduced cohesion of corneocytes.

The pathogenic mechanisms underlying the biphasic response of adult mouse epidermis to the arotinoid ethyl ester remain unknown. A reduced uptake or enhanced metabolism of the arotinoid in or by the keratinocytes during the third and fourth week of drug administration seems unlikely, since other non-cytotoxic phenomena due to the arotinoid action occurred in this period. An arotinoid-induced alteration of lysosomal membranes, leading to a reduced susceptibility to the drug and to a decline in lysosomal enzyme release during the second phase, may be one of the most plausible hypotheses for explaining the occurrence of morphological alterations in the adult mouse epidermis in two phases. Finally, the question remains to be answered of whether the biphasic response of epidermal proliferation may also be related to labilization of lysosomal membranes and release of proteolytic and hydrolytic enzymes, as reported by Lazarus & Hatcher (1975) using vitamin A. It is thought that gap junctions play an important role in intercellular communication and growth control in normal and transformed epithelia (Weinstein et al 1976, McNutt 1977). However, it is not yet known whether the increase in the numbers of gap junctions observed at all arotinoid doses in this study is of primary importance for the action of this compound on epidermal proliferation, or simply represents a secondary phenomenon.

One of the most important features of mouse epidermis after arotinoid administration was the increase in the number of keratinization markers, such as tonofilaments, desmosomes and keratohyaline granules, which possibly reflects an arotinoid-induced enhancement of epidermal differentiation. This possibility is supported by the results of a recent collaborative study with the Ann Arbor group on the effects of arotinoid on keratinocyte protein fractions of adult mouse epidermis (Grekin et al 1983). Arotinoid ethyl ester was found to enhance the formation of keratohyaline macroaggregates and covalently cross-linked fibrous proteins, but it induced a significant decrease in the fraction of non-cross-linked fibrous proteins.

Neonatal mouse keratinocytes

Retinoids have been reported to affect cyclic AMP levels (Aso et al 1976) and to increase cyclic AMP-dependent protein kinase phosphorylation in various cellular systems (Ludwig et al 1980, Plet & Crain 1981). However, Grekin et al (1982) were unable to detect any effects of different synthetic retinoids on cyclic AMP-dependent protein kinase activity and protein phosphorylation in keratinocytes. This observation is compatible with our finding that arotinoid ethyl ester failed to stimulate the cyclic AMP content of neonatal mouse keratinocytes grown in the presence of the drug for six days. Since cyclic AMP stimulates keratinocyte proliferation (Green 1978, Marcelo 1979) and retinoids, on the other hand, inhibit the proliferation of keratinocytes (Madison et al 1981), it seems unlikely that stimulation of cyclic AMP is implicated in the mechanism of retinoid action. Treatment of our keratinocyte cultures with arotinoid ethyl ester alone did not affect the intracellular concentrations of cyclic AMP. However, when keratinocytes were exposed to arotinoid after stimulation with cholera toxin plus MIX their cyclic AMP content and proliferation rate both decreased significantly. To test the possibility that arotinoid ethyl ester inhibited keratinocyte proliferation at a step other than cyclic AMP synthesis, we added 8-bromo cyclic AMP to the medium of keratinocyte cultures which were then exposed to arotinoid. Arotinoid ethyl ester was able to inhibit the stimulation of keratinocyte proliferation by 8-bromo cyclic AMP, which intracellularly acts as cyclic AMP. Therefore, although arotinoid ethyl ester cannot directly affect the keratinocyte cyclic AMP level *in vitro*, it may modify its response to various pharmacological agents and alter keratinocyte functions in which cyclic AMP is involved.

REFERENCES

Aso K, Rabinowitz J, Farber EM 1976 The role of prostaglandin E, cyclic AMP and cyclic GMP in the proliferation of guinea-pig ear skin stimulated by topical application of vitamin A acid. J Invest Dermatol 67:231-234

Banerjee SD, Cohen RH, Bernfield MR 1977 Basal lamina of embryonic salivary epithelia. Production by the epithelium and role in maintaining lobular morphology. J Cell Biol 73:445-463

Bollag W 1981 Arotinoids. A new class of retinoids with activities in oncology and dermatology. Cancer Chemother Pharmacol 7:27-29

Braun-Falco O, Christophers E 1969 Psoriasiforme Epidermis-Reaktion der Meerschweinchenhaut durch örtliche Vitamin A-Säure-Applikation. Arch Klin Exp Dermatol 234:70-86

Gerstein W 1971 Cell proliferation in human fetal epidermis. J Invest Dermatol 57:262-265

Gonzalez LF, Krawczyk WS, Wilgram GF 1976 Ultrastructural observations on the enzymatic activity of keratinosomes. J Ultrastruct Res 55:203-223

Green H 1978 Cyclic AMP in relation to proliferation of the epidermal cell: a new view. Cell 15:801-811

Grekin R, Duell EA, Voorhees JJ 1982 Effects of retinoic acid analogs on epidermal cell proliferation, protein phosphorylation and protein kinase activity. Clin Res 30:587

Grekin R, Tsambaos D, Duell EA, Voorhees JJ 1983 Effects of the retinoid Ro 13-6298 on the epidermal differentiation in mice. J Invest Dermatol 80:356

Hardy HH 1967 Responses in embryonic mouse skin to excess vitamin A in organotypic cultures from the trunk, upper lip and lower jaw. Exp Cell Res 46:367-384

Hardy HH 1974 Epithelial–mesenchymal interactions in vitro altered by vitamin A and some implications. In Vitro (Rockville) 10:358

Hardy HH, Bellows CG 1978 The stability of vitamin A-induced metaplasia of mouse vibrissa follicles in vitro. J Invest Dermatol 71:236-241

Hardy HH, Sonstergard KS, Sweeny PR 1973 Light and electron microscopic studies of the reprogramming of epidermis and vibrissa follicles by excess vitamin A in organ culture. In Vitro (Rockville) 8:405

Hardy HH, Sweeny PR, Bellows CG 1978 The effects of vitamin A on the epidermis of the fetal mouse in organ culture—an ultrastructural study. J Ultrastruct Res 64:246-260

Herken R, Schultz-Ehrenburg U 1981 Autoradiographic investigations on the cell kinetics of epidermis and periderm of limb buds from mouse embryos in vitro. Br J Dermatol 104:277-284

Jarrett A 1980 The action of vitamin A on adult epidermis and dermis. In: Jarrett A (ed) The physiology and pathophysiology of the skin. Academic Press, London, vol 6:2059-2091

Jarrett A, Wrench R, Mahmoud BL 1978 Effects of retinyl acetate on epidermal proliferation and differentiation. Induced enzyme reactions in the epidermis. Clin Exp Dermatol 3:173-181

Lazarus GS, Hatcher VB 1975 Lysosomes and the skin. In: Dingle JT, Dean RT (eds) Lysosomes in biology and pathology. North-Holland Publishing Co, Amsterdam, p 111

Lazarus GS, Hatcher VB, Levine N 1975 Lysosomes and the skin. J Invest Dermatol 65:259-271

Lugwig WK, Lowry B, Niles PM 1980 Retinoic acid increases cyclic AMP-dependent kinase activity in murine melanoma cells. J Biol Chem 255:5999-6002

Loeliger P, Bollag W, Mayer H 1980 Arotinoids, a new class of highly active retinoids. Eur J Med Chem Chim Ther 15:9-15

Madison K, Tong PS, Marcelo CL, Voorhees JJ 1981 Ro 10-9359 retinoid inhibits both in vitro epidermal cell proliferation and differentiation. In: Orfanos CE et al (eds) Retinoids. Advances in basic research and therapy. Springer-Verlag, Berlin, p 161

Marcelo CL 1979 Differential effects of cAMP and cGMP on in vitro epidermal cell growth. Exp Cell Res 120:201-210

Marcelo CL, Kim YG, Kaine JL, Voorhees JJ 1978 Stratification, specialization and proliferation of primary keratinocyte cultures. J Cell Biol 79:356-370

Marks R, Pearse AD, Hashimoto T, Barton S 1984 Overview of mode of action of retinoids. In: Cunliffe WJ, Miller AJ (eds) Retinoid therapy. MTP Press Ltd, Lancaster, p 91-99

McNutt NS 1977 Freeze-fracture techniques and application to the structural analysis of the mammalian plasma membrane. Cell Surf Rev 3:75-126

New DAT 1963 Effects of vitamin A on cultures of skin and buccal epithelium of the embryonic rat and mouse. Br J Dermatol 75:320-325

Peck GL, Elias PM, Wetzler B 1977 Effects of retinoic acid on embryonic chick skin. J Invest Dermatol 69:463-476

Plet A, Crain D 1981 Effect of retinoic acid treatment of F9 embryonal carcinoma cells on the activity and distribution of cAMP dependent protein kinase. J Biol Chem 257:889-893

Schultz-Ehrenburg U 1975 Differentiation of the epidermis in the limb bud culture. In: Neubert D, Merker HJ (eds) New approaches to the evaluation of abnormal embryonic development. Georg Thieme, Stuttgart, p 213-226

Stadler R, Marcelo CL, Voorhees JJ, Orfanos CE 1984 Effect of a new retinoid, arotinoid Ro 13-6298 on the in vitro keratinocyte proliferation and differentiation. Acta Dermato-Venereol 64:405-411

Tsambaos D 1984 Effekte von Arotinoidäthylester auf die Embryogenese, Proliferation und Differenzierung der Epidermis und auf die Talgdrüsenaktivität. Habilitationsschrift, Freie Universität Berlin

Weinstein RS, Merk FB, Alroy J 1976 The structure and function of intercellular junctions in cancer. Adv Cancer Res 23:23-32

Wessels NK 1963 Effects of extra-epithelial factors on the incorporation of thymidine by embryonic epidermis. Exp Cell Res 30:36-55

Zimmermann B 1978 The development of alkaline phosphatase activity in limb buds of mouse embryos in vitro and its relation to chondrogenesis. Anat Embryol 153:95-106

DISCUSSION

Peck: Have you noticed any specific effects of TTNPB ethyl ester on desmosomes?

Tsambaos: I don't think that there are arotinoid effects on desmosomes that can be regarded as specific. In general, epidermal changes occurring with this compound can be also seen after application of other agents, for example after topical benzoyl peroxide.

Peck: Elias & Williams (1984) studied the effects of retinoids on epidermis from normal, uninvolved skin of patients being treated for disease. In serial sections desmosomes appeared to be breaking off from the cell and lying free in the intracellular space, rather than just being stretched from the cell. There was no attachment of the desmosomes to the membrane. Holes in the cell membrane were found which could have been desmosome attachment sites.

Tsambaos: I cannot confirm this. We have not seen such effects even with high arotinoid doses. It is possible, however, that their occurrence may be related to the pH of the fixative.

Peck: There is controversy about the function of the keratinosomes and about whether the extracellular material that you see after retinoid treatment represents mucus or cell debris. Williams & Elias (1981) used a variety of staining techniques but could not identify this material as mucin, and they found no intracellular membrane-bound mucus granules. So the material could well represent cell debris from ruptured cells; the stimulation of its production may not be a specific retinoid effect.

Tsambaos: I agree. We have seen the same effects with orally or topically administered agents other than retinoids. This granular material is probably not synthesized by the keratinocytes but originates from cytotoxic processes induced by the retinoids. With low retinoid doses the granular material is present in much smaller amounts or is completely absent.

Hicks: With electron microscopy there is a great danger of attributing everything one sees in the treatment group to the specific action of the compound being tested, whereas many of the effects may be epiphenomena associated with a general process. In your experiments you have two concomitant phenomena, toxicity and increased cell turnover, and a lot of the substructural changes you describe in your retinoid-treated group are general symptoms of increased cell turnover and/or toxicity, for example the gaps in the basal lamina. It is well established that if you make any epithelium turn over more rapidly than usual, you get direct contact between the basal cells and the supporting mesenchyme through gaps in the basal lamina. It has nothing to do with the retinoid *per se*. The problem is that it is often quite difficult to distinguish between specific drug-related changes in substructure and general phenomena.

Tsambaos: Gaps in the basal lamina have been found after treatment of embryonic epithelia with various retinoids (Peck et al 1977, Hardy et al 1978, Tsambaos et al 1984). Since they occur even with very low retinoid doses I don't think we can say that they always represent toxic phenomena. In contrast to

Hardy et al (1978), I believe that the retinoid-induced gaps in the basal lamina are not of primary importance for the action of these compounds on embryonic epithelia.

Hicks: But you will see gaps in the basal lamina in any regenerating epithelium, irrespective of the agent that has caused the regeneration.

Tsambaos: Yes, but in that case they are rarely seen and much less pronounced than those found with retinoids.

Turton: We have carried out many experiments in rats and mice with various retinoids including retinamides, 13-*cis*-retinoic acid, etretinate and retinyl acetate. However, we have not seen the changes in our female F344 rats or B6D2F1 mice that are typical of the epithelial or mucocutaneous side-effects produced in humans in response to etretinate or isotretinoin therapy. We are not even certain that we see unequivocal hair loss, which is often reported to occur in rats and mice given retinoids. Why is it that nude or hairless mice must be used to obtain some of the epithelial changes seen in humans?

Tsambaos: I don't think that you have to use such animals. We were able to observe arotinoid effects similar to those presented here using C57BL mice, which are not hairless.

Moon: Perhaps you are not using a high enough concentration of retinoid, Dr Turton.

Dawson: How do you administer your retinoids?

Turton: In the diet.

Tsambaos: We had the same problem with dietary administration, because we did not protect the retinoid-containing diet from light. A significant part of the retinoid put into the diet may be inactivated or altered in the presence of air and light.

Turton: The retinoid is not inactivated.

Moon: We can show this by analysing the diets by high performance liquid chromatography (HPLC). It depends on the retinoid and how it is formulated of course.

Sporn: One can stabilize retinoids, especially retinamides, by putting them in an oily vehicle with antioxidants. One can then recover more than 90% of the original material, as shown by HPLC analysis.

Turton: We know our retinoids are stable in the diet. The beadlet preparations we have used from Roche (13-*cis*-retinoic acid, retinyl acetate and beta-carotene) are very stable. We also know that etretinate and the six retinoids we have tested when mixed into the diet with oil and antioxidants are stable. The absence of epithelial side-effects in our trials may be associated with the mode of retinoid administration and the resulting plasma levels. Perhaps when feeding retinoid in the diet there is not the high peak of retinoid in the plasma that might be obtained with intragastric, intravenous, subcutaneous or intraperitoneal administration. I think our experience in not seeing these side-effects is similar to that of Dick Moon, who has tested many more compounds than we have by dietary administration.

Tsambaos: Fritsch et al (1981) have put etretinate into the diet and have reported effects very similar to those we observed after application of this drug by gastric tube.

Turton: Yes, but with all the retinamides we have tested we do not see such epithelial changes.

Dawson: We see more pronounced toxic effects when we administer retinoids in an aqueous suspension of propylene glycol and Cremophor EL® (method of Hixson & Denine, 1979) rather than in arachis oil. So the vehicle may determine whether or not one sees marked effects.

REFERENCES

Elias P, Williams M 1984 Retinoid effects on epidermal structure, differentiation and permeability. Dermatologica 169:230

Fritsch PO, Pohlin G, Längle V, Elias PM 1981 Response of epidermal cell proliferation to orally administered aromatic retinoid. I. Invest Dermatol 77:287-293

Hardy HH, Sweeny PR, Bellows CG 1978 The effects of vitamin A on the epidermis of the fetal mouse in organ culture—an ultrastructural study. J Ultrastruct Res 64:246-260

Hixson EJ, Denine EP 1979 Comparative subacute toxicity of retinyl acetate and three synthetic retinamides in swiss mice. J Natl Cancer Inst 63:1359-1364

Peck GL, Elias PM, Wetzler B 1977 Effects of retinoic acid on embryonic chick skin. J Invest Dermatol 69:463-476

Tsambaos D, Zimmermann B, Orfanos CE 1984 Effects of retinoids on chondrogenesis and epidermogenesis in vitro. In: Cunliffe WJ, Miller AJ (eds) Retinoid therapy. MTP Press, Lancaster, p 119-133

Williams M, Elias P 1981 Nature of skin fragility in patients receiving retinoids for systemic effect. Arch Dermatol 117:611-619

Immunostimulation by retinoic acid

GUNTHER DENNERT

University of Southern California Comprehensive Cancer Center, Los Angeles, CA 90033, USA

Abstract. Retinoids have been shown to inhibit tumour growth in several model systems. In this paper evidence that immune effectors are important for this effect is discussed. Injection of retinoic acid (RA) into mice before challenge with allogeneic or syngeneic tumour cells results in a strong increase in cell-mediated cytotoxicity specific for the respective tumour. This stimulation appears to be due to effects taking place before or during the induction phase rather than the effector phase of cell-mediated cytolysis. The effector cells responsible for cytotoxicity express the Thy 1 antigen, are H-2 specific and are therefore T killer cells. The induction of T cell-mediated cytotoxicity requires the participation of the lymphokine interleukin 2 (IL-2). The possibility was tested that RA directly or indirectly influences the production of IL-2 and thereby stimulates the induction of T killer cells. Results indeed show that RA-injected mice display an increased capacity to produce IL-2 upon stimulation of their splenocytes in a mixed lymphocyte reaction. It appears therefore that RA has an effect on T cells that are destined to produce IL-2 upon antigenic challenge. Since IL-2 plays a role not only in the induction of specific cytotoxic T cells but also in the induction of natural killer (NK) cells, RA was also tested in a model system in which NK cells appear to play an important protective role. Results showed that split-dose irradiated mice that lose their NK activity and subsequently develop leukaemia can be protected from leukaemogenesis either by reconstitution with NK cells or by injection with RA. The question of whether this effect is due to stimulation of immune effectors or is a direct effect on the preleukaemic cells is discussed.

1985 Retinoids, differentiation and disease. Pitman, London (Ciba Foundation Symposium 113) p 117–131

The tumour-suppressive effects of vitamin A and several of its derivatives are now well documented in *in vivo* tumour model systems (for reviews see Lotan 1980, Dennert 1984). There are two principal mechanisms by which retinoids may suppress tumour growth. Retinoids may act either by suppressing carcinogenesis or the growth of transplanted tumours at the level of the tumour cell (Lotan 1980), or by stimulating immune effectors that are able to recognize and kill the tumour cell (Dennert 1984). Although *in vitro* the two possibilities can be easily distinguished and independently studied, the *in vivo* situation is much less amenable to analysis since difficult manipulations of the tumour-bearing host are required. The aim of this paper is to analyse the

117

evidence that the immune system stimulated by retinoids can exert inhibitory effects on tumour growth. This will be done by examining a number of different *in vivo* and *in vitro* tumour model systems.

Retinoids may suppress tumorigenesis and the growth of transplantable tumours: evidence for immune-mediated effects

In the majority of transplantable tumours studied so far, retinoids have not been shown to have inhibitory effects on tumour growth (Bollag 1972, Felix et al 1976, Baron et al 1981, Patek et al 1979, Meltzer & Cohen 1974, Kurata & Micksche 1977, Pavelic et al 1980). But in a selected number of tumour models, very significant inhibition of tumour growth has been demonstrated. For example, Deneufbourg (1979) showed that etretinate inhibits the growth of a syngeneic sarcoma in C57BL/6 mice, and Tannock et al (1972) observed that mice that have received retinyl palmitate require a reduced dose of local irradiation for tumour suppression. Moreover, Felix et al (1975, 1976) reported that growth of the S91 melanoma in BALB/c mice could be inhibited by retinoic acid. There are several pieces of evidence in these model systems that suggest that retinoids stimulate tumour rejection by activating immune effectors rather than by directly affecting the tumour cell itself. For instance, in Deneufbourg's experiments (Table 1) mice that received retinoid not only survived the initial tumour inoculation but were able to show a

TABLE 1 Model systems in which tumour growth is inhibited by retinoids: evidence for participation of immune effectors

Tumour	Tumour–host relationships	Retinoid	Evidence for participation of immune effectors	Reference
Sarcoma	Syngeneic	Etretinate	Second-set rejection of tumour in retinoid-injected hosts	Deneufbourg (1979)
Fibrosarcoma	Syngeneic	Retinyl palmitate	No effects in immuno-compromised irradiated hosts	Tannock et al (1972)
Melanoma S91	Allogeneic	Retinyl palmitate	No effects in allogeneic hosts or in ALS-treated immunocompromised hosts	Felix et al (1975, 1976)
Melanoma S91	Allogeneic	Retinoic acid	No effect in thymectomized hosts	Patek et al (1979)
Fibrosarcoma L33	Syngeneic	Retinoic acid	No effect in thymectomized hosts	Patek et al (1979)

ALS, anti-lymphocyte serum.

second-set tumour rejection, indicating the presence of immunological memory, probably stimulated by the prior injection of retinoid. Tannock et al (1972) reported that mice that were immunocompromised by whole-body irradiation could not be protected by injection of retinoid, whereas normal mice were protected. This suggests that a radiosensitive component plays a role in the retinoid-mediated effect. Similarly, Felix et al (1975, 1976) reported that animals immunocompromised by injection of anti-lymphocyte serum (ALS) did not show any retinoid-dependent inhibition of tumour growth. They also observed that although the retinoid-dependent tumour suppression was reproducible in allogeneic hosts, it did not take place in semiallogeneic hosts. This suggests that allogeneic histocompatibility antigens are recognized during tumour growth suppression and that the immune effectors responsible for tumour suppression are stimulated by the retinoid injection. In support of this conclusion, the experiments by Patek et al (1979) showed that, in thymectomized mice, retinoic acid has no effect in the allogeneic S91 model or in the syngeneic L33 model. If retinoic acid suppresses tumour growth by direct inhibition of tumour cell proliferation, then similar suppressive effects would be expected in normal and thymectomized hosts.

There is, however, a general dilemma in these experiments. One could postulate that although the tumour is amenable to immune attack, the immune response is not sufficiently strong to eliminate it or even to suppress its growth noticeably; however, in conjunction with the retinoid, which has inhibitory effects on the growth of the tumour, the immune-mediated effectors may be able to suppress tumour growth. In this case, tumour inhibition by the retinoid might be interpreted to be due to stimulation of immune effectors because elimination of the immune effectors would eliminate the retinoid-induced suppression. There are similar difficulties in interpreting tumour induction experiments. In a recent series of experiments, C57BL/6 mice were immunocompromised by four weekly doses of 200 rad ^{60}Co irradiation (split-dose irradiation). Such mice develop radiation leukaemia five to eight months after the leukaemogenic irradiation (Kaplan & Brown 1952). One of the main long-term defects in these mice is the absence of natural killer (NK) activity (Warner & Dennert 1982). It was therefore hypothesized that the NK defect might lead to a lack of immunosurveillance and subsequent leukaemia development. In support of this, injection of split-dose irradiated mice with cloned NK cells shortly after the last radiation dose was found to increase dramatically the rate of survival of the mice (Table 2). Injection of retinoic acid (RA) into irradiated mice also significantly decreased death from thymic leukaemia (Table 2). One might conclude that RA stimulates the recovery of NK activity and thereby suppresses leukaemogenesis. But another plausible explanation would be that RA directly inhibits

TABLE 2 Suppression of radiation leukaemia by natural killer cells and retinoic acid

Injection[a]	Time of injection (weeks[b])	Deaths from leukaemia/ total no. animals[c] (%)
No injection		11/13 (85)
NK cells	0	3/14 (21)
NK cells	4	4/10 (40)
NK cells	8	5/10 (50)
NK cells	12	7/10 (70)
Methylcellulose	0	14/20 (70)
RA	0	9/26 (35)
RA	8	9/16 (56)

[a] Injections into mice of 10 g. NK B61B10 cells: 2×10^6 injected as a single dose i.v.; RA: 100 μg in methylcellulose injected daily i.p. for a total of five days.
[b] After last radiation dose.
[c] C57BL/6 mice received four weekly doses of 200 rad ^{60}Co irradiation and were observed for 10 months after the last irradiation. Animals that died during this period were autopsied and examined for thymic leukaemias (Warner & Dennert 1982).
RA, retinoic acid; NK cells, natural killer cells.

the transition from preleukaemic to leukaemic cells. The dilemma of *in vivo* tumour growth experiments, therefore, is that tumours may be subject to multiple influences and that, in the animal, these influences cannot be easily separated from each other.

Stimulation of cell-mediated immunity by retinoids

Another way to explore whether retinoids have effects on the immune system is to assay directly the effects of retinoids on the activity of immune effectors, rather than on the growth of the tumour or skin graft. In the past, immunostimulatory effects of retinoids on humoral immunity have been reported by various groups (see review by Dennert 1984). For the purpose of this discussion, it may be more relevant to examine the effects of retinoids on cell-mediated immunity since this effector arm of the immune system is the more important one for graft rejection. Injection of retinoids into cancer patients has been reported to stimulate delayed-type skin reactions to various antigens (Micksche et al 1977). Moreover, Glaser & Lotan (1979) showed that mice injected with RA, and subsequently transplanted with a syngeneic tumour transformed by simian virus 40, generate effector cells in their spleens. These effector cells, if injected with the respective tumour cells into normal mice, are able to inhibit tumour growth. Since this effect is specific for the tumour to which the effector cells are sensitized and since it can be transferred by cells displaying the T-cell antigen Thy 1, the response to retinoids is probably caused by stimulation of the immune system. This reasoning is somewhat indirect, but there are more direct ways of showing that retinoids do in fact stimulate cell-mediated immunity. C57BL/6 mice injected with 25–100 μg RA per day for five days and subsequently challenged with the allogeneic tumour S194 were tested for cell-mediated cytotoxicity seven days after injection of the tumour. Results in Table 3 show that cytotoxicity to S194 was strongly stimulated. The effector cells responsible for

TABLE 3 Retinoic acid stimulates the induction of cell-mediated cytotoxicity: involvement of Thy 1+ effector cells

	RA dose (μg)	Treatment of splenocytes	% Cytotoxicity on S194
Experiment I	0	—	10
	25	—	52
	100	—	38
	300	—	12
Experiment II	0	—	16
	25	—	28
	25	Anti-Thy 1.2[a]	1

C57BL/6 (H-2^b) mice (10 g) were injected with retinoic acid (RA) in corn oil, or with vehicle only, daily for five days and were subsequently challenged with 10^7 cells of the allogeneic tumour S194 (H-2^d). Splenocytes from the mice were tested for cell-mediated cytotoxicity towards S194 cells seven days after injection of the tumour; assay length was 4 h and attacker to target cell ratio was 100:1. Results are from two independent experiments.
[a]Splenocytes were treated with anti-Thy 1.2 and complement before the cytotoxicity assay.

this effect are Thy 1.2$^+$ cells since treatment of splenocytes with anti-Thy 1.2 and complement before the cytotoxicity assay eliminated effector cell activity. There are two cell types that could be responsible for this effect, both expressing the Thy 1 antigen. One is the specific cytotoxic T cell (T killer) and the other the NK cell. Both can be distinguished by their specificity patterns as well as their cell-surface markers. In a cytotoxicity assay (Table 4) the

TABLE 4 Retinoic acid stimulates specific cell-mediated cytotoxicity: evidence for T killer cells

RA dose (μg)	% Cytotoxicity			
	S194 (H-2^d)	P815 (H-2^d)	BALB/c (H-2^d)	C$_3$H (H-2^k)
0	12	11	10	1
25	32	36	21	1

C57BL/6 (H-2^b) mice (10 g) were injected with retinoic acid (RA) in corn oil, or with vehicle only, daily for five days and were subsequently challenged with 10^7 cells of the allogeneic tumour S194 (H-2^d). Splenocytes from the mice were tested for cell-mediated cytotoxicity seven days after injection of the tumour; assay length was 4 h and attacker to target cell ratio was 100:1. BALB/c and C$_3$H cells were splenocytes cultured with lipopolysaccharide for two days.

effector cells caused lysis of the H-2^d targets S194, P815 and BALB/c splenocytes but not of H-2^k splenocytes. This shows that cytotoxicity is specific. Moreover, although P815 and BALB/c splenocytes are NK-resistant targets, they were lysed by the effector cells. This strongly suggests that the stimulated effector cells are indeed cytotoxic T cells and not NK cells.

The demonstration that RA stimulates T killer cells in an allogeneic response does not preclude the participation of NK cells in other systems, for instance syngeneic ones. The effects of RA were therefore tested in syngeneic tumour models. BALB/c mice were injected with 25 μg RA per day for five days and their spleen cells harvested and challenged *in vitro* with syngeneic S194 cells. The results (Table 5) reveal that cytotoxicity is stimulated in the

TABLE 5 Retinoic acid stimulates cell-mediated cytotoxicity to syngeneic tumours

Mouse strain	Tumour challenge in vitro	RA dose (μg)	% Cytotoxicity S194 (H-2^d)	S49 (H-2^d)	EL$_4$ (H-2^b)
BALB/c (H-2^d)	S194 (H-2^d)	0	27	4	—
BALB/c (H-2^d)	S194 (H-2^d)	25	58	11	—
C57BL/6 (H-2^b)	EL$_4$ (H-2^b)	0	—	—	7
C57BL/6 (H-2^b)	EL$_4$ (H-2^b)	25	—	—	61

Mice (10 g) were injected with retinoic acid (RA) in corn oil, or with vehicle only, daily for five days. On day 8 splenocytes were harvested and challenged *in vitro* with syngeneic tumours S194 (BALB/c) or EL$_4$ (C57BL/6). After five days the splenocytes were tested for cell-mediated cytotoxicity on syngeneic and allogeneic targets. Assay length was 8 h and attacker to target cell ratio was 100:1.

presence of RA and that it is specific for the tumour S194 as there is little cytotoxicity on the BALB/c tumour S49 and BALB/c splenocytes. Very similar results are seen when C57BL/6 mice are injected with RA and their spleen cells subsequently challenged *in vitro* with syngeneic EL$_4$ tumour cells: there is a strong stimulation of cytotoxicity for EL$_4$ targets (Table 5). These results show that RA stimulates cell-mediated cytotoxicity to syngeneic tumours and that the cytotoxicity induced is specific and unlikely to be due to NK cells since EL$_4$ is an NK-resistant target. The observations in the syngeneic systems concur with the conclusion drawn from the allogeneic model that the RA-stimulated effector cells are indeed T cells.

The finding that RA injected into C57BL/6 mice *in vivo* stimulated cell-mediated cytotoxicity to EL$_4$ quite strongly in splenocytes *in vitro* opened the possibility that the growth of this tumour might be inhibitable *in vivo*. Groups of six mice were injected with RA for five days and then challenged with 10^4 EL$_4$ cells. Drug injection was continued on a daily basis for one week and three times a week thereafter. The results (Fig. 1) show that in the mice that received RA before tumour challenge, the appearance of palpable tumours was delayed, but in the animals that received RA only after tumour injection there was no suppression of tumour growth or appearance. There are two important points to this result. First, tumour growth is only inhibited when RA is injected before the tumour challenge. This suggests that the effect of the drug is not on the tumour itself but rather on other cells which,

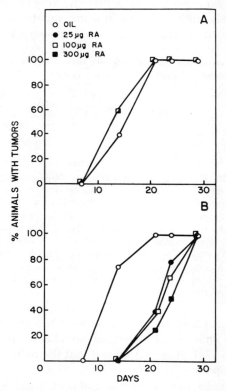

FIG. 1. Effect of retinoic acid on the growth of EL_4 tumours *in vivo*. Groups of six C57BL/6 mice were (A) given no treatment for five days or (B) injected daily with oil (vehicle) or retinoic acid (RA) for five days. On day 8 all animals were challenged with 10^4 EL_4 cells, and RA or oil was then injected into both groups of animals (A and B) daily for one week and three times a week thereafter.

upon tumour challenge, inhibit tumour growth. Second, although tumour growth or appearance is inhibited by prior RA injection, the animals are not cured of the tumours. At early times, i.e. day 21, only a few of the mice that had received prior injection of RA had tumours, whereas all other mice had palpable tumours; by day 30 however, mice in all groups had progressively growing tumours. Therefore, the ability of RA to induce rather high cytolytic activity to EL$_4$ *in vitro* does not suffice to cause rejection of this tumour *in vivo*.

Mechanism of action of retinoic acid

The experiments so far described show that injecting animals with RA before antigenic challenge is sufficient, and in some situations even necessary, to cause stimulation of the immune response. This suggests that RA acts in or

TABLE 6 Splenocytes from retinoic acid-injected mice have an increased capacity to produce IL-2 in mixed lymphocyte cultures and show elevated cytotoxicity

RA dose (μg)	IFN assay (IU/ml)	IL-2 assay on CTLL-2 (c.p.m. ± SEM)	% Cytotoxicity on S194 (H-2^d)
0	12	100 ± 20	12
25	12	290 ± 35	45
100	8	250 ± 15	55

C57BL/6 (H-2^b) mice (10 g) were injected with 25 μg or 100 μg retinoic acid (RA) daily for 14 days. Splenocytes were harvested two days after the last dose of RA and were cultured at a density of 10^7 cells/ml with BALB/c (H-2^d) stimulator cells (5 × 10^6/ml) in RPMI 1640 containing 5% fetal bovine serum. Supernatants were harvested on days 1–4 of the culture and assayed for activity on the IL-2-dependent cell line CTLL-2 by a standard proliferation assay. Data given for IL-2 activity are for day 2 of culture only and are expressed as c.p.m. ± SEM for CTLL-2 proliferation. The same supernatants were tested for interferon (IFN) activity by a standard virus production assay. The data for the day 2 supernatants are given and expressed as international units (IU) per ml. Cytotoxicity was assayed on day 5 at an attacker to target cell ratio of 60:1; assay length 4 h.

before the induction phase of cell-mediated cytotoxicity. Previous experiments have shown that RA has indeed no effect during target cytolysis (Dennert & Lotan 1978). It is well known that the induction of T killer cells requires the participation of lymphokines, which are often produced by accessory or helper cells. One lymphokine of central importance is interleukin 2 (IL-2) and another which may play a role is interferon (IFN). We tested the hypothesis that RA may act by enhancing the ability of the transplant recipient to produce IL-2 or IFN and so stimulate the cell-mediated cytotoxic response. RA (25 μg or 100 μg per mouse) was injected daily into C57BL/6 mice for 14 days and splenocytes from the mice were subsequently cultured with BALB/c stimulator cells. Cell supernatants were harvested and tested for IL-2 with the IL-2-dependent cell line CTLL-2. The results (Table 6) show that splenocytes from mice injected with RA not only have the capacity to generate cytotoxic effector cells but also have increased IL-2 activity in their tissue-culture supernatants. In contrast, no elevated IFN activity was noted.

Conclusion

There is now conclusive evidence that retinoic acid, and probably several other retinoids (Lotan & Dennert 1979), stimulate cell-mediated cytotoxicity *in vivo* and *in vitro* in both syngeneic and allogeneic tumour model systems. Since the killer cell stimulated is Thy-1$^+$ and H-2 specific it is most likely to be a T killer cell. The mechanism by which T killer cells are stimulated remains obscure. Experiments clearly show however that the effector phase of cytolysis is not stimulated. Rather, RA seems to influence the immune system before it is confronted with the antigenic stimulus, since injection of the drug before tumour transplantation often produces good stimulatory effects. A possible mechanism by which RA may stimulate the induction phase of T killer cells is indicated by the finding that injection of the drug stimulates the secretion of IL-2 in splenocytes cultured in a mixed lymphocyte culture. It remains to be determined, however, whether RA has direct or indirect effects on those T cells responsible for IL-2 secretion. If in fact RA acts via stimulation of IL-2 secretion, effects on other lymphocytes should be expected. For one, helper T cells require IL-2 for proliferation. Although in our own experiments we found little effect of RA on the humoral response or on priming of helper T cells, stimulatory effects of retinoids on antibody responses have been reported by various laboratories (see review by Dennert 1984). Another cell type that requires IL-2 for its proliferation and action is the NK cell (Dennert 1980). One might therefore expect that RA could have stimulatory effects on NK activity. In our own experiments in which we tested this hypothesis with RA-injected mice and assay of either NK cytolytic activity *in vitro* or natural resistance to bone marrow grafts *in vivo*, variable results were obtained which did not lead to firm conclusions (G. Dennert & J.F. Warner, unpublished work 1984). However, in *in vitro* experiments, Goldfarb & Herberman (1981) observed stimulatory effects of RA on NK cytolytic activity. It may therefore be that RA and other retinoids have a rather broad effect on various IL-2-dependent immune functions. The answer to the question of whether tumour suppression by retinoids is primarily due to the immunoenhancing capacity of these compounds or whether other mechanisms are involved awaits further experimentation and perhaps vitamin A analogues with stronger effects.

Acknowledgements

I thank Carol Gay Anderson for expert technical assistance. This work was supported by the National Institutes of Health via grants CA 39501 and CA 37706 and a grant by the American Cancer Society IM 284.

REFERENCES

Baron S, Kleyn KM, Russell JK, Blalock JE 1981 Retinoic acid: enhancement of a tumor and inhibition of interferon's antitumor action. J Natl Cancer Inst 67:95-97

Bollag W 1972 Prophylaxis of chemically induced benign and malignant epithelial tumors by vitamin A acid. Eur J Cancer 8:689-693

Deneufbourg JM 1979 Anti-tumor effect of an aromatic retinoic acid analog in a mouse syngeneic transplantable sarcoma. Biomed Express (Paris) 31:122-123

Dennert G 1980 Cloned lines of natural killer cells. Nature (Lond) 287:47-49

Dennert G 1984 Retinoids and the immune system: immunostimulation by vitamin A. In: Sporn MB et al (eds) The retinoids. Academic Press, Orlando, vol 2:373-388

Dennert G, Lotan R 1978 Effects of retinoic acid on the immune system: stimulation of T killer cell induction. Eur J Immunol 8:23-29

Felix EL, Loyd B, Cohn MH 1975 Inhibition of the growth and development of a transplantable murine melanoma by vitamin A. Science (Wash DC) 189:886-887

Felix EL, Cohn MH, Loyd B 1976 Immune and toxic antitumor effects of systemic and intralesional vitamin A. J Surg Res 21:307-312

Glaser M, Lotan R 1979 Augmentation of specific tumor immunity against syngeneic SV40 induced sarcoma in mice by retinoic acid. Cell Immunol 45:175-181

Goldfarb RH, Herberman RB 1981 Natural killer cell reactivity: regulatory interactions among phorbol ester, interferon, cholera toxin, and retinoic acid. J Immunol 126:2129-2135

Kaplan HS, Brown MB 1952 A quantitative dose–response study of lymphoid-tumor development in irradiated C57 black mice. J Natl Cancer Inst 13:185-208

Kurata T, Micksche M 1977 Suppressed tumor growth and metastasis by vitamin A + BCG in Lewis lung tumor bearing mice. Oncology (Basel) 34:212-215

Lotan R 1980 Effects of vitamin A and its analogs (retinoids) on normal and neoplastic cells. Biochim Biophys Acta 605:33-91

Lotan R, Dennert G 1979 Stimulatory effects of vitamin A analogs on induction of cell-mediated cytotoxicity *in vivo*. Cancer Res 39:55-58

Meltzer MS, Cohen BE 1974 Tumor suppression by *Mycobacterium bovis* (strain BCG) enhanced by vitamin A. J Natl Cancer Inst 53:585-587

Micksche M, Cerni C, Kokorn O, Titshcer R, Wrba H 1977 Stimulation of immune response in lung cancer patients by vitamin A therapy. Oncology (Basel) 34:234-238

Patek PQ, Collins JL, Yogeeswaran G, Dennert G 1979 Anti-tumor potential of retinoic acid: stimulation of immune mediated effectors. Int J Cancer 24:624-628

Pavelic ZP, Dave S, Bialkowski S, Priore RL, Greco WR 1980 Anti-tumor activity of *Corynebacterium parvum* and retinyl palmitate used in combination on the Lewis lung carcinoma. Cancer Res 40:4617-4621

Tannock IF, Suite HD, Marshall N 1972 Vitamin A and the radiation response of experimental tumors: an immune-mediated effect. J Natl Cancer Inst 48:731-741

Warner JF, Dennert G 1982 Effects of a cloned cell line with NK activity on bone marrow transplants, tumour development and metastasis *in vivo*. Nature (Lond) 300:31-34

DISCUSSION

*Malkovský:*Vittorio Colizzi and I have obtained very similar results, but in an entirely different system, that of antibacterial immunity. We injected mice with about 30 million cells of *Mycobacterium bovis*, strain Bacille Calmette-

Guérin (BCG cells). These mice are unable to respond to the purified protein derivative (PPD) of mycobacteria, and if you expose their lymphocytes to PPD they do not produce interleukin 2 (IL-2). We used two groups of BCG-injected mice, one group fed with a diet containing retinyl acetate and the second with a conventional diet. Fig. 1 shows that in cells taken from control mice we could detect little IL-2 production by splenocytes in response to stimulation with

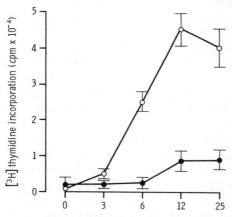

Dilution of IL-2 containing supernatants (%)

FIG. 1. (*Malkovský*) Stimulation of interleukin 2 production by purified protein derivative (PPD) in spleen cells from control (closed circles) or retinyl acetate-fed (open circles) mice infected with Bacille Calmette-Guérin (BCG). Sex- and age-matched CBA mice (9–12 weeks old) were maintained on a conventional diet (Spratts Laboratory Diet 1, Spillers) *ad libitum*, either with or without supplementary retinyl acetate (0.5 g per kg conventional diet) in the form of stable gelatinized beadlets (Roche) as described previously (Malkovský et al 1983a,b). The retinyl acetate diet was commenced six weeks before mice received an intravenous injection of 3×10^7 viable BCG cells (kindly provided by Glaxo, Middlesex, England) in 0.5 ml phosphate-buffered saline. Spleen cells were taken 14 days after the injection of BCG and incubated as single-cell suspensions depleted of erythrocytes (Malkovský et al 1982) in 24-well flat-bottomed tissue culture plates (Costar). Each well contained 10^6 cells in 1 ml RPMI 1640 medium (Flow) supplemented with 2 mM-L-glutamine (Sigma), 40 µg ml⁻¹ gentamycin, 10% heat-inactivated fetal calf serum (Flow) and 20 µg ml⁻¹ PPD (Serum Institute, Copenhagen, Denmark). After a 48-hour incubation at 37° C in a humidified atmosphere of 95% air and 5% CO_2, the supernatants were collected and tested for the presence of IL-2 with an IL-2-dependent T cell line (CTLL), as described by Colizzi (1984). Samples of 10^4 CTLL cells in 0.2 ml culture medium were incubated in triplicate in 96-well round-bottomed tissue culture plates (Falcon) in the presence of serial dilutions of IL-2-containing supernatants. Six hours before harvesting, 1 µCi of [*methyl*-³H]thymidine (specific activity 5 Ci mmol⁻¹, The Radiochemical Centre, Amersham, England) was added to each well. CTLL cells were harvested after 24 hours with a multicell cell harvester and thymidine incorporation was determined by liquid scintillation counting (Malkovský et al 1982). Radioactivity was expressed as counts per minute (c.p.m.) of triplicate cultures. Spleen cells from BCG-infected mice on the conventional diet released little IL-2, whereas spleen cells from BCG-infected mice fed retinyl acetate produced high levels of IL-2.

PPD. We assayed the IL-2 activity in the supernatant of PPD-stimulated splenocytes using CTLL cells, which are highly sensitive to IL-2 and die in its absence. Splenocytes from mice fed a diet supplemented with retinyl acetate produced a lot of IL-2 in response to PPD stimulation (Fig. 1). This observation together with recent results from our laboratory (Malkovský et al 1984a, 1985) supports Dr Dennert's suggestion that enhancement of IL-2 production may be one of the main effects of retinoids in the immune system. We also did a limiting dilution analysis and found that in retinyl acetate-fed mice the number of IL-2-producing cells was increased, so it is not only the absolute level of production of IL-2 that is affected.

Sporn: Is it known whether the induction of IL-2 represents *de novo* protein synthesis?

Dennert: You don't find IL-2 in cells that are not induced.

Sporn: Is there any evidence that retinoids control expression of the IL-2 gene? Has the IL-2 gene been cloned?

Dennert: The gene has been cloned but I doubt that retinoids directly affect its expression. You cannot induce IL-2 production with retinoic acid *per se*, i.e. you cannot stimulate it with retinoids *in vitro*. Retinoids seem to do something to the cells *before* they are stimulated to produce IL-2; either, as Dr Malkovský suggests, we induce more cells that are able to produce IL-2, or cells change so that they can produce more IL-2 in response to an appropriate stimulus. This happens *in vivo*; we inject the retinoid, wait for a week and then when we induce the splenocytes we find increased IL-2 production. We would not see the same effect if we stimulated normal splenocytes *in vitro* in the presence of retinoic acid.

Lotan: I have looked at a continuously proliferating cytotoxic cell line from Dr Y. Kaufmann, and found that retinoic acid actually inhibits IL-2 production and secretion (Y. Kaufmann & R. Lotan, unpublished work).

Dennert: But this cell line is really not very well understood. It is a hybridoma, not a clone of T killer cells, and it has a number of very strange properties. We cannot expect to understand the effect of retinoic acid on these cells.

Hartmann: Your hypothesis is that retinoids have some effect on IL-2 production. Since the production of IL-2 depends on IL-1 release, has anybody looked at IL-1 production by macrophages or keratinocytes?

Dennert: No.

Malkovský: I would like to stress one very important point. You have to treat animals for a certain time before you can detect immunostimulation by retinoids (Malkovský et al 1984b). We are now doing a very detailed study of the time course of retinoid effects and the dose–response relationship using retinyl acetate. We find that it is best to maintain animals on a diet supplemented with retinyl acetate for at least three weeks before we start to estimate their immunoreactivity.

Dennert: If you inject retinoic acid, a period of five days or a week is usually sufficient.

Malkovský: Yes, but injecting a substance is different from feeding a substance, and effects of injections usually appear faster. The point is that there is a delay between retinoid encounter with the immune system and its effect. T helper cells or some other cell type might have to be generated before the effect is manifested.

Hartmann: In your *in vivo* study, Dr Dennert, you did not pretreat all your animals with retinoic acid. The group that received retinoid at the same time as (but not before) the tumour inoculum showed no beneficial effect (Fig. 1, p 123). This might mean that retinoids *must* be given before tumour challenge. It also could mean that the inoculum was too high to be challenged successfully by retinoids.

Dennert: No. The inoculum we used was the only dose of tumour cells that gave any effect.

Koeffler: You suggested that retinoids might stimulate helper T cells, but your results show no increase in interferon production. How do you reconcile an increase in IL-2 production with no change in interferon?

Dennert: Why not? We really don't know what the role of interferon is in these systems. In splenocytes from mice treated with retinoid we get increased IL-2 production, and since we know that IL-2 is a product of helper T cells, we think that retinoids may have a direct or indirect effect on this type of cell.

Hicks: This is yet another example of retinoids having a stimulatory effect on a cell system. Here you see cell proliferation in the immune system. What you measure in terms of the production of a particular hormone or antibody is just an indication of which cells have been stimulated to proliferate; it is not a direct effect of the retinoid.

Eccles: Many of our results are consistent with Dr Dennert's. We have looked at the effects of etretinate on the growth and metastasis of about 15 transplantable syngeneic tumours (Eccles 1985). In all cases where we get a tumour-inhibitory effect *in vivo* it is T cell dependent. Effects are not seen in congenitally athymic nude mice; they are abolished by thymectomy plus irradiation with cytosine arabinoside priming, and they are also abrogated by cyclosporin A which interferes with the response to IL-2 of certain subsets of T lymphocytes. But I think we are still no nearer to knowing what the target for retinoids is and what kills the tumour cells; they may not be the same thing. We haven't yet considered the possible importance of antigen-presenting cells. It is interesting that *in vitro* one cannot stimulate IL-2 production with retinoids, so it might be worth looking at the role of accessory cells, for example with a bisphosphonate such as APD [3-amino-1-hydroxypropylidene-1,1-bis-phosphonate] which is supposed to inhibit accessory cell function. The retinoid *target* cell might be operative at a very early phase of the immune response and the *effector* cell could be a mononuclear phagocyte activated later in a T cell-dependent fashion. It might be that neither the target cell nor the effector cell is a helper T cell, but that both cell types are T helper dependent.

Dennert: How did you demonstrate that T cells are important in your system? Did you do direct cytolytic assays?

Eccles: No, because *in vitro* the tumours are resistant to natural killer cells and also T cells. Although these tumour cells can be controlled by retinoids *in vivo* we've yet to find a single cell type that is cytotoxic *in vitro*. They are all solid sarcomas and carcinomas, which can't be killed easily *in vitro*. Lymphoid cells are a different matter altogether.

Dennert: That is why we use them.

Eccles: We don't use them because we are interested in studying solid tumour growth and metastasis. Our evidence for T cell involvement in retinoid effects is simply that in T cell-deprived animals all the anti-tumour effects are abrogated. Also, the direct cytotoxic effects of retinoids *in vitro* do not correlate with *in vivo* tumour sensitivity.

Dennert: With thymectomized animals or nude mice you can also exclude antibody-dependent cellular cytotoxicity because you should not get a humoral response.

Turton: We have recently looked at the effects of feeding retinyl acetate, 13-*cis*-retinoic acid and other retinoids to skin-grafted mice (Hunt et al 1983, Turton et al 1983). There was a gross enlargement of the spleen and lymph nodes; in some mice the lymph nodes were 13 times their normal weight. The experiments demonstrated that skin grafts were rejected faster in animals fed retinoids, but in one or two experiments we found a lack of uniformity in the enhacement of skin graft rejection, and the rates of rejection were not always related to the retinoid dose level.

Dennert: What system was it? Male/female or allogeneic?

Malkovský: We used H-Y differences: male skin grafted onto female mice. The dose of retinyl acetate that is effective in accelerating skin graft rejection is lower than the effective dose in the tumour systems. We don't know why, and in fact if you use a higher dose in the skin graft system the effect is not so pronounced.

Turton: It seems odd that we generally obtained a gross proliferation of the lymphomedullary organs which was related to retinoid dose level, but that this response was not always correlated with a dose-related enhancement of skin graft rejection.

Malkovský: Perhaps the retinoid doses, which are high, could stimulate some kind of suppressor cell.

Koeffler: Patients with Sézary syndrome (a T-lymphocyte leukaemia) can apparently be treated with high doses of retinoids and their neoplastic cells are helper T lymphocytes. Do you see this as just a toxic effect of retinoids?

Dennert: Yes. The difference between retinoid doses that are stimulatory and those that are inhibitory in the immune system is small. To get stimulation you have to stay in a very narrow dose range: in our case 25–100 µg per 10 g mouse. With higher doses you get inhibitory effects.

Sherman: Your responses to 25 µg and to 100 µg retinoic acid are very similar (Table 3, p 121). What do lower doses do? Can you actually see a dose–response relationship?

Dennert: Yes we can, but there is a lot of variability in these experiments. We always titrate between 5 µg and 300 µg retinoic acid. With low doses we see much smaller effects than with 100 µg, but with the higher doses we again see small effects, because of toxicity.

Sherman: What is the lowest dose that will give you an effect?

Dennert: About 10 µg.

REFERENCES

Colizzi V 1984 *In vivo* and *in vitro* administration of interleukin-2-containing preparation reverses T-cell unresponsiveness in *Mycobacteruium bovis* BCG-infected mice. Infect Immun 45:25-28

Eccles SA 1985 Effects of retinoids on growth and dissemination of tumours: immunological considerations. Biochem Pharmacol, in press

Hunt R, Palmer L, Medawar PB, Turton J, Gwynne J, Hicks RM 1983 Immunopotentiation by retinoids in rats and mice. Proc Nutr Soc 42:13A

Malkovský M, Asherson GL, Stockinger B, Watkins MC 1982 Nonspecific inhibitor released by T acceptor cells reduces the production of interleukin-2. Nature (Lond) 300:652-655

Malkovský M, Doré C, Hunt R, Palmer L, Chandler P, Medawar PB 1983a Enhancement of specific antitumor immunity in mice fed a diet enriched in vitamin A acetate. Proc Natl Acad Sci USA 80:6322-6326

Malkovský M, Edwards AJ, Hunt R, Palmer L, Medawar PB 1983b T-cell mediated enhancement of host-versus-graft reactivity in mice fed a diet enriched in vitamin A acetate. Nature (Lond) 302:338-340

Malkovský M, Medawar PB, Hunt R, Palmer L, Doré C 1984a A diet enriched in vitamin A acetate or *in vivo* administration of interleukin-2 can counteract a tolerogenic stimulus. Proc R Soc Lond B Biol Sci 220:439-445

Malkovský M, Hunt R, Palmer L, Doré C, Medawar PB 1984b Retinyl acetate-mediated augmentation of resistance to a transplantable 3-methylcholanthrene-induced fibrosarcoma: the dose response and time course. Transplantation (Baltimore) 38:158-161

Malkovský M, Medawar PB, Thatcher DR et al 1985 Acquired immunological tolerance of foreign cells is impaired by recombinant interleukin 2 or vitamin A acetate. Proc Natl Acad Sci USA 82:536-538

Turton JA, Hicks RM, Gwynne J, Hunt R, Palmer L, Medawar PB 1983 Skeletal development as an index of retinoid toxicity. Proc Nutr Soc 42:12A

Retinoids and the control of pattern in regenerating limbs

MALCOLM MADEN

National Institute for Medical Research, The Ridgeway, Mill Hill, London NW7 1AA, UK

Abstract. It has recently been discovered that, as well as having effects on cell division and differentiation, retinoids induce dramatic changes in the development of pattern in limbs. Local application of retinoic acid to the anterior side of chick limb buds causes anteroposterior mirror-imaging such that the limb has six digits instead of three. In *Rana* limb buds retinoids induce changes in both the anteroposterior and proximodistal axes. In regenerating axolotl limbs their effect is primarily on the proximodistal axis. These proximodistal effects result in the regeneration of a complete limb from distal amputation levels. Concentration effects, time effects and the relative efficacy of various retinoids have been established. Cellular changes observed include a stimulation of epidermal mucopolysaccharide production, inhibition of cell division, induction of cartilage matrix breakdown and a stimulation of fibronectin production by mesodermal cells. The relevance of each of these changes to pattern effects has been determined. Initial experiments on the cellular location of radiolabelled retinoic acid are described. It thus seems that retinoids can change the determination of developing cells, and once we know the molecular basis of retinoid action then we should also know how developing cells become specified to form particular cell types.

1985 Retinoids, differentiation and disease. Pitman, London (Ciba Foundation Symposium 113) p 132–155

Once the molecular mechanisms of retinoid action have been uncovered we will be able to prevent and cure certain types of cancer and understand many details of the control of cell proliferation and cell differentiation. It is my opinion that knowledge of retinoid mechanisms will also provide an answer to the central problem of developmental biology, namely, how does a single cell, the fertilized egg, become a highly organized and extremely complex adult?

It is the process of pattern formation which turns the fertilized egg into a recognizable adult rather than a large ball of cells or a tumour. Pattern formation must therefore involve a complex series of cell interactions, the molecular basis of which, either in the embryo or part of the embryo such as the limb, is totally unknown. An investigation into their nature presents

132

enormous problems right from the start—which molecules are involved: proteins, glycoproteins, glycolipids? Where do we look for differences: in the extracellular matrix, on the cell surface, in the cytoplasm? Is the technology available to detect such small and presumably quantitative differences between single, or at least small numbers of, cells?

A short cut to these answers would be provided if we could change pattern formation at will. Fortunately, we can now do so with retinoids. This is why their effects on the developing and regenerating limb are so important, because they allow us for the first time to change the pattern of a developing system in a precise and controlled manner. We can then look at the biochemical alterations undergone during retinoid administration and therefore at the alterations undergone in the pattern-forming process. This will then tell us which molecules to look at, in which places and at what time.

The effects of retinoids on limbs have been investigated in three different systems: in the developing chick limb, the developing amphibian limb and the regenerating amphibian limb. I will begin by describing the morphological effects of retinoids on these systems and then consider in more detail the cellular effects of retinoids on the regenerating amphibian limb.

Retinoids and chick limb development

When a piece of filter paper or newsprint soaked in retinoic acid is grafted into the anterior side of the chick limb bud, mirror-imaged duplications in the anteroposterior axis are produced (Tickle et al 1982, Summerbell 1983). Thus instead of the normal three-digit limb (Fig. 1a), a six-digit double posterior limb develops (Fig. 1b) with the digital sequence of 432234. If the retinoic acid is applied to the posterior side instead of the anterior then either normal limbs or defective limbs (similar to that shown in Fig. 1d) arise.

The effect is concentration dependent—at lower concentrations a lesser degree of duplication occurs. For example a five- or four-digit limb rather than a six-digit limb may develop (Fig. 1c). At the other end of the scale, if the concentration becomes too high, then development is inhibited and defective limbs arise (Fig. 1d). The effect is also stage dependent—at later stages of development a lesser degree of duplication occurs, and at still later stages only normal limbs develop.

These results have been explained in two ways. The similarities between the effects of retinoic acid and the effects of grafting the zone of polarizing activity (a region of the posterior limb bud which when grafted to the anterior side produces six-digit double posterior limbs and is thought to act by the production of a diffusible morphogen) led Tickle et al (1982) to suggest that retinoic acid was either a morphogen itself or stimulated anterior cells to

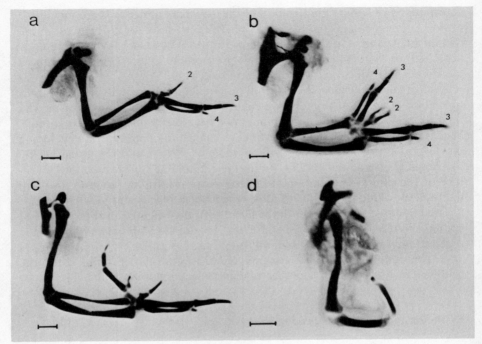

FIG. 1. The effects of applying retinoic acid locally to the anterior margin of the chick limb bud with a paper carrier. (a) Normal limb with digits 2, 3 and 4; (b) full anteroposterior duplication with a digital sequence of 4 3 2 2 3 4 produced after a medium dose of retinoic acid; (c) at lower doses less complete duplications appear—this limb has digits 3 2 2 3 4 (from left to right); (d) at high doses reductions of the limb appear. Alcian green stain; cleared whole mounts. Bar = 1 mm.

produce the normal limb morphogen. Another explanation (Summerbell 1983) is based on the premise that the normal limb pattern in the antero-posterior axis is generated by a reaction–diffusion system (Gierer & Mein-hardt 1972). If it is assumed that retinoic acid binds to the inhibitor, the experimental behaviour can be precisely reproduced in simulations.

Retinoids and amphibian limb development

When retinoids are applied to the developing limb buds of toads (Niazi & Saxena 1978) or frogs (Maden 1983a) not only are effects on the antero-posterior axis obtained, as described above for chicks, but effects on the proximodistal axis too. In these experiments the hindlimb buds of the animals are amputated through the shank when still at an early developmental stage and are then placed in solutions of various strengths of retinyl palmitate for various lengths of time. Instead of regenerating just the foot as might be expected (Fig. 2a), the limb buds produce extra segments in the proximodistal

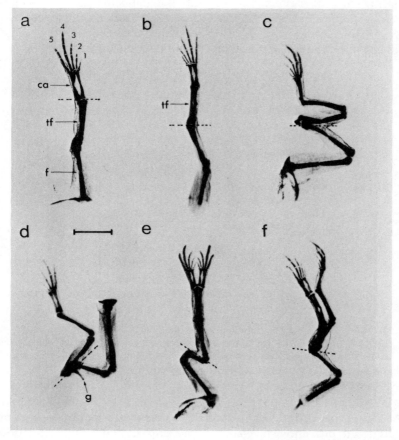

FIG. 2. The effects of systemic application of retinyl palmitate on hindlimb regeneration in *Rana temporaria*. Broken lines mark the amputation plane. (a) Control limb showing perfect regeneration of the foot; f = femur, tf = tibia and fibula, ca = calcaneum and astragalus, 1 2 3 4 5 = digits. (b–f) Limbs treated with increasing concentrations of retinyl palmitate (30–120 mg/ml) for increasing times (2–9 days) after amputation. (b) A complete extra tibia and extra fibula are present (tf). (c) A complete limb has developed from the amputation plane beginning with the femur. (d) A complete limb and pelvic girdle (g) have regenerated. (e) A complete limb has regenerated and it is bifurcated at the foot level to form a mirror-image duplicate. Note that the digital sequence is 5 4 3 2 1 1 2 3 4 5, i.e. the anterior faces are adjacent as in the chick limb shown in Fig. 1b. (f) A complete limb, bifurcated at the top, has developed with a complete pelvic girdle. The duplicated limbs have their anterior faces adjacent. Victoria blue stain; cleared whole mounts. Bar = 1 mm.

axis (Fig. 2b–d). For example, a limb with two tibiae and fibulae in tandem can be produced (Fig. 2b), or one with a whole limb regenerating from the shank level (Fig. 2c), and even one with a whole limb plus the girdle (Fig. 2d). As in the chick, this effect is concentration dependent—as the concentration increases, the level of the limb from which regeneration commences becomes more proximal (Fig. 3). It is also time dependent (Fig. 3)—extra tissue is produced if the animals are in retinyl palmitate for a longer time.

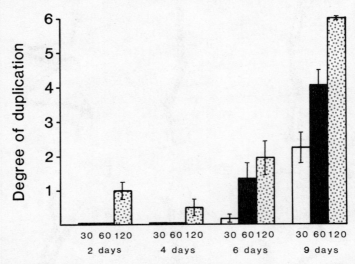

FIG. 3. The effects of systemically applied retinyl palmitate on the regeneration of *Rana temporaria* hindlimbs: dependence on duration of exposure (2, 4, 6 or 9 days) and concentration (30, 60 or 120 mg/1). The increasing proximalization of the level from which regeneration commences is scored as the degree of limb duplication, from 0 for a control limb (Fig. 2a) to 6 for a double complete limb (Fig. 2f). See Maden (1983a) for details. Bars mark standard errors.

At higher concentrations of vitamin A, anteroposterior duplications appear in addition to proximodistal changes. The limbs have two feet in mirror-image symmetry as well as being proximodistally elongated (Fig. 2e,f). The point of bifurcation of the mirror-imaging can vary from the foot level all the way back to the girdle. The latter results in the most remarkable cases of complete pairs of limbs, including the pelvic girdle, growing from the shanks of the original limbs (Fig. 2f).

A point of great interest here is that the structure of these duplicated limbs is identical to that of duplicated limbs obtained with chicks, i.e. their anterior sides are adjacent, so they form double posterior limbs (Maden 1983a). This is despite the fact that retinoids were administered locally to chicks, but systemically to *Rana*. Thus, during anteroposterior duplication, retinoids seem to act specifically on the anterior cells of the limb, causing them to become more posterior in their determination. In the case of proximodistal effects, retinoids seem to act on all the cells of the limb causing them to become more proximal in their determination.

An important principle of retinoid action is that it is the precise state in which the cells are at the time of administration that determines the effect retinoids will have. The information for the nature of the response is not in the retinoid itself (Roberts & Sporn 1984). This point is well demonstrated in a recent experiment of Niazi & Ratnasamy (1984). They took *Bufo melano-sticus* tadpoles at the stage when their hindlimbs had nearly completed development and only the digits had yet to form. They amputated the left limbs through the shank, but let the right limbs continue development untouched and immersed the animals in retinyl palmitate. The left limbs, which had been damaged and were regenerating, produced proximodistal and anteroposterior duplications. The right limbs, however, which were developing, produced defective limbs with digits missing.

Retinoids and limb regeneration

When the limbs of axolotls or newts are amputated, the animals have the remarkable ability to regenerate exactly what has been removed (Fig. 4a). But when retinoids are administered externally (Maden 1982), by intra-peritoneal injection (Thoms & Stocum 1984) or by local implantation of silastin blocks (Maden et al 1984), then extra elements are regenerated in the proximodistal axis (Fig. 4b–f). Thus, after amputation of a limb through the radius and ulna, retinyl palmitate treatment can induce an axolotl to produce a new limb with a complete extra forearm segment in tandem (Fig. 4b); even a complete limb can be made to grow from the amputation plane (Fig. 4d). Reduplication also occurs after amputation through other levels of the forelimb, such as the hand (Fig. 4e), and in the hindlimbs as well (Fig. 4f).

The effects of different retinoids were investigated (Fig. 5) and the order of efficacy was found to be retinoic acid > retinol ≫ retinyl palmitate > retinyl acetate with retinoic acid being at least 20 times more potent than retinyl acetate. With other retinoid effects it has been found that analogues with a free terminal carboxyl group are invariably the most potent (Sporn & Roberts 1984), and this is certainly true in the case of pattern formation.

Proximodistal duplication of regenerating limbs is a concentration-dependent (Fig. 6a) and time-dependent (Fig. 6b) phenomenon. That is, as the concentration of retinoid is increased or the time for which it is administered is increased, the level from which regeneration commences becomes more proximal (towards the shoulder). The data in Fig. 6b show the effects of two

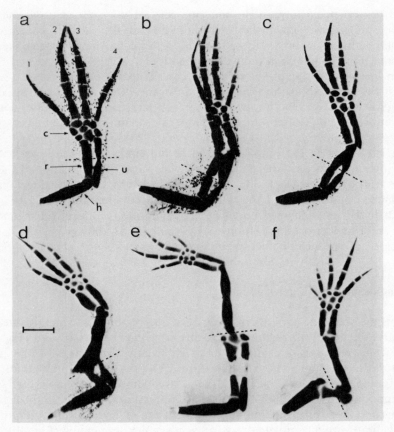

FIG. 4. Regeneration of axolotl limbs treated systemically with retinyl palmitate; concentrations of retinoid used increase from a–d. Broken lines mark the amputation plane. (a) Control forelimb amputated through the radius and ulna. The parts of the limb that were removed have been replaced; h = humerus, r = radius, u = ulna, 1 2 3 4 = digits. (b) Retinoid-treated forelimb with an extra radius and ulna in tandem. (c) Regenerate which begins at the distal humerus level, producing an elbow joint then a lower limb. (d) A complete limb beginning with a humerus has regenerated from the amputation plane. (e) Forelimb amputated through the carpals leaving three carpals behind; a complete limb has regenerated in tandem. (f) Hindlimb amputated through the tibia and fibula; a complete limb has regenerated from that level. Victoria blue stain; cleared whole mounts. Bar = 1 mm.

different retinoids—retinoic acid, which is the more potent, induced a more rapid rise in the degree of duplication. For maximal effects, retinoic acid should be administered for four to nine days after amputation and retinyl palmitate given for 12–18 days. A further point of interest here is that if the animals are left in retinoids for too long (> 10 days in 0.05 mM-retinoic acid,

> 20 days in 0.6 mM-retinyl palmitate) then regeneration is permanently inhibited. The limbs will never regenerate unless reamputated higher up. The inhibition of regeneration is caused by inhibition of cell division (see below).

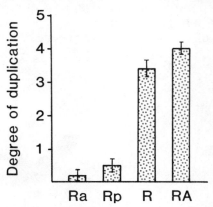

FIG. 5. The effect of four different retinoids applied systemically to axolotl forelimbs amputated through the radius and ulna. Each retinoid was administered for 15 days at a concentration of 0.12 mM. The degree of duplication is related to the proximalization of the level from which regeneration commences and ranges from 0 (control, Fig. 4a) to 5 (complete limb regenerating, Fig. 4d). See Maden (1983b) for details. Ra = retinyl acetate, Rp = retinyl palmitate, R = retinol, RA = retinoic acid. Retinoic acid is the most potent. Bars mark standard errors.

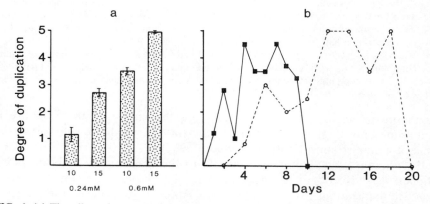

FIG. 6. (a) The effect of two concentrations of systemically applied retinyl palmitate on axolotl forelimb regenerates; 10, 15 = 10 days, 15 days of administration. Bars mark standard errors. (b) The effect of the duration of administration of two retinoids on axolotl forelimb regenerates. Squares, solid lines = 0.05 mM-retinoic acid; circles, broken lines = 0.6 mM-retinyl palmitate. Vertical axis: degree of duplication. Retinoic acid has a much more immediate effect than retinyl palmitate. Both cause an inhibition of regeneration if administered for too long. In both (a) and (b) the degree of duplication is measured about four weeks after retinoid administration, when limb regeneration is completed.

The converse experiment to this was performed by amputating limbs and allowing various periods of time to elapse *before* administering retinoids. The results revealed (Fig. 7a) that blastemal cells are only capable of being proximally respecified during the first eight days of regeneration. After this stage, development has progressed too far and teratogenic effects are manifest (Fig. 7b)—digits and carpals are missing or fused. This emphasizes

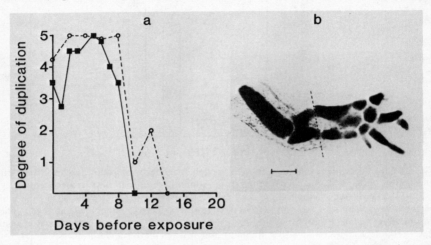

FIG. 7. (a) The effect on axolotl forelimb regenerates of increasing the time delay between amputation and systemic administration of retinoids. Squares, solid lines = 0.05 mM-retinoic acid; circles, broken lines = 0.6 mM-retinyl palmitate. The retinoids must be applied before 8–12 days of regeneration have passed for full duplication to occur. After that time only teratogenic effects appear, such as missing digits as shown in (b). (b) Victoria blue whole mount of a limb amputated through the radius and ulna (broken line) and treated with 0.6 mM-retinyl palmitate from 10 days after amputation for 15 days. Digits are missing and there are no pattern duplications. Bar = 1 mm.

again the principle of retinoid action that the nature of the response is determined not by the retinoid but by the state in which the cells are at the time of administration (Roberts & Sporn 1984).

By administering retinoic acid locally to the blastema in a silastin block implant we have obtained an estimate of the absolute amount needed to produce duplications (Maden et al 1984). About 16 µg is needed to induce a complete limb from the amputation plane (Fig. 4d).

The cellular effects of retinoids on regenerating limbs

The cellular effects of retinoids are many and varied. Among them are the induction of mucous metaplasia in epithelium (Fell & Mellanby 1953), the induction of cartilage matrix breakdown (Fell & Mellanby 1952), the inhibition of cell division (Lotan 1980) and the alteration of cell-surface and matrix glycoproteins (Roberts & Sporn 1984). Each of these has been found to occur in the regenerating limb (Maden 1983b).

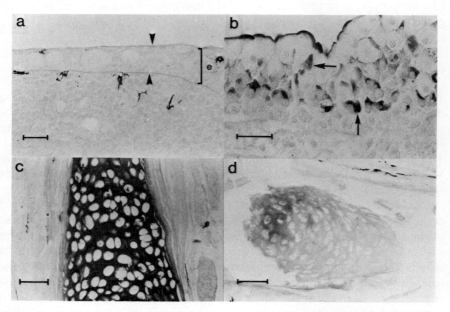

FIG. 8. Alcian blue staining of axolotl limb sections at pH 2.5. (a) Control epidermis (e) over a regenerate showing that only the outer mucus layer of the epidermis (arrow) and the basement membrane on the inner surface (arrow) stain. (b) Epidermis over a regenerate treated systemically with 0.12 mM-retinoic acid for 12 days. Alcian blue material is also present within and between cells in the lower layers of the epidermis (arrows). (c) Section through the cartilage of a limb showing the dark matrix staining with Alcian blue. (d) Cartilage after 12 days of systemic treatment with 0.12 mM-retinoic acid showing the virtual absence of staining, caused by loss of proteoglycans from the matrix. Bar = 100 μm.

Epidermal effects

Only the surface mucus layer of the normal axolotl epidermis stains with Alcian blue (Fig. 8a), but after retinoic acid treatment Alcian blue-staining material begins to appear within and between the epidermal cells over the regenerate (regenerating limb) (Fig. 8b). This is presumably a mucopolysaccharide as has been found in other systems (Bellows & Hardy 1977, Yuspa & Harris 1974).

We can then ask the question: is this epidermal change responsible for the pattern effects? To test this hypothesis, epidermis was exchanged between retinoid-treated and untreated limbs (Maden 1984). When treated epidermis was grafted onto untreated mesoderm, the regenerates were normal (e.g. Fig. 4a). When untreated epidermis was grafted onto treated mesoderm the regenerates were proximodistally duplicated (e.g. Fig. 4b). Thus stimulation of mucopolysaccharide production in the epidermis is not responsible for the pattern alterations.

Cartilage effects

The induction of cartilage matrix breakdown by retinoic acid can also be detected by Alcian blue staining. Normal cartilage stains deep blue (Fig. 8c), but after eight days of treatment to the regenerating limb the stain begins to show patchy areas, and after 12 days the cartilage hardly stains at all (Fig. 8d). This is the same effect that vitamin A has on cartilage cultured *in vitro* (Fell & Mellanby 1952) and is caused by the release of proteoglycans from the matrix (Dingle et al 1972).

Could this be the cause of pattern alterations? Perhaps pattern formation is controlled by a concentration gradient of proteoglycan, and as more than normal is released after retinoid treatment, blastemal cells become more proximalized. This is very unlikely for two reasons. Firstly, if all the cartilage is removed from an axolotl limb and the limb is then amputated through the forearm, pattern formation distal to the amputation plane, including the regeneration of the distal cartilage elements, is perfectly normal. Secondly, if extra cartilage is grafted into the limb (for example an extra radius and ulna) and part of the limb is amputated, pattern formation in the regenerate is again normal (M. Maden, unpublished work 1983). Clearly, if the organization of the regenerate is controlled by a gradient of proteoglycans, both these operations would be expected to disturb pattern formation, but they do not.

Inhibition of cell division

It seems paradoxical that a treatment which stimulates extra regeneration should, at the same time, inhibit cell division. But this is precisely what occurs. The labelling indices and mitotic indices of retinoid-treated vs. control limbs are shown in Fig. 9. The concentration of retinyl palmitate used in these experiments was the same as that needed to produce maximal pattern effects. During the period of retinoid treatment, cell division in the blastemas becomes inhibited (Fig. 9), the cells aggregate together (see below) and only when retinoid treatment is stopped do the cells start dividing again so that regeneration can commence. The treatment must be terminated before day

10 when 0.05 mM-retinoic acid is used or before day 20 when 0.6 mM-retinyl
palmitate is used or cell division becomes permanently inhibited and the limbs
will not regenerate at all.

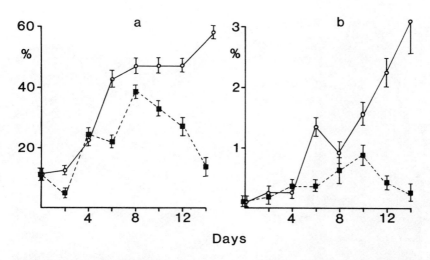

FIG. 9. (a) Labelling indices of control (circles, solid lines) and retinoid-treated (squares, broken
lines) regenerating axolotl limbs sampled every day after amputation. Labelling index = % of
cells in DNA synthesis at the time of [³H]thymidine administration. Experimental limbs were
systemically treated with 0.6 mM-retinyl palmitate throughout the experiment. The initial
stimulation of DNA synthesis after amputation occurs in both control and experimental limbs,
but after eight days an inhibition by the retinoid begins to take effect. (b) Mitotic indices of
control (circles, solid lines) and retinoid-treated (squares, broken lines) regenerates. Mitotic
index = % of cells in mitosis at the time of fixation. Same experimental conditions as (a). Mitosis
is severely inhibited when control limbs start to divide rapidly after day 8. Bars mark standard
errors.

This inhibition of cell division, which is a well-recognized effect of retinoids
on a wide variety of cell types (Lotan 1980), could be responsible for pattern
effects. It is possible to devise a theory of pattern formation based on cell
cycles as ticks of a clock; any delay could cause a proximalization of the cells.
This theory can be tested by applying other cell-cycle inhibitors to limbs, and
the easiest way of doing this is to denervate the limbs. Denervated axolotl
limbs do not regenerate until the nerves grow back and it is thought that the
nerves supply a neurotrophic factor, possibly a peptide, which is a mitogen
for blastemal cells (Singer 1978). After denervation, limbs give curves of
labelling indices and mitotic indices (Maden 1978) that are identical to those
of vitamin A-treated limbs (Fig. 9). Thus we can denervate a limb and induce

a delay of the same period that is observed after retinoid treatment. When this is done no pattern abnormalities arise and the limbs are perfectly normal (Maden 1983c). Thus cell cycle delays *per se* are not the cause of pattern effects.

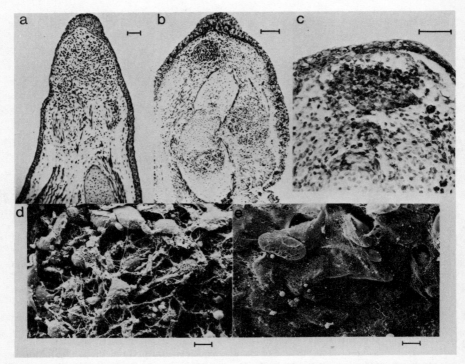

FIG. 10. (a) Section through a normal cone-stage blastema from a regenerating axolotl limb full of rapidly dividing mesenchymal cells evenly spread out underneath the thickened epithelial cap at the top. Bar = 100 μm. (b) Section through a regenerate treated with retinoid for 15 days. The blastemal cells have clumped together into a tight ball. Bar = 100 μm. (c) Retinoid-treated regenerating limb. Section treated with rabbit anti-fibronectin antibody showing dark staining in the cell aggregate. Bar = 100 μm. (d) A scanning electron micrograph of the mesoderm at the tip of a normal blastema. The epidermis has been stripped off, revealing a matrix composed of thick strands. Bar = 10 μm. (e) A scanning electron micrograph of the mesoderm of a limb treated with retinoic acid for six days. The matrix has changed extensively; the network is thicker than normal but is composed of thinner strands. Bar = 10 μm.

Extracellular matrix changes

The histology of retinoid-treated limbs reveals a surprising behaviour of blastemal cells. A normal blastema is composed of uniformly dispersed, lightly staining mesenchymal cells (Fig. 10a), derived from the dedifferentiation of stump tissues—cartilage, muscle, connective tissue, dermis etc. But after two weeks of retinoid treatment, the blastemal cells which are not dividing (see above) clump together into a ball (Fig. 10b). This is a unique feature of retinoid treatment, rather than being caused by the inhibition of cell division, because neither denervation nor X-irradiation of limbs (both these treatments also inhibit cell division) produces this behaviour.

A more detailed study of this phenomenon has revealed that the retinoid-induced cell clumping is due to the protein fibronectin (S. Keeble, unpublished work 1984) because the spaces between the cells stain heavily with a rabbit antibody to fibronectin (Fig. 10c). The heavy staining could be caused by an induction of fibronectin synthesis as has been observed in various cell types (Bolmer & Wolf 1982, Jetten et al 1979) or by an increase in the ability of cells to bind fibronectin to the cell surface (Hassell et al 1979, Roberts & Sporn 1984).

These matrix changes in the blastema have also been observed with the scanning electron microscope (Fig. 10d,e). Such wide-spread alterations are presumably due in part to fibronectin, but could also be caused by retinoid-induced synthesis of other macromolecules such as collagen and proteoglycans (Roberts & Sporn 1984). We are currently testing the hypothesis that the matrix is responsible for the pattern effects by administering fibronectin to regenerating limbs.

Discussion

I have summarized the recently discovered effects of retinoids on limb development and regeneration, and have tried to emphasize that the principles of retinoid action which have been established by research on other cell types also apply to the limb. These principles are:

(1) The effects of retinoids are concentration dependent. If the concentration is too low no pattern alterations occur, but as the concentration is raised pattern changes are manifest to an increasing extent. At very high concentrations, development is abolished altogether.

(2) Pattern changes increase as the time for which retinoids are administered is increased up to a point beyond which development is abolished.

(3) Retinoic acid is the most potent of four retinoids tested in causing pattern changes.

(4) The cellular context in which retinoids are administered is important. In one cell state (when the limbs are undamaged or have passed a certain stage) retinoids have teratogenic effects and less structure than normal develops, but in another cell state (during the period when pattern changes

are occurring) retinoids have 'super-developmental' effects and more structure than normal develops.

(5) Retinoids stimulate mucopolysaccharide production in the epidermis of the limb.

(6) Retinoids cause a loss of metachromasia in the cartilage of the limb by releasing proteoglycans from the matrix.

(7) Retinoids inhibit cell division in a concentration-dependent manner.

(8) Retinoids cause extracellular matrix changes and stimulate fibronectin production (and/or sequestration) in the mesodermal cells.

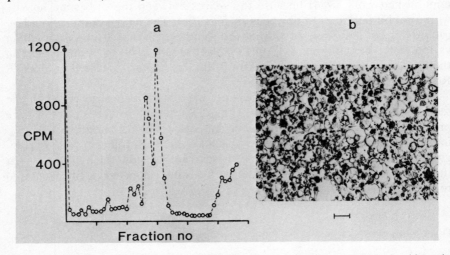

FIG. 11. (a) Distribution of [³H]retinoic acid in fractions obtained after centrifugation (through a 15–60% sucrose gradient) of homogenized blastemas from regenerating axolotl limbs treated for 24 h with [³H]retinoic acid. The peak on the left is in the pellet, the one in the middle is in a membrane fraction (see b) and the one on the right is probably free retinoic acid and/or retinoic acid-binding proteins. (b) An electron micrograph of the central peak which consists of membrane vesicles, a few Golgi and autophagic vacuoles. Bar = 1 μm.

Thus the changes in specification that retinoids induce in the cells of the limb do not take place in isolation from this multitude of other, well-characterized effects, but in addition to them. This gives us confidence that the biochemical data on the mechanisms of action of retinoids will also be relevant to the developing and regenerating limb.

Are any of the cellular effects described above responsible for pattern formation changes? So far the answer is no, so it seems that some other pathway must play the crucial role. Currently we are investigating the cellular site of action of retinoic acid. When tritiated retinoic acid is administered to the regenerating axolotl limb or the chick limb bud for 24 h and the tissue is then homogenized and spun through a 15–60% sucrose gradient, three peaks of radioactivity appear (Fig. 11a). The one at the top of the gradient is probably free retinoic acid and/or retinoic acid-binding proteins. The peak in the middle is in a fraction which consists of membranes (Fig. 11b) and the peak at the bottom is in a fraction containing nuclei and/or matrix debris. The retinoic acid can be removed from the membrane fraction only by organic extraction and it thus resembles the vitamin A localized in microsomes in pigment epithelium (Berman et al 1979). Further work will establish the biological significance of this result.

Thus the developing and regenerating limb provides a further system for research into the biochemistry of retinoid action. This is, I believe, the most promising research direction for solving the central problem of developmental biology, namely what the molecular basis of pattern formation is.

Acknowledgements

I wish to thank Stella Keeble, Katriye Mustafa and Liz Hurst for research assistance and Hoffmann-La Roche, Basel for generous gifts of radiolabelled retinoic acid.

REFERENCES

Bellows CG, Hardy MH 1977 Histochemical evidence of mucosubstances in the metaplastic epidermis and hair follicles produced *in vitro* in the presence of excess vitamin A. Anat Rec 187:257-272

Berman ER, Segal N, Feeney L 1979 Subcellular distribution of free and esterified forms of vitamin A in the pigment epithelium of the retina and in liver. Biochim Biophys Acta 572:167-177

Bolmer SD, Wolf G 1982 Stimulation of fibronectin production by retinoic acid in mouse skin tumors. Cancer Res 42:4465-4472

Dingle JT, Fell HB, Goodman DS 1972 The effect of retinol and of retinol-binding protein on embryonic skeletal tissue in organ culture. J Cell Sci 11:393-402

Fell HB, Mellanby E 1952 The effect of hypervitaminosis A on embryonic limb-bones cultivated *in vitro*. J Physiol (Lond) 116:320-349

Fell HB, Mellanby E 1953 Metaplasia produced in cultures of chick ectoderm by high vitamin A. J Physiol (Lond) 119:470-488

Gierer A, Meinhardt H 1972 A theory of biological pattern formation. Kybernetik 12:30-39

Hassell JR, Pennypacker JP, Kleinman HK, Pratt RM, Yamada KM 1979 Enhanced cellular fibronectin accumulation in chondrocytes treated with vitamin A. Cell 17:821-826

Jetten AM, Jetten MER, Shapiro SS, Poon JP 1979 Characterisation of the action of retinoids on mouse fibroblast cell lines. Exp Cell Res 119:289-299

Lotan R 1980 Effects of vitamin A and its analogues (retinoids) on normal and neoplastic cells. Biochim Biophys Acta 605:33-91

Maden M 1978 Neurotrophic control of the cell cycle during amphibian limb regeneration. J Embryol Exp Morphol 48:169-175

Maden M 1982 Vitamin A and pattern formation in the regenerating limb. Nature (Lond) 295:672-675

Maden M 1983a The effects of vitamin A on the developing limbs of *Rana temporaria*. Dev Biol 98:409-416

Maden M 1983b The effect of vitamin A on the regenerating amphibian limb. J Embryol Exp Morphol 77:273-295

Maden M 1983c Vitamin A and the control of pattern in regenerating limbs. In: Fallon JF, Caplan AI (eds) Limb development and regeneration, Part A. Alan Liss, New York (Progress in clinical and biological research series, vol 110A) p 445-454

Maden M 1984 Does vitamin A act on pattern formation via the epidermis or mesoderm? J Exp Zool 230:387-392

Maden M, Keeble S, Cox RA 1984 The characteristics of local application of retinoic acid to the regenerating axolotl limb. Wilhelm Roux's Arch Dev Biol, in press

Niazi IA, Ratnasamy CS 1984 Regeneration of whole limbs in toad tadpoles treated with retinol palmitate after the wound-healing stage. J Exp Zool 230:501-505

Niazi IA, Saxena S 1978 Abnormal hind limb regeneration in tadpoles of the toad, *Bufo andersonii*, exposed to excess vitamin A. Folia Biol (Crakow) 26:3-8

Roberts AB, Sporn MB 1984 Cellular biology and biochemistry of the retinoids. In: Sporn MB et al (eds) The retinoids. Academic Press, Orlando, vol 2:209-286

Singer M 1978 On the nature of the neurotrophic phenomenon in Urodele limb regeneration. Am Zool 18:829-841

Sporn MB, Roberts AB 1984 Biological methods of analysis and assay of retinoids—relationships between structure and activity. In: Sporn MB et al (eds) The retinoids. Academic Press, Orlando, vol 1:235-279

Summerbell D 1983 The effect of local application of retinoic acid to the anterior margin of the developing chick limb. J Embryol Exp Morphol 78:269-289

Thoms SD, Stocum DL 1984 Retinoic acid-induced pattern duplication in regenerating urodele limbs. Dev Biol 103:319-328

Tickle C, Alberts B, Wolpert L, Lee J 1982 Local application of retinoic acid to the limb bud mimics the action of the polarizing region. Nature (Lond) 296:564-566

Yuspa SH, Harris CC 1974 Altered differentiation of mouse epidermal cells treated with retinyl acetate *in vivo*. Exp Cell Res 86:95-105

DISCUSSION

Tickle: Malcolm Maden has described our experiments on the effects of retinoids on the developing chick wing and I would like to explain why we always use a local application of retinoic acid in this system. In previous work on the development of the limb, a signalling region called the polarizing region has been identified (Saunders & Gasseling 1968, Tickle et al 1975). It is a region of cells at the posterior margin of the developing limb and one can show its signalling ability by transplanting it to the anterior margin of a second wing bud. This wing bud now has polarizing tissue at both posterior and anterior margins and mirror-image duplications arise. In the normal limb the pattern of digits is 2 3 4, but with a grafted polarizing region at the anterior margin the pattern of digits is now 4 3 2 2 3 4. The extra structures arise in response to a signal from the polarizing region tissue. If one looks at the effects of grafting polarizing tissue to different positions in the limb, one finds a correlation between the pattern of digits and distance from the polarizing region. The rule is that the cells closest to the polarizing region form digit 4, cells that are a bit further away form digit 3, and cells further away still form digit 2. We have put forward the idea that the signal from the polarizing region is a diffusible morphogen, and that a concentration gradient of morphogen is set up across the limb (Tickle et al 1975). Cells exposed to a high concentration would form digit 4 whereas those exposed to a low concentration would form digit 2. This would explain the distance-dependent effects that we see when we graft in another polarizing region.

We wanted to identify the morphogen, and the approach we took was to apply extracts of the polarizing region cells and other chemicals that we thought might be important (such as prostaglandins and cyclic AMP) to the anterior margin of the developing wing. All the compounds that we looked at and all the extracts gave negative results. However, when we heard that retinoic acid could inhibit cell–cell communication (Pitts et al 1981), we tried applying the retinoid to the developing wing. To our surprise mirror-image duplications resulted, so for the first time we had found a chemical that mimics the action of the natural signalling region (Tickle et al 1982).

In our initial experiments we soaked a small piece of DEAE paper, which is positively charged, in the retinoic acid solution and then implanted it into the developing limb. Since then Bruce Alberts and Gregor Eichele have developed much more efficient ways of applying retinoic acid (Eichele et al 1984). They investigated the release properties of a number of potential carriers by soaking the carriers in radioactively labelled retinoic acid and then measuring the release of the retinoid into tissue culture medium. The most effective carrier was AG1-X2, which is a positively charged ion-exchange bead. AG1-X2 beads release large amounts of retinoic acid over a prolonged time period, that continues for at least a day in $vitro$. Furthermore the amount of retinoic acid released is proportional to the concentration of retinoic acid in which the beads are soaked. Other types of beads, for example a more highly cross-linked positively charged ion-exchange bead and a non-charged bead, release smaller amounts of retinoid and the time course of release is not as prolonged. AG1-X2 beads should be extremely useful for locally applying retinoic acid to tissues where one needs controlled and prolonged release.

We have used these beads to repeat our initial experiments on the chick wing in more detail, and we see a very clear dose–response relationship (Tickle et al 1985). As we raise the concentration of retinoic acid in which the beads are soaked from 0.005 mg/ml to 0.01 mg/ml we get the sequential addition of digits; first an extra digit 2, then additional digits 3 and 2 and finally a full reduplicated wing (digit pattern 4 3 2 2 3 4). As we increase the concentration beyond 0.01

mg/ml we get progressive thinning of the symmetrical limb (digit patterns 4 3 3 4, 4 3 4 and just a single symmetrical digit 4) and finally at high concentrations (1–10 mg/ml) truncated wings develop. We can quantitate these digit patterns in two ways. One measure is the '% respecification' that involves scoring wing patterns according to which additional digits form. The second measure is simply the mean number of digits. Using these measures we can plot dose-response curves. When % respecification is plotted against the concentration of *all-trans*-retinoic acid in the soaking solution the curve has a bell shape. First, as the concentration of retinoic acid increases, the % respecification rises, while at high concentrations it falls because we get limb truncations that score zero. The curve for the mean number of digits plotted against *all-trans*-retinoic acid concentration again initially rises—to a peak of about five digits per wing—but then drops with very high concentrations of retinoic acid.

Gregor Eichele has worked out the amount of retinoic acid in the limb bud after application of a retinoic acid-soaked bead. He soaked beads in radioactively labelled retinoic acid, implanted them into wing buds and left them there for 14 hours, which is the minimum time required to produce effects. He then removed the beads, extracted the tissue of the bud with n-hexane and measured the radioactivity. The limb bud cells metabolize the released retinoic acid and he found by high performance liquid chromatography analysis that after 14 hours only 9% of the radioactivity in the bud was in the form of *all-trans*-retinoic acid. Taking this into account, he calculated the amount of retinoic acid in the bud and found that this is directly proportional to the concentration of the retinoid in the soaking solution (Tickle et al 1985). This proportionality means that our dose–response curves are true response curves. From these curves, we can estimate that the amount of *all-trans*-retinoic acid that is required in the bud to specify additional digits is 1–25 nM. Since these concentrations are within the physiological range, we think it possible that retinoids could be natural morphogenetic signalling molecules. In fact Gregor Eichele has recently found, by measuring the radioactivity in different slices across a wing bud after implanting a bead soaked in radiolabelled retinoic acid, that a gradient of retinoic acid, which stabilizes very quickly, is established across the bud (Tickle et al 1985).

Yuspa: Are you saying that vitamin A gradients could be the natural way that limb development is controlled in the chick?

Tickle: It is possible that retinoids could be morphogens, but of course there are other explanations. There are retinoids present in the limb, and Gregor Eichele is beginning to look at this.

Yuspa: It would be leaving a lot to chance to control limb development with a substance that must be obtained from an exogenous source.

Sporn: All sorts of vertebrate eggs are very rich in retinoid material. In vitamin A-deficient quail eggs the whole developmental programme comes to a halt, but it is quite likely that there is enough retinol in normal eggs to execute this programme.

Yuspa: Have the effects of retinoids on limb development been tested in mammals?

Maden: No, it is assumed that retinoids are teratogens in limb development in mammals. Grafting experiments have not been done because very few people work on mammalian limbs.

Lotan: Can you make newts or axolotls vitamin A deficient to see whether they need vitamin A for normal development?

Maden: I would imagine that it is extremely difficult, and I don't know what the endogenous levels of retinoids are in the system. The only way of looking at the natural function of retinoids in these animals would be to administer an antagonist or antibody during development.

Sporn: There is no good way of antagonizing the effect of any retinoid at present; an antagonist is very badly needed. Retinoids themselves are small molecules and are not antigenic so we have no antibodies either, and it is therefore not easy to create a functional lack of active retinoid in a tissue.

Moon: The compounds that both of you have used are all highly teratogenic in animals. Have you tried any compounds that are less toxic, for example retinamides like *N*-(4-hydroxyphenyl)retinamide (HPR) which has little or no teratogenicity in animals? Is there a relationship between the teratogenicity of a compound and its effects on the development of limbs?

Tickle: We have done some preliminary experiments to test HPR. One of the problems is that, although AG1-X2 beads release acids extremely well, we have no good way of releasing amides at the moment. With local applications of HPR, in two out of four cases wings with an additional digit 2 developed, but there may be a problem with release (J. Lee, unpublished work).

Moon: What concentration did you use?

Tickle: The concentration in the soaking solution was very high, but we would have to find out how HPR loads onto the particular beads that we used, how much is released and so on before we could say what the concentration was in the bud.

Hicks: Were any of the other retinoids that you tried active?

Tickle: Etretinate was not active, but we need to develop better ways of applying these compounds.

Sporn: The amides may act as 'pro-drugs' that have to be converted back to *all-trans*-retinoic acid before they can have a biological effect, and these embryonic tissues may be devoid of the appropriate metabolizing enzymes. F9 cells are low in both esterase and amidase activity.

Wald: Do you know the reasons for the different potencies of the retinoids that you have studied, Dr Maden?

Maden: No.

Sporn: The simplest hypothesis is that there is no esterase. Retinyl palmitate and retinyl acetate were relatively inactive compared with retinol. In many other systems retinyl acetate is as potent as retinol. Have either of you used any of the retinoidal benzoic acid derivatives?

Tickle: We have used TTNPB [(*E*)-4-[2-(5,6,7,8-tetrahydro-5,5,8,8-tetramethyl-2-naphthalenyl)-1-propenyl]benzoic acid] and we find that it is 10–100 times more active than retinoic acid (J. Lee, unpublished work).

Strickland: Can you change pattern formation by putting the retinoid-soaked bead in a different position on the limb?

Tickle: Yes, there is a position-dependent effect that correlates very well with the effect of grafting polarizing tissue to different positions. If you put the bead in a posterior position you have no effect on the pattern of the wing, so this is the same as grafting the polarizing region to the posterior margin. If you put the bead at the apex of the wing bud, with low retinoid concentrations, that give complete reduplications when administered anteriorly, you get additional structures. These patterns are similar to those that form in response to polarizing region grafts at the apex of the bud. If you increase the concentration of retinoic acid applied at the apex you get truncated limbs; in fact the limbs may be obliterated completely. This can be understood in terms of the effect on the apical ectodermal ridge, which is essential for outgrowth.

Wald: Have you studied the effects of any retinoids on the neural tube?

Tickle: We have been using the beads to apply retinoic acid to several parts of the embryo other than the limb. We have looked at the developing spine, at a rather later stage than neural tube formation, but in one or two cases we have seen spina bifida in the chick (S. Weale & S.E. Wedden, unpublished work).

Lotan: Dr Maden, I was surprised to see loss of metachromasia with such low concentrations of retinoid. I understood that cartilage resorption was a toxic effect of high doses of retinoids and was caused by the release of cathepsin B-like enzymes. How could this occur with low concentrations of retinoic acid?

Maden: I was equally surprised. I thought that loss of metachromasia was just a generalized toxic effect but I see it with retinoid concentrations that are low and that cause pattern changes. I don't have any explanation.

Tsambaos: You have shown retinoid-induced changes in the limb epithelium. What influence do these changes exert on the pattern of limb regeneration?

Maden: None. You can take treated epidermis and graft it onto an untreated limb and it has no effect on the pattern. The epidermis does not play a role in pattern formation. The retinoid-induced changes of importance for pattern take place in the mesodermal cells.

Yuspa: How important are the growth-inhibitory effects? If you gave another growth-inhibitory substance, for example cytosine arabinoside, for five days and then withdrew it would you get the same response?

Maden: All the cellular effects that I described (growth inhibition, cartilage breakdown, changes in fibronectin) have been independently tested for a possible role in the effect on pattern. I have not tested cytosine arabinoside itself; instead I have looked at this by denervating a limb. A normal limb can regenerate only if it has its full nerve supply, which is thought to provide a mitogen, possibly a protein. If you denervate the limb and stop the supply of mitogen for the same length of time that you normally administer retinoid and then allow the nerve to grow back, the cells begin to divide again but you don't get any pattern effects. So the effects of retinoids on limb regeneration do not occur via the inhibition of cell division.

Sherman: Dr Tickle, do you have to administer retinoic acid for a lengthy period of time before you get growth and proliferation, as Malcolm Maden does? In view of your observation of the rapid metabolism of retinoic acid, one might not expect to see a delay in proliferation.

Tickle: Our time course is completely different from Malcolm's. We have to apply retinoic acid for only 14–20 h to get the full effect.

Sherman: So the two systems are quite different in terms of kinetics and the time required to elicit a response?

Maden: Yes. I have to administer retinoid locally for a minimum of two or three days.

Sherman: Do you keep on using fresh retinoic acid implants?

Maden: No, we use a single implantation.

Sporn: There is a big difference between the two systems in terms of their kinetics. One is a very rapidly changing chick embryo in its early stages of development. The other is a mature axolotl; you are cutting off the whole limb and regeneration will take many days.

Sherman: Yes, but what I'm trying to find out is whether, in Malcolm Maden's case, we are looking at a triggering effect or at a response to a constant and long-acting administration of retinoid.

Maden: I don't know. We have to administer retinoid for days, but then metabolism and the whole process of regeneration are much slower than in the chick.

Sherman: But if the *metabolism of the retinoid* is very quick, as it is in the chick, then the role of the retinoid in regeneration could be restricted to initiation. Under these circumstances one might expect regeneration to be induced even when retinoid is given only for short periods.

Maden: I can't explain the mechanisms. I would certainly like to understand the biochemistry of retinoid action in developing and regenerating limbs, but do you think that there is any future in studying this in the limb system itself or should one use cell culture or some other method?

Lotan: Limb development and regeneration are processes that take time; there is a progression of events so you really cannot hold the cells at any one point, ask what happens to fibronectin or any specific parameter at that point and isolate a macromolecule. I don't know how many cells you could isolate for assay. How would you pick single cell types out of the mixture of different types? It would be quite difficult to use your cells for tissue culture, and in any case what cells would you choose and what would be the effect of isolating them from interactions with other cells? You may not be able to follow the process that you see in the intact limb.

Sporn: Can the blastema be cultured?

Maden: It can be cultured as a whole but not as individual cells.

Sporn: What happens if you try to disaggregate blastemas in culture?

Maden: They can be disaggregated, but we have no medium for keeping the cells alive. One would change to a system that could be cultured easily, like chick or *Xenopus* cells.

Koeffler: Could any of the cells be immortalized by a virus?

Tickle: I don't think it has been tried with limb cells. Chick cells are very peculiar in that they do not form immortal cell lines for some reason, but one might be able to use mouse tissue.

Lotan: Lewis et al (1978) looked at mesenchymal cells from the limb bud of the mouse, but all they could show in culture was an inhibition of differentiation and maintenance of fibronectin.

Tickle: But their system is totally different from ours. They were using whole limbs and were looking in micromass cultures at cartilage differentiation. We find that if we take chick limb buds at late stages and implant beads loaded with retinoic acid in a region where the cells are just about to differentiate into cartilage, we can produce malformations of a phocomelic type. Similar effects have been reported in mammals when pregnant females have been fed retinoic acid (Kochhar 1977). These effects are rather different from the limb reduplication effects; they are effects on cartilage differentiation. Cottrill (see Archer et al 1984) has been studying micromass cultures of specific parts of the chick limb, for instance the progress zone, which is a region of undifferentiated cells at the tip of the limb bud. It may be much more relevant for the pattern effects to apply retinoic acid to these cultures than to the ones that Pennypacker used (Lewis et al 1978).

Sporn: I find all this work very exciting. In Malcolm Maden's system, we are dealing with the question of regeneration and I am reminded of the old idea of a relationship between wound healing and cancer (Haddow 1972). In both these, there is a recalling of processes that were functionally useful at some early time in embryonic development. In limb regeneration the blastema, which is a mass of mesenchyme, is important. The two growth factors that are found in particularly high concentrations in blood platelets are platelet-derived growth factor (PDGF) and transforming growth factor β (TGF-β), and these two synergize very strongly in promoting fibroblastic proliferation. You don't find PDGF receptors on cells that are not of mesenchymal origin, like epithelial cells. These two peptides are extremely important for turning on the wound healing response. Malcolm Maden showed that retinoic acid suppresses the proliferative response and I found this particularly exciting because there is a parallel in Anita Roberts' work with *myc*-transfected cells. In soft agar a combination of PDGF and TGF-β is enough to induce these cells to grow in an anchorage-independent manner, but retinoic acid can totally abolish this regenerative response (A. Roberts, unpublished work). Something like that might be going on in the limb. Both growth factors are active at 10^{-10}–10^{-12} M, and it would be very interesting to see what application of these would do *in vivo* to the limb bud of the chick or the regenerating limb of the axolotl. Once you have described a phenomenon you will probably have to go into a cell culture system to look at mechanisms, but at least you can establish *in vivo* whether the growth factors have anything to do with the responses that you see. Retinoids may change the receptors for these growth factors or the sensitivity to growth factors. They may affect sialyl transferases. Whatever the mechanism, it would be very useful to know what factors can modify the response of either the chick or the amphibian limb to retinoids.

REFERENCES

Archer CW, Cottrill CP, Rooney P 1984 Cellular aspects of cartilage differentiation and morphogenesis. In: Kemp RB, Hinchliffe JR (eds) Matrices and cell differentiation. Alan R Liss, New York, p 409-426

Eichele G, Tickle C, Alberts BM 1984 Microcontrolled release of biologically active compounds in chick embryos: beads of 200-μm diameter for the local release of retinoids. Anal Biochem 142:542-555

Haddow A 1972 Molecular repair, wound healing and carcinogenesis: tumour production a possible overhealing? Adv Cancer Res 16:181-234

Kochhar DM 1977 Cellular basis of congenital limb deformity induced in mice by vitamin A. In: Bergsma D, Lenz W (eds) Morphogenesis and malformation of the limb. Alan R. Liss, New York (Birth defects: original article series) vol 13:111-153

Lewis CA, Pratt RM, Pennypacker JP, Hassell JR 1978 Inhibition of limb chondrogenesis in vitro by vitamin A: alterations in cell surface characteristics. Dev Biol 64:31-47

Pitts JD, Burk RR, Murphy JP 1981 Retinoic acid blocks junctional communication between animal cells. Cell Biol Int Rep 5 Suppl A:45

Saunders JW, Gasseling MT 1968 Ectodermal–mesenchymal interactions in the origin of wing symmetry. In: Fleishmajer R, Billingham RE (eds) Epithelial–mesenchymal interactions. Williams & Wilkins, Baltimore, p 78-97

Tickle C, Summerbell D, Wolpert L 1975 Positional signalling and specification of digits in chick limb morphogenesis. Nature (Lond) 254:199-202

Tickle C, Lee J, Eichele G 1985 A quantitative analysis of the effect of all-*trans*-retinoic acid on the pattern of chick wing development. Dev Biol, in press

Tickle C, Alberts B, Wolpert L, Lee J 1982 Local application of retinoic acid to the limb bud mimics the action of the polarizing region. Nature (Lond) 296:564-565

Retinoids and mammary gland differentiation

R. C. MOON, R. G. MEHTA and D. L. McCORMICK

Laboratory of Pathophysiology, Life Sciences Research, IIT Research Institute, Chicago, Illinois 60616, USA

Abstract. Certain retinoids serve as effective chemopreventive agents against breast cancer. The effective retinoids are also antiproliferative agents for the mammary gland both *in vivo* and *in vitro*. *N*-(4-hydroxyphenyl)retinamide (HPR) can inhibit the occurrence of hyperplastic alveolar nodules in C3H mice *in vivo* and 7,12-dimethyl-benz[*a*]anthracene-induced nodule-like alveolar lesions *in vitro*. Moreover, HPR can also inhibit the phorbol ester-induced promotion of hyperplastic alveolar nodule development *in vitro*. HPR is metabolized by the mammary gland *in vitro* and one of the metabolites competes for the cytosolic retinoic acid-binding protein although the metabolite is not *all-trans*-retinoic acid.

1985 Retinoids, differentiation and disease. Pitman, London (Ciba Foundation Symposium 113) p 156–167

Experimental data from a number of laboratories have demonstrated that administration of non-toxic pharmacological doses of natural and synthetic retinoids can suppress the chemical induction of tumours in several epithelial tissues (Moon et al 1983, Moon & Itri 1984). In the rat mammary gland, initial studies of anticarcinogenesis were performed with the natural retinoid retinyl acetate (Moon et al 1976). Although retinyl acetate was found to be highly effective in mammary cancer chemoprevention, the potential for hepatic toxicity associated with chronic, high-dose administration of natural retinoids has prompted the development of synthetic analogues of these compounds. The goal in the design of such synthetic retinoids is to retain or increase anticarcinogenic activity while decreasing toxicity.

Modification of the basic vitamin A structure has produced retinoids with increased target organ specificity; these compounds maintain anticarcinogenic efficacy while toxicity is reduced. Over the past several years, many natural and synthetic retinoids have been evaluated in our laboratory for activity against chemically induced mammary carcinogenesis. Retinyl acetate, retinyl methyl ether, axerophthene, *N*-(4-hydroxyphenyl)-*all-trans*-retinamide and *N*-(4-hydroxyphenyl)-13-*cis*-retinamide all have significant inhibitory activity in experimental models for breast cancer, but several other retinoids are ineffective (Moon et al 1983). Of the active retinoids, *N*-(4-hydroxyphenyl)-*all-trans*-retinamide (HPR) is currently the compound that best combines anticarcinogenic activity with reduced toxicity.

Although retinoids can inhibit mammary cancer induction, no retinoid has yet been developed that is totally effective, i.e. that reduces mammary cancer incidence to zero. However, significant increases in chemopreventive activity can be achieved when retinoid administration is combined with hormonal manipulation (McCormick et al 1982). The synergistic activity of retinoids and bilateral ovariectomy in mammary cancer inhibition suggests an interaction between hormones and retinoids in the control of differentiation and proliferation of the mammary parenchyma. Studies of such interactions may be of value in understanding the mechanism(s) of retinoid action in the inhibition of cancer induction.

Effects of retinoids on mammary gland development *in vivo*

HPR appears to be the retinoid that best combines anticarcinogenic efficacy in the rodent mammary gland with reduced toxicity. Although HPR has received extensive study in carcinogen-treated animals, much less is known about the effects of HPR on the normal (non-carcinogen-treated) mammary gland *in vivo*. Some insight into the problem has been gained through examination of wholemount preparations of the mammary glands of animals maintained on a diet supplemented with HPR.

Fig. 1 shows that the mammary glands of rats receiving HPR exhibit primarily a bare ductal system, with little branching and few end buds. By contrast, extensive ductal branching with many lobuloalveolar structures is evident in the control rats receiving a placebo diet. The antiproliferative effect of HPR does not appear to be a result of an altered hormonal secretion, since serum prolactin levels (Welsch et al 1980) and oestrous cycles (Moon et al 1979) are not altered in retinoid-treated rats. The effect of HPR on the mammary epithelium seems to be truly an antiproliferative response, and not a result of terminal differentiation of the ductal epithelium, because retinoid-treated glands may be induced to proliferate as a result of pregnancy. However, milk secretion is somewhat impaired (R.C. Moon, unpublished work 1984). Although retinoids may inhibit the normal cyclic proliferation of the mammary epithelium, this inhibition may be overriden by intense hormonal stimulation such as that accompanying pregnancy. The anti-proliferative effect of HPR is dose dependent; the glands of animals fed higher retinoid doses are less proliferative than those fed the lower dose of the compound. Similar results have been obtained with retinyl acetate.

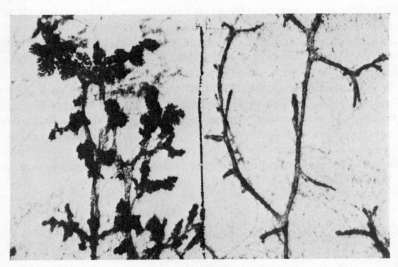

FIG. 1. Representative areas of mammary gland wholemounts from Sprague-Dawley female rats given i.v. injections of 0.8% NaCl solution at 50 and 57 days of age and fed either control diet (left) or diet supplemented with N-(4-hydroxyphenyl)retinamide (782 mg/kg diet) (right). The animals were killed 182 days after the first i.v. injection. The abdominal pair of glands was removed, fixed and stained with alum carmine (× 20). [Taken with permission from Moon et al (1979).]

Although retinyl acetate is a highly effective inhibitor of mammary carcinogenesis in rats, this retinoid is not effective against mammary tumour induction in mice. Experiments conducted in both the C3H/A^{vy} (Maiorana & Gullino 1980) and GR (Welsch et al 1981) strains of mice showed dietary supplementation with retinyl acetate to be ineffective: retinyl acetate had no effect on tumour induction in the C3H/A^{vy} strain and enhanced the occurrence of mammary tumours in GR animals. Consistent with these results, normal mammary glands from GR mice exhibited extensive alveolar development as a result of retinoid exposure; this extensive mammary growth was not seen in untreated controls (Welsch et al 1981). Thus it appears that the effect of individual retinoids on the mammary epithelium is species specific; furthermore, the role of retinoids as anticarcinogenic agents may be related to their antiproliferative effects on the mammary epithelium.

Effects of retinoids on mammary gland development *in vitro*

Mouse mammary glands can be induced to undergo morphological and functional differentiation in organ culture by the addition of appropriate mammogenic hormones to the culture medium. The procedure for organ culture of mammary glands has been described previously (Mehta & Banerjee 1975). Mammary glands incubated with insulin and prolactin exhibit extensive end-bud development, with proliferation of some end buds into alveolar structures (Fig. 2A). Glands do not differentiate into such structures in the absence of prolactin. The addition of either HPR or *all-trans*-retinoic acid to the culture medium at a concentration of 1 µM inhibits prolactin-induced differentiation of the glands (Fig. 2B) (Mehta et al 1983). By contrast, retinyl acetate is ineffective at this concentration and is toxic at higher concentrations. Administration of high concentrations of retinyl acetate allows maintenance of prolactin-induced alveolar structures but causes extreme ductal dilatation (Fig. 2C). These *in vitro* data are consistent with the published *in vivo* data: HPR is a highly effective inhibitor of mouse mammary tumorigenesis, but retinyl acetate has no chemopreventive activity in mouse systems (Welsch et al 1983).

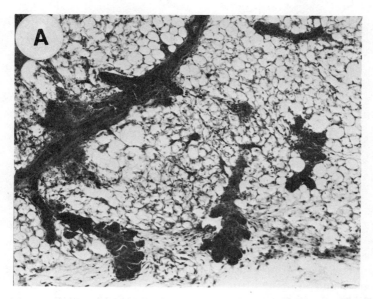

FIG. 2. (*ctd on p. 160*) Histological sections (× 80) of mouse mammary glands incubated *in vitro* with insulin and prolactin with or without retinoid. (A) no retinoid; (B) 1µM-*N*-(4-hydroxyphenyl)retinamide; (C) 1µM-retinyl acetate. [Taken with permission from Mehta et al (1983).]

FIG.2. (*ctd*)

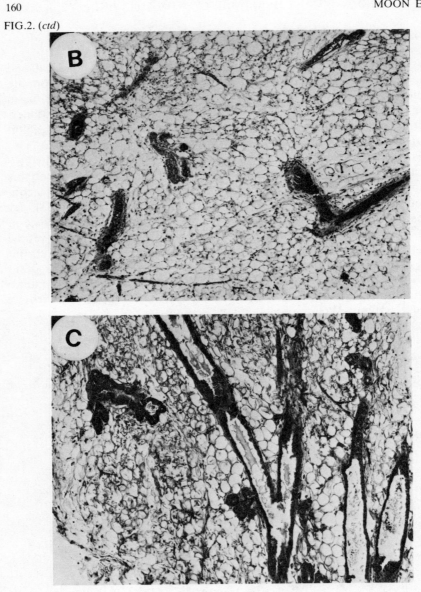

Telang & Sarkar (1983) showed that treatment of organ cultures of adult mammary glands obtained from R111 or C57BL mice with HPR (5–10 μM) resulted in a significant reduction in the number of spontaneous mammary alveolar lesions. However, the development and maintenance of hormone-induced alveolar structures in glands were unaffected by the presence of HPR during the development period. The different effects of HPR on mammary alveolar lesions and on normal lobuloalveolar development suggest that the retinoid may preferentially induce regression of abnormal structures, since these structures were affected by the retinoid while normal lobuloalveolar development was not. Consistent with these findings is our observation that the induction of lobuloalveolar development by the simultaneous addition of insulin, prolactin, aldosterone and hydrocortisone to mammary cultures of immature BALB/c mice is also not affected by HPR (1 μM) (R.G. Mehta et al, unpublished work 1984). Thus, it appears that retinoid modulation of differentiation and/or proliferation of the mammary epithelium *in vitro* depends on the stage of differentiation at which retinoid treatment is initiated, and that such modulation may be counteracted by intense hormonal stimulation.

Effects of retinoids on hyperplastic alveolar nodule development in C3H mice *in vivo*

Hyperplastic alveolar nodules (HAN) in mammary tumour virus-positive C3H mice are considered to be preneoplastic lesions because they develop into tumours when transplanted into syngeneic mice. By contrast, carcinogen-induced HAN in rats do not appear to be preneoplastic (Sinha & Dao 1977). Thus, to evaluate the effects of retinoids on the development of alveolar preneoplastic lesions, a mouse model is required.

The results of a study to determine the effect of retinoids on HAN development in C3H mice are given in Table 1. In this experiment, dietary administration of non-toxic levels of HPR inhibited the development of HAN in a dose-related manner, but retinyl acetate had no effect. In a subsequent study, tumour incidence and number in nulliparous C3H mice were significantly reduced by HPR; however, the retinoid had no effect on mammary tumorigenesis in multiparous animals (Welsch et al 1983). The lack of activity of retinyl acetate in inhibiting HAN development in C3H mice correlates with the lack of an effect of this retinoid on mammary tumorigenesis in the C3H/A^{vy} strain of mice. These results may indicate that the effectiveness of a retinoid in the inhibition of HAN development can be correlated with its activity in inhibiting tumorigenesis.

Effect of retinoids on the development of nodule-like alveolar lesions *in vitro*

The mammary gland organ culture system has also been utilized successfully to evaluate the effects of chemopreventive or therapeutic drugs on the

TABLE 1 Effect of retinoids on the number of hyperplastic alveolar nodules in abdominal–inguinal mammary glands of virgin female C3H/He mice

Group	Diet[a]	No. of mice	No. of hyperplastic alveolar nodules >0.5 mm (mean ± SE)	Mean initial body weight (g)	Mean final body weight (g)	Mice surviving at 20 months (%)
1	Basal[b]	35	4.06 ± 8.87	20.6	31.2	70
2	Basal + retinoid solvent	37	3.64 ± 0.60	20.5	31.0	74
3	Basal + HPR (194 mg/kg diet)	35	2.43 ± 0.46	20.3	30.0	70
4	Basal + HPR (391 mg/kg diet)	31	1.84 ± 0.61*	19.0	31.2	62
5	Basal + retinyl acetate (82 mg/kg diet)	15	3.86 ± 0.46	21.1	27.3	60

* $P < 0.05$ compared with Group 1 (basal diet) and Group 2 (basal diet plus retinoid solvent).

[a] Animals received diets from six weeks of age to 20–24 months of age.

[b] Basal diet, Wayne Lab Meal.

HPR, N-(4-hydroxyphenyl)retinamide.

[Taken with permission from Moon et al (1983).]

development of transformed parenchyma. The protocol for the development of such lesions in organ culture was developed by Banerjee and his colleagues (Lin et al 1976). Mice are pretreated with ovarian steroids for nine days and their mammary glands are then dissected out and incubated with growth-promoting hormones *in vitro* (insulin, prolactin, aldosterone and hydro-cortisone) for 10 days. On day 3 of the incubation period, the glands are exposed to 7,12-dimethylbenz[a]anthracene (DMBA, 2 µg/ml) for 24 h. The glands are incubated from day 10 onwards in medium supplemented with insulin only for three weeks. This allows the regression of all the normal untransformed alveolar structures to the ductal stage; however, some alveolar areas remain unregressed. These unregressed areas are termed nodule-like alveolar lesions (NLAL). Transplantation of cells prepared from these nodule-bearing glands to syngeneic host results in the development of adenocarcinomas (Telang et al 1979).

Dickens et al (1979) showed that the addition of retinylidene dimedone (10^{-6} M) to the medium inhibited the development of DMBA-induced NLAL in this system. Since then we have observed that *all-trans*-retinoic acid or HPR can also significantly inhibit the induction of NLAL by DMBA. Chatterjee & Banerjee (1982) showed that the reduction by HPR of NLAL occurrence depended on the hormone combination used during the growth-promoting phase. The effect of DMBA on nodulogenesis was similar when either oestrogen, progesterone and growth hormone or aldosterone and hydrocortisone were used with prolactin as growth-promoting hormones (Table 2A). The presence of ovarian steroids in the medium, however, tended to reduce the efficacy of HPR in inhibiting DMBA-induced NLAL *in vitro*: HPR reduced the number of NLAL by 68% in glands incubated with adrenocorticoids, but only by 15% in glands incubated with oestradiol plus progesterone (Table 2B). These results are consistent with our finding that HPR was more effective against ovarian hormone-independent tumours *in vivo* than against tumours that are ovarian hormone dependent (McCormick et al 1982).

More recently, Som et al (1984) have observed that β-carotene (10^{-6} M) also inhibits the development of NLAL; the carotenoid is most effective when present during the growth phase (days 0–10) of the culture process. This effect is apparently not due to retinol; however, the metabolic conversion of β-carotene to a compound other than retinol can not be ruled out.

Effect of retinoids on phorbol ester-induced promotion of nodule-like alveolar lesions in organ culture

As noted above, retinoids can effectively inhibit the development of DMBA-induced NLAL in culture. This culture procedure was extended to determine

TABLE 2A Induction of nodule-like alveolar lesions by 7,12-dimethylbenz[*a*]anthracene in mammary glands exposed to different combinations of hormones in organ culture

		Glands with NLAL	
Treatment	Hormones	Number	%
DMBA	I, Prl, F, A	27/34	79
DMSO	I, Prl, F, A	0/34	0
DMBA	I, Prl, GH, E, P	28/40	70
DMSO	I, Prl, GH, E, P	0/28	0

Mammary glands were cultured in the presence of the listed hormones for 10 days and exposed to 7.8 μM-DMBA (Sigma Chemical Company, St. Louis, MO, USA) dissolved in DMSO for 24 h between the third and fourth days of culture. The concentration of DMSO in the medium was 0.1%.

NLAL, nodule-like alveolar lesions; DMBA, 7,12-dimethylbenz[*a*]anthracene; DMSO, dimethylsulphoxide; I, insulin; Prl, prolactin; GH, growth hormone; A, aldosterone; E, 17β-oestradiol; F, hydrocortisone; P, progesterone.

[Taken with permission from Chatterjee & Banerjee (1982).]

TABLE 2B Influence of hormones on retinoid inhibition of the transformation of mammary glands exposed to 7,12-dimethylbenz[*a*]anthracene in organ culture

		Glands with NLAL		Inhibition of
Treatment	Hormones	Number	%	NLAL (%)[a]
HPR	I, Prl, F, A	11/44	25	68
HPR	I, Prl, F, A + E, P	16/24	67	15
HPR	I, Prl, F, A + E	19/38	50	37
HPR	I, Prl, F, A + P	11/29	38	52
HPR	I, Prl, GH, E, P	22/40	55	21

Mammary glands were exposed to 7.8 μM-DMBA dissolved in DMSO for 24 h between the third and fourth days of culture. For days 4–10 the glands were incubated in DMBA-free fresh medium containing 10^{-6} M-HPR in DMSO and the hormone combinations indicated. The concentration of DMSO in the medium was 0.1%.

[a] Compared to control cultures not treated with HPR, as shown in Table 2A.

NLAL, nodule-like alveolar lesions; DMBA, 7,12-dimethylbenz[*a*]anthracene; DMSO, dimethylsulphoxide; I, insulin; Prl, prolactin; GH, growth hormone; A, aldosterone; E, 17β-oestradiol; F, hydrocortisone; P, progesterone; HPR, *N*-(4-hydroxyphenyl)retinamide.

[Taken with permission from Chatterjee & Banerjee (1982).]

whether 12-*O*-tetradecanoylphorbol-13-acetate (TPA) promotes the occurrence of DMBA-induced NLAL, and, if so, whether retinoids can inhibit such promotion. DMBA-initiated glands were exposed to TPA (25 ng/ml) for a number of days during the 24-day culture period. As indicated in Table 3, glands exposed to TPA for days 9–14 of the culture period showed an increased incidence of DMBA-induced NLAL and an increase in the number of nodules per gland. Other regimens were less effective. The addition of HPR to the medium with TPA resulted in a reduction in NLAL incidence from 80% to 30%, and a decrease in the number of lesions per gland. These data clearly indicate that the retinoid can indeed suppress TPA-induced promotion of NLAL development, but it is not known whether enhancement

TABLE 3 Influence of TPA and HPR on 7,12-dimethylbenz[a]anthracene-induced nodule-like alveolar lesions in the mouse mammary gland *in vitro*

Treatment			Glands with lesions		Number of lesions
DMBA[a]	TPA[b]	HPR[c]	Number	%	per gland
—	—	—	3/20	15	0.3
Day 3	—	—	17/28	60	1.6
Day 3	Days 4–9	—	6/10	60	1.1
Day 3	Days 9–14	—	27/32	84	3.4
Day 3	Days 14–21	—	8/10	80	2.2
Day 3	Days 19–24	—	5/10	50	1.2
Day 3	Days 9–14	Days 4–9	8/14	57	1.5
Day 3	Days 9–14	Days 9–12	4/12	40	1.3
Day 3	Days 9–14	Days 4–12	3/10	30	0.6

[a] DMBA, 7,12-dimethylbenz[a]anthracene, at $2\,\mu g/ml$ for 24 h.
[b] TPA, 12-O-tetradecanoylphorbol-13-acetate, at $25\,ng/ml$.
[c] HPR, N-(4-hydroxyphenyl)retinamide, at $1 \times 10^{-6}\,M$.

by TPA of NLAL development is accompanied by an increase in ornithine decarboxylase activity as in skin tumour promotion, or whether the retinoid can inhibit this enzyme activity in the mammary gland.

Retinoid metabolism in mammary gland differentiation

Over the years, several hypotheses have been proposed for the mechanism of retinoid action; however, none has been universally accepted. Most experimental evidence indicates that retinoic acid, rather than retinol, is the active natural component in the control of epithelial differentiation (Sporn & Roberts 1983, Breitman et al 1980). The binding of retinoic acid to a specific cytosolic protein (CRABP) has been suggested to be a requisite step for retinoid action (Chytil & Ong 1984, Mehta et al 1980, 1982). However, recent unpublished data from our laboratory indicate that the absolute concentration of CRABP in a cell may be of less significance in retinoid action than once believed.

We have more recently shown that HPR is metabolized by the mammary gland under organ culture conditions. High performance liquid chromatography (HPLC) analysis of extracts of glands treated with HPR for 10 days in culture yielded three principal metabolites. Two of these metabolites have been identified as 13-*cis*-HPR and *N*-(4-methoxyphenyl)retinamide. The third, more polar, metabolite remains to be identified. A similar metabolic pattern was also observed in the liver and mammary glands of the rats treated with HPR for five days *in vivo*.

CRABP is normally detectable in cultured mammary glands. Although *all-trans*-retinoic acid competes for these binding sites, HPR does not. Interestingly, however, the polar metabolite of HPR isolated by HPLC, which is not *all-trans*-retinoic acid, 13-*cis*-retinoic acid or 5,6-epoxyretinoic acid, does compete for the CRABP. The other two metabolites [13-*cis*-HPR and *N*-(4-methoxyphenyl)retinamide], like the parent HPR, do not compete for these binding sites (R.G. Mehta et al, unpublished work). These results strongly suggest that the mammary gland *per se* is capable of metabolizing retinoids, and this could in turn account for the specificity of retinoids for mammary glands prone to carcinogenesis. If the unidentified metabolite of HPR does prove to have biological activity, it may be possible to determine conclusively whether or not CRABP plays a regulatory role in the action of retinoids on mammary epithelium.

REFERENCES

Breitman TR, Selonick SE, Collins SJ 1980 Induction of differentiation of human promyelocytic leukemia cell line (HL-60) by retinoic acid. Proc Natl Acad Sci USA 77:2936-2940

Chatterjee M, Banerjee MR 1982 Influence of hormones on *N*-(4-hydroxyphenyl)retinamide inhibition of 7,12-dimethylbenz(a)anthracene transformation of mammary cells in organ culture. Cancer Lett 16:239-245

Chytil F, Ong DE 1984 Cellular retinoid-binding proteins. In: Sporn MB et al (eds) The retinoids. Academic Press, Orlando, vol 2:89-123

Dickens MS, Custer RP, Sorof S 1979 Retinoid prevents mammary gland transformation by carcinogenic hydrocarbon in whole-organ culture. Proc Natl Acad Sci USA 76:5891-5895

Lin FK, Banerjee MR, Grump LR 1976 Cell cycle-related hormone carcinogen interaction during chemical carcinogen induction of nodule-like mammary lesions in organ culture. Cancer Res 36:1607-1614

Maiorana A, Gullino PM 1980 Effect of retinyl acetate on the incidence of mammary carcinomas and hepatomas in mice. J Natl Cancer Inst 64:655-664

McCormick DL, Mehta RG, Thompson CA, Dinger N, Caldwell JA, Moon RC 1982 Enhanced inhibition of mammary carcinogenesis by combined treatment with *N*-(4-hydroxyphenyl)retinamide and ovariectomy. Cancer Res 42:508-512

Mehta RG, Banerjee MR 1975 Action of growth promoting hormones on macromolecular biosynthesis during lobulo-alveolar development of the entire mammary gland in organ culture. Acta Endocrinol 80:501-516

Mehta RG, Cerny WL, Moon RC 1980 Distribution of retinoic acid-binding proteins in normal and neoplastic mammary tissues. Cancer Res 40:47-49

Mehta RG, Cerny WL, Moon RC 1982 Nuclear interactions of retinoic acid-binding proteins in chemically induced mammary adenocarcinoma. Biochem J 208:731-736

Mehta RG, Cerny WL, Moon RC 1983 Retinoids inhibit prolactin-induced development of the mammary gland *in vitro*. Carcinogenesis (Lond) 4:23-26

Moon RC, Grubbs CJ, Sporn MB 1976 Inhibition of 7,12-dimethylbenz(a)anthracene-induced mammary carcinogenesis by retinyl acetate. Cancer Res 36:2626-2630

Moon RC, Thompson HJ, Becci PJ et al 1979 N-(4-Hydroxyphenyl)retinamide, a new retinoid for prevention of breast cancer in the rat. Cancer Res 39:1339-1346

Moon RC, McCormick DL, Mehta RG 1983 Inhibition of carcinogenesis by retinoids. Cancer Res (suppl) 43:2469s-2475s

Moon RC, Itri LM 1984 Retinoids and cancer. In: Sporn MB et al (eds) The retinoids. Academic Press, Orlando, vol 2:327-371

Sinha D, Dao TL 1977 Hyperplastic alveolar nodules of the rat mammary gland. Tumor-producing capability *in vivo* and *in vitro*. Cancer Lett 2:153-160

Som S, Chatterjee M, Banerjee MR 1984 β-Carotene inhibition of 7,12-dimethylbenz[a]anthracene-induced transformation of murine mammary cells *in vitro*. Carcinogenesis (Lond) 5:937-940

Sporn MB, Roberts AB 1983 Role of retinoids in differentiation and carcinogenesis. Cancer Res 43:3034-3040

Telang NT, Banerjee MR, Iyer AP, Kundu AB 1979 Neoplastic transformation of epithelial cells in whole mammary gland *in vitro*. Proc Natl Acad Sci USA 76:5886-5890

Telang NT, Sarkar NH 1983 Long-term survival of adult mouse mammary glands in culture and their response to a retinoid. Cancer Res 43:4891-4900

Welsch CW, Brown CK, Goodrich-Smith M, Chiusano J, Moon RC 1980 Synergistic effect of chronic prolactin suppression and retinoid treatment in the prophylaxis of N-methyl-N-nitrosourea-induced mammary tumorigenesis in female Sprague-Dawley rats. Cancer Res 40:3095-3098

Welsch CW, Goodrich-Smith M, Brown CK, Crowe N 1981 Enhancement by retinyl acetate of hormone-induced mammary tumorigenesis in female GR/A mice. J Natl Cancer Inst 67:935-938

Welsch CW, DeHoog JV, Moon RC 1983 Inhibition of mammary tumorigenesis in nulliparous C3H mice by chronic feeding of the synthetic retinoid, N-(4-hydroxyphenyl)-retinamide. Carcinogenesis (Lond) 4:1185-1187

Discussion of this paper appears on p 182

Modulation of carcinogenesis in the urinary bladder by retinoids

R. M. HICKS, J. A. TURTON, J. CHOWANIEC, C. N. TOMLINSON, J. GWYNNE, K. NANDRA, E. CHRYSOSTOMOU and M. PEDRICK

School of Pathology, Middlesex Hospital Medical School, Riding House Street, London W1P 7LD, UK

Abstract. Bladder cancer has a 70% recurrence rate within five years and a high associated mortality. It commonly occurs in one or both of two predominant growth/behaviour patterns: either well-differentiated, relatively benign exophytic papillary lesions, or flat, poorly differentiated invasive carcinoma usually arising from carcinoma-in-situ. We have used the F344 rat treated with *N*-butyl-*N*-(4-hydroxybutyl)nitrosamine (BBN) as a model for the papillary disease, and the BBN-treated B6D2F1 mouse for flat, invasive bladder carcinoma. In the rat, carcinogenesis is a multistage process and several retinoids will delay or even halt the development of bladder cancer. Inhibition of carcinogenesis is not complete, but there is a consistent reduction in the time-related incidence of papillomas and carcinomas and a concomitant improvement in the overall differentiation of the urothelium. In the BBN/mouse model, retinoids also have anticarcinogenic activity but interpretation of the results is more complicated. Unlike the F344 rat, the B6D2F1 mouse has a non-uniform response to BBN; not all mice develop bladder cancer even after treatment with very high doses of BBN and in those that do, more than one mechanism of carcinogenesis may be involved. Individual retinoids differ markedly in their ability to modulate bladder carcinogenesis in rodents; the behaviour of one analogue cannot be predicted automatically from data obtained with another. Combined data from rodent trials in this and other laboratories have identified *N*-(4-hydroxyphenyl)retinamide (HPR) as the most anticarcinogenic retinoid tested so far for the rodent bladder. It is also less toxic in rodents and better tolerated in humans than either 13-*cis*-retinoic acid or etretinate, two retinoids currently used in dermatological practice. A prophylactic chemopreventive trial of HPR in bladder cancer patients starting in 1985 will be centred on the Middlesex Hospital, London.

1985 Retinoids, differentiation and disease. Pitman, London (Ciba Foundation Symposium 113) p 168–190

Bladder cancer is a disease of world-wide distribution and, though it has been epidemiologically linked to cigarette smoking in Europe and the USA, the aetiology of the majority of cases is still unknown. It is largely a disease of multifocal primaries with a 70% recurrence rate within five years and a high associated mortality; prognosis for the bladder cancer patient is poor. Until now, treatment has been aimed at destruction of the tumour and its metastases. Although this is of prime value for the rescue of the cancer patient, the advent of retinoid therapy may permit a new preventive approach to bladder cancer which should complement traditional treatments and benefit the patient. Experimental studies have shown that cancer of the rat bladder develops by a multistage process (Hicks et al 1978, Hicks 1980), each stage of which can be influenced by different factors. If the same is true for humans, the rate of progression of the disease may well be open to manipulation. If the length of the latent period between induction and cancer growth could be increased, this would offer a significant and useful additional method of controlling bladder cancer in populations identified to be at risk.

Vitamin A has been known to be potentially anticarcinogenic for some years (e.g. Bollag 1972). Epidemiological studies have also suggested an inverse correlation between blood retinol levels and cancer risk (Kark 1980, Wald et al 1980), though recently this has been questioned (Willett et al 1984). However, the prophylactic use of natural vitamin A (retinol, *all-trans*-retinoic acid, retinyl palmitate) cannot be advocated for cancer therapy, because high doses of these compounds cause systemic toxic effects without significantly raising blood retinol levels which are under homeostatic control. Synthetic retinoids have been developed that are less toxic than natural vitamin A and yet retain biological activity (Sporn et al 1979), and several of these inhibit the development of chemically induced cancer in various organs in experimental animals (Sporn & Newton 1979, Moon & McCormick 1982). In particular, the incidence of rat or mouse bladder carcinoma, induced either by *N*-methyl-*N*-nitrosourea (MNU) or by *N*-butyl-*N*-(4-hydroxy-butyl)nitrosamine (BBN), can be reduced by post-carcinogen feeding with 13-*cis*-retinoic acid (Grubbs et al 1977, Becci et al 1978, 1981, Sporn et al 1977, Hicks et al 1982), *all-trans*- and 13-*cis*-*N*-ethylretinamide (Thompson et al 1981, Hicks et al 1982, Moon & McCormick 1982), *N*-(2-hydroxy-ethyl)retinamide (Thompson et al 1981), *all-trans*- and 13-*cis*-4-hydroxy-phenylretinamide (Moon & McCormick 1982) and 2-hydroxypropyl-retinamide (Moon & McCormick 1982).

The consistency of rodent data from different laboratories supports the thesis that retinoid treatment may prove therapeutic for human bladder cancer patients. However, if retinoids are to be of value in the management of human bladder cancer, it is important to consider the pathology of the disease and the limitations of currently available therapies. Human bladder cancer behaves as if at least two types of disease process are involved; it usually presents either as well-differentiated, slow-growing, exophytic papillary/nodular lesions or as flat carcinoma-in-situ, although both conditions may occur in the same bladder. The papillary disease, if uncomplicated by carcinoma-in-situ, is usually controllable by local resection or by radiotherapy for many years, even though it may recur with increasing frequency. The flat carcinoma-in-situ, by contrast, is usually a field change affecting large areas of the urothelium; it can progress rapidly to aggressive, poorly differentiated invasive carcinomas which fail to respond to radiation, chemotherapy or local

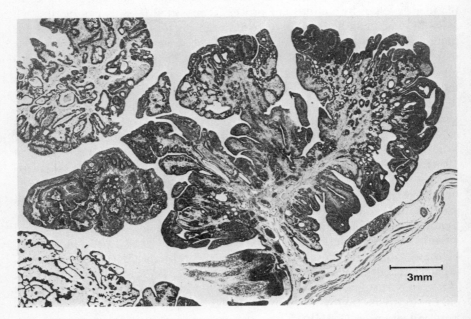

FIG. 1. Typical papillary lesion in the bladder of an F344 rat 52 weeks after treating with 600 mg *N*-butyl-*N*-(4-hydroxybutyl)nitrosamine. Although such tumours may become large enough to kill the animal by urethral obstruction, in this experimental bladder cancer model they seldom invade the thickness of the bladder wall or metastasize to distant sites. Wax section stained with haematoxylin and eosin.

surgery. Although it would be best to prevent bladder cancer altogether, diversion of the disease into the less aggressive, better-differentiated papillary form would also be useful, as would be the development of any treatment capable of halting the further progression of pre-existing invasive disease. Any of these actions would be anticarcinogenic, although very different physiological processes might be involved.

Rodent models now exist for both forms of bladder cancer. Most investigators have used rat models in which the tumours produced are predominantly well-differentiated, slow-growing, papillary/nodular lesions (P/N hyperplasia) which, although they frequently progress and invade the lamina propria (P1 invasion), only rarely extend through the muscle (P2) to the peritoneal surface of the bladder (P3) or produce distant metastases (P4). The MNU-treated Wistar rat (Hicks & Wakefield 1972) and the BBN-treated F344 rat (Ito et al 1969) both provide good models for the exophytic papillary

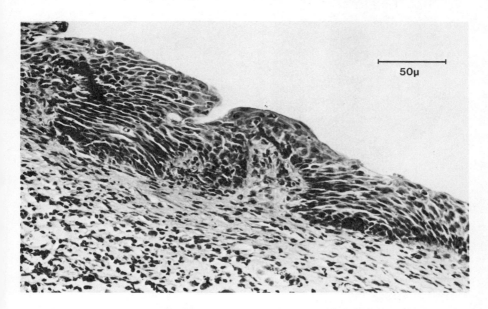

FIG. 2. An area of carcinoma-in-situ in the urothelium of a B6D2F1 mouse treated 19 weeks previously with 30 mg N-butyl-N-(4-hydroxybutyl)nitrosamine. There is gross dysplasia, loss of normal cell polarity and differentiation, and streaming of basal-type cells within the thickness of the urothelium. The animal had been fed tetrazol-5-ylretinamide but the retinoid failed to prevent tumour development. Wax section stained with haematoxylin and eosin.

form of the human disease (Fig. 1). By contrast, the BBN-treated B6D2F1 hybrid mouse, a model developed by Becci et al (1981), develops both extensive carcinoma-in-situ (Fig. 2) and invasive, poorly differentiated transitional-cell carcinoma (Fig. 3), conditions closely resembling the lesions of the aggressive disease in humans. The effects of using retinoids as anticarcinogenic agents in the F344 rat and in the B6D2F1 mouse bladder cancer model systems are considered here, and the possible implications of these studies for the therapeutic use of retinoids in human bladder cancer patients are discussed.

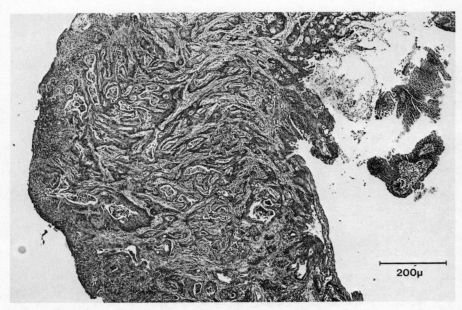

FIG. 3. An aggressively invasive P3 carcinoma arising from wide-spread carcinoma-in-situ in a B6D2F1 mouse treated 19 weeks previously with 30 mg N-butyl-N-(4-hydroxybutyl)nitrosamine. This poorly differentiated tumour filled and thickened the bladder wall and overgrew its peritoneal surface. In places it was also exophytic with papillary extensions into the bladder lumen. The animal had been fed tetrazol-5-ylretinamide but the retinoid failed to prevent tumour development. Wax section stained with haematoxylin and eosin.

The effects of retinoid-containing diets on bladder cancer development in BBN-treated F344 rats

The most frequently used measure of the anticarcinogenic activity of retinoids in experimental animal models has been the percentage reduction in the incidence of carcinoma of the urinary bladder at a single point in time in a retinoid-treated group by comparison with that in placebo-fed controls, e.g. Grubbs et al (1977). Measured thus, several retinoids inhibit carcinogenesis in the BBN/F344 rat model. However, in long-term feeding studies with 13-*cis*-retinoic acid (isotretinoin, CRA) and N-ethylretinamide (NER), we found that if the retinoid-fed animals were maintained for long enough, the retinoid restraint was overcome, the bladder cancer incidence eventually reached the same level as in controls and the retinoid-treated animals also died of bladder cancer. The experimental details and results of these experiments have been published elsewhere (Hicks et al 1982). In the early stages of the experiment the retinoids undoubtedly produced a reduction in time-related bladder cancer incidence; this was attributable entirely to prolongation of the latent period between exposure to the carcinogen and the appearance of histologically detectable tumours. The exact time of assessment of tumour incidence in such systems is thus a crucial factor in the evaluation of retinoid efficacy and, in any particular experiment, the estimate of inhibition could vary from 100% to zero depending on the time at which the animals were killed and the estimate was made (see Fig. 4).

What is noteworthy, however, is that the average life-span of the retinoid-treated animals was extended by that length of time by which the retinoid extended the latent period, i.e. for any particular dose of carcinogen, the carcinogenic response of the urothelium was shifted by some weeks along the time axis.

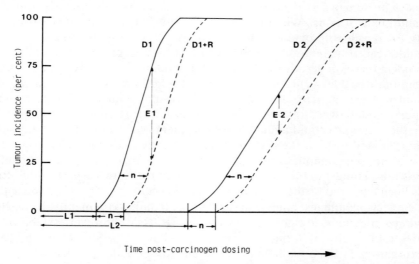

FIG. 4. The effect of a retinoid on the response of the urothelium to two different doses of N-butyl-N-(4-hydroxybutyl)nitrosamine (BBN), based on results obtained with the F344/BBN rat model. The latent periods, L1 and L2, before tumours develop are inversely proportional to the carcinogen doses, D1 and D2, and the slope of the response curve is more acute with the higher than with the lower dose. Inclusion of a retinoid, R, in the diet extends the latent periods by the time n, thus shifting the incidence curves by that period along the time axis. Depending on the time at which the experiment is terminated, retinoid inhibition of tumour incidence may appear to be zero (tumour incidence has reached 100% in both carcinogen-only and in retinoid-fed groups), or to be total [experiment terminated between L1 and (L1 + n), or between L2 and (L2 + n)], or it may reach constant levels, E1 and E2, between the times (L1 + n) or (L2 + n) and the times at which tumour incidences in the carcinogen-only groups reach 100%. If the time n is a constant for any particular dose of retinoid and is not related to carcinogen dose, then retinoid inhibition of tumour incidence will appear more effective after the higher than after the lower dose of carcinogen, because E1 > E2.

The analysis of tissue changes in the bladders of BBN-treated animals killed at different time intervals over a two-year period confirmed that CRA and NER lengthened the latent period of symptom-free urothelial hyperplasia but that once the restraint was overcome, the subsequent rate of tumour growth and progression was the same as in controls. However, since at any point in time the tumours in retinoid-fed animals were 'younger' than those in placebo-fed controls, they were both smaller and better differentiated. Nevertheless, as time progressed, the retinoids did not prevent the development of all the usual markers of neoplastic growth, nor indeed did they prevent the development of areas of squamous metaplasia within the tumours.

Extrapolation from these results suggests that retinoids should be capable of reducing the age-related prevalence of the papillary/nodular form of cancer in humans by delaying, perhaps for some years, the appearance of symptomatic lesions. When it is remembered that the individual at highest risk for developing bladder cancer is the patient who has already had one bladder cancer diagnosed and removed and that, in general, bladder cancer occurs in the sixth decade of life or later, providing an extra five years or so of symptom-free life would be a very useful additional way of managing the bladder cancer patient. The experimental results obtained with rat bladder cancer models were' thus valuable and encouraging, but did not indicate whether retinoids could affect the development of the more invasive, poorly differentiated form of the disease which currently is so difficult to control.

Another point emphasized by these studies was that retinoid toxicity in the rat is not manifest in the same way as it is in humans; this aspect of the work has been explored further by Turton et al (this volume). The most noticeable side-effects in humans, namely dryness of the mouth, skin fragility, headaches and eye infections, are not observed in rats. Rats do, however, exhibit other measurable signs of retinoid toxicity including reduction in weight gain, disturbances in bone modelling, and haematological disturbances. Thus, rat models may give an indication of the *likelihood* of experiencing some toxic side-effect from retinoid therapy for bladder cancer, but cannot be relied upon to give qualitative guidance on the way such toxicity may be expressed.

The effect of retinoids on bladder cancer development in BBN-treated B6D2F1 mice

Moon and his co-workers have assessed the relative inhibitory effects of several retinoids on the per cent incidence of bladder cancer in the BBN-treated B6D2F1 hybrid male mouse. Relatively high total doses of BBN were used, for example 180 mg or 90 mg per mouse (Becci et al 1981, Thompson et al 1981) or 90 mg and 60 mg per mouse (Moon et al 1982), and the per cent inhibition of tumour development was assessed after 4.5–5.5 months on the diet. The most extensively investigated retinoid was CRA but it is apparent from these studies that some other analogues are more effective and others considerably less effective than this compound at inhibiting bladder cancer development in the male mouse; about half the retinoids tested were without any inhibitory effect on bladder carcinogenesis (Moon & McCormick 1982, Moon et al 1982).

With the carcinogen-treated B6D2F1 mouse model, as with the rat model, we aimed to investigate the long-term effects of retinoids. We used females and, to permit longer-term survival of the carcinogen-treated animals, we used lower total doses of BBN: 30 mg and 15 mg per mouse. The higher of these doses still proved to be powerfully carcinogenic and, because of the number of deaths from bladder cancer, the experiments with 30 mg BBN had to be terminated after five months in order to retain sufficient numbers of animals per group to achieve meaningful results. The experimental protocol and detailed histological and toxicological findings from this study are presented elsewhere (Hicks et al 1985).

TABLE 1 **Effect of retinoid-containing diets on the state of the urothelium in B6D2F1 mice pretreated with 30 mg BBN**

Diet[a]	No. of animals	State of urothelium (% of animals affected; total no. affected in parentheses)			
		Normal	Simple hyperplasia	P/N hyperplasia[b]	Neoplasia (cis[c], P1, P2, P3)
Placebo	49	12 (6)	14 (7)	6 (3)	67 (33)
HPR	52	25 (13)	25 (13)	25 (13)	25 (13)
TZ	50	16 (8)	12 (6)	6 (3)	66 (33)
BR	52	10 (5)	13 (7)	21 (11)	56 (29)

[a] Retinoids at a concentration of 1.0 mmol/kg diet.
[b] Papillary/nodular hyperplasia.
[c] Carcinoma-in-situ.
BBN, N-butyl-N-(4-hydroxybutyl)nitrosamine; HPR, N-(4-hydroxyphenyl)retinamide; TZ, tetrazol-5-ylretinamide; BR, N-butylretinamide.

With 30 mg BBN and retinoids at a concentration of 1.0 mmol/kg diet, we confirmed the report (Moon et al 1982) that N-(4-hydroxyphenyl)retinamide (HPR) was the most effective retinoid tested for protecting the mouse bladder against the carcinogenic effect of BBN (Table 1). It had a marked effect both on total tumour incidence and on the differentiation of the urothelium. When this retinoid was fed after carcinogen treatment, twice the number of animals in the HPR-fed group as in the placebo-fed group retained normal bladders, and the number of animals with carcinoma-in-situ or invasive tumours was reduced by more than 60%, from 67% in the placebo-fed group to 25% in the HPR-fed group (Table 1 and Fig. 5). There was a highly significant reduction in the number of animals with dysplastic lesions and only a single carcinoma-in-situ was found in the HPR-fed group by comparison with 13 (26%) in the placebo-fed BBN-treated mice (difference, $P = 0.01$) (Fig. 5a). Eight of the

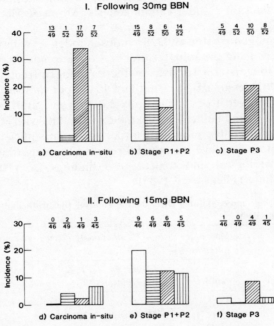

FIG. 5. The effects of retinoids on the stage of bladder cancers induced in B6D2F1 mice by pretreatment with (I) 30 mg N-butyl-N-(4-hydroxybutyl)nitrosamine (BBN) or (II) 15 mg BBN. ☐ , Placebo-fed control group; ▤ , N-(4-hydroxyphenyl)retinamide at 1.0 mmol/kg diet; ▨ , tetrazol-5-ylretinamide at 1.0 mmol/kg diet; ▥ , N-butylretinamide at 1.0 mmol/kg diet. The histograms show the percentage incidences of carcinoma-in-situ, P1 + P2 cancers and P3 carcinomas in the various groups. Actual numbers of tumour-bearing animals and group sizes are shown above each column.

12 invasive carcinomas which developed arose from nodular hyperplasias and were moderately well differentiated, and in only one animal was there a focus of squamous metaplasia within an area of nodular hyperplasia. However, maintaining the mice on dietary HPR apparently had no effect on the development of large and highly aggressive P3 carcinomas, four of which were found in this group (Fig. 5c).

By comparison, N-butylretinamide (BR), given after 30 mg BBN, was relatively ineffective. It did not markedly alter the appearance of the urothelium and had only a very slight inhibitory effect on the numbers of bladder carcinomas found up to five months after completing carcinogen dosing (Table 1). Like BR, tetrazol-5-ylretinamide (TZ) afforded little protection for the carcinogen-treated mouse bladder (Table 1). Neither retinoid prevented the development of aggressive P3 carcinomas in some of the animals treated with 30 mg BBN (Fig. 5c).

Halving the total dose of BBN from 30 mg to 15 mg reduced the overall incidence of invasive carcinomas in the mouse bladder up to 56 weeks from 41% to 21%. Far fewer urothelial dysplasias developed, no carcinoma-in-situ was found and there were more bladders with normal or mildly hyperplastic urothelia (Table 2). HPR was anticarcinogenic against 30 mg BBN and, on

TABLE 2 The dose-related response of the B6D2F1 mouse urothelium to the carcinogen BBN

		State of urothelium (% of animals affected; total no. affected in parentheses)					
Dose of BBN (mg)	No. of animals	Normal	Simple hyperplasia	P/N hyperplasia[a]	Carcinoma-in-situ	Invasive P1, P2	Invasive P3
15	46	17 (8)	46 (21)	15 (7)	0 (0)	19 (9)	2 (1)
30	49	12 (6)	14 (7)	6 (3)	26 (13)	30.5 (15)	10.5 (5)

[a] Papillary/nodular hyperplasia.
BBN, N-butyl-N-(4-hydroxybutyl)nitrosamine.

first consideration, might be expected to be even more effective against the lower dose of 15 mg. On the other hand, our previous experience with the rat model suggested that any one dose of retinoid would be *less* effective against low than against high doses of BBN (Fig. 4). This also proved to be the case with the mouse model.

The effect of retinoid diets on the development of bladder cancer in mice treated with 15 mg BBN is shown in Table 3. The protective effects were less than in mice given 30 mg BBN, although the HPR-treated animals did have fewer carcinomas and more normal bladders than did the placebo-fed group. This trend might have achieved significance had much larger group sizes been used but, because of the small absolute numbers of tumours produced by 15 mg BBN, anything less than total inhibition of carcinogenesis would fail to be statistically significant with the numbers of animals used for these studies. A further confounding factor is the non-uniform response of this hybrid mouse to the carcinogen BBN. Unlike the F344 rat, in which there is a direct dose–response relationship between BBN and the incidence of bladder cancer, some individual B6D2F1 mice are 'resistant' to BBN and fail to develop bladder cancer even after treatment with very high doses (Becci et al 1981, R.C. Moon, personal communication). This further complicates the interpretation of results obtained with low dose levels of BBN when the absolute numbers of tumours produced are small.

TABLE 3 The effect of retinoid-containing diets on the urothelium of B6D2F1 mice pretreated with 15 mg BBN

Diet[a]	No. of animals	State of urothelium (% of animals affected; total no. affected in parentheses)			
		Normal	Simple hyperplasia	P/N hyperplasia[b]	Neoplasia (cis[c], P1, P2, P3)
Placebo	46	17 (8)	46 (2)	15 (7)	22 (10)
HPR	49	45 (22)	27 (13)	12 (6)	16 (8)
TZ	49	41 (20)	27 (13)	10 (5)	22 (11)
BR	45	24 (11)	47 (21)	9 (4)	20 (9)

[a] Retinoids at a concentration of 1.0 mmol/kg diet.
[b] Papillary/nodular hyperplasia.
[c] Carcinoma-in-situ.
BBN, N-butyl-N-(4-hydroxybutyl)nitrosamine; HPR, N-(4-hydroxyphenyl)retinamide; TZ, tetrazol-5-ylretinamide; BR, N-butylretinamide.

As before, neither BR nor TZ prevented the development of aggressive P3 carcinomas in a few of these animals (Fig. 5f). These results are difficult to interpret without understanding the particular mechanisms involved in BBN-induced carcinogenesis in this hybrid-mouse bladder. The fact that none of the retinoids tested prevented the very rapid development of a few P3 carcinomas suggests that two different tumour cell populations with different susceptibilities to retinoids may develop in this model. As discussed elsewhere, more than one pathway of carcinogenesis may well coexist in the bladder after exposure to a carcinogen (Hicks 1983).

In the rat bladder, carcinogenesis is known to be a multistage process involving initiation, promotion and propagation (Hicks 1980, 1983, 1984). The tumours characteristically pass through a phase of clonal expansion of premalignant cells (benign tumours) before malignant conversion takes place. Cancers developing via this process should be susceptible to retinoid intervention, for retinoids have demonstrable antipromoting activity in several animal organ systems (Slaga et al 1980, Yuspa 1983); indeed, the results obtained with the BBN-treated F344 rat model support this postulate. By contrast, the invasive carcinomas which rapidly develop in the mouse bladder after even very low doses of the carcinogen BBN may arise by a different process which involves neither promotion nor the formation of benign tumours composed of premalignant cells. For example, in the mouse bladder BBN could produce a mutation in a cellular proto-oncogene which would then permit immediate transformation and expression of the malignant phenotype [cf. the ability of the mutant Ha-*ras*-1 gene from the T24 human bladder cell line to induce complete malignant transformation without the necessity for cooperation with any second 'immortalizing' oncogene (Spandidos & Wilkie 1984)]. If, as appears to be the case in the rat model, retinoids are unable to affect the subsequent development of fully committed cancer cells, i.e. those which have undergone complete malignant transformation, then they would be unlikely to prevent the rapid growth of cancers developing by such a direct process, irrespective of whether gene activation was the result of a high or of a low dose of BBN.

Relevance of retinoid data obtained with rodent models to the management of human bladder cancer

The extrapolation of data from inbred strains of rodents to a genetically variable human population is fraught with numerous and well-recognized problems. However, there is no currently available economically or scientifically viable *in vivo* alternative to the rodent models if the metabolism, toxicity and organotropism of new compounds and the responses of the whole animal are to be investigated. Models such as those discussed here, in which the human disease condition is accurately reproduced, give a better indication of the potential therapeutic value of new compounds than can *in vitro* studies with human, other animal or bacterial cell systems.

The aetiology of most human bladder cancers is still not understood. Recent advances have been made in describing the molecular biology of bladder carcinogenesis where this involves the Ha-*ras* oncogene, but it must be remembered that evidence for activation of this oncogene could be detected in only about 10% of randomly selected human urinary tract cancers surveyed immediately after surgery (Fujita et al 1984). The remaining neoplasms could well have developed via different pathways, involving more than one developmental stage, the production of more than one tumour cell population and/or cooperation between other oncogenes not identified in the study. Thus, completely transformed urothelial cells may develop by several different mechanisms, some of which, though not all, may be expected to be open to modulation by retinoids such as HPR. In this respect, the BBN-treated B6D2F1 mouse appears to model the human condition more nearly than does the BBN/F344 rat.

Since selected retinoids consistently show anticarcinogenic activity in rodent bladder models, their therapeutic potential in human bladder cancer patients should be investigated. If human bladder cancer responds to HPR in a comparable fashion to urothelial cancer in the rat and mouse, this retinoid could reduce significantly the number and rate of recurrences of papillary urothelial cancers in the high-risk group of patients who have had one such tumour diagnosed and resected. It might also improve the overall differentiation of the urothelium and permit lower-grade tumours to develop which are more easily controlled by current treatment modalities. Preliminary clinical trials with another retinoid, etretinate, provide positive support for these suggestions (Alfthan et al 1983, Studer et al 1984). This compound has been in clinical use for some years and is the retinoid of choice for treatment of psoriasis (in contrast to CRA which is better for acnes). The selection of etretinate for these preliminary bladder cancer trials was doubtless influenced more by considerations of clinical availability than by any evidence of particular suitability or organotropism for the urinary bladder. Observations on the effect of etretinate in rodents are conflicting; this retinoid was reported to inhibit promotion in mouse forestomach epithelium (Wagner et al 1983) but in rat bladder it was only effective if given before exposure to the carcinogen, i.e. it modulated initiation of carcinogenesis (Murasaki et al 1980). Furthermore, at effective dose levels in the mouse it was also toxic; a daily dose of 4.59 mg/kg body weight had to be discontinued because of severe intoxication (Wagner et al 1983). In our laboratory, rats were found to tolerate only about 4.0 mg/kg body weight etretinate by comparison with more than 70 mg/kg body weight of other retinoids including NER and HPR. In rodent trials, both here and elsewhere, HPR has been the most effectively anticarcinogenic retinoid tested so far. In Phase 1 human trials also, HPR has proved notably less toxic than either CRA or etretinate and up to 450 mg per day (higher dose levels are now under investigation) are well tolerated and without significant side-effects. It should therefore be possible to use this retinoid for chemoprevention in bladder cancer patients at doses at least 10 times higher than has been possible so far with retinoids available for dermatological practice.

Bladder cancer is an unpleasant and dangerously life-threatening disease. Any measure that shows promise of improving the prognosis for some patients, even if not for all individuals, is worthy of further investigation. It is hoped that a clinical trial of HPR in bladder cancer patients will be started in 1985, based at the Middlesex Hospital and associated institutes.

Acknowledgements

This work was supported by federal funds from the American Department of Health and Human Services under contract numbers NO1 CP75938 and NO1 CP05602-56. The contents of this publication do not necessarily reflect the views or policies of the Department of Health and Human Services, nor does the mention of trade names, commercial products or organizations imply endorsement by the US Government.

REFERENCES

Alfthan O, Tarkkanen J, Grohen P, Heinonen E, Pyrhonen S, Saila K 1983 Tigason (etretinate) in prevention of recurrence of superficial bladder tumours. Eur Urol 9:6-9

Becci PJ, Thompson HJ, Grubbs CJ et al 1978 Inhibitory effect of 13-*cis*-retinoic acid on urinary bladder carcinogenesis induced in C57BL/6 mice by *N*-butyl-*N*-(4-hydroxybutyl)nitrosamine. Cancer Res 38:4463-4466

Becci PJ, Thompson HJ, Strum JM, Brown CC, Sporn MB, Moon RC 1981 *N*-Butyl-*N*-(4-hydroxybutyl)nitrosamine-induced urinary bladder cancer in C57BL/6× DBA/2F$_1$ mice as a useful model for study of chemoprevention of cancer with retinoids. Cancer Res 41:927-932

Bollag W 1972 Prophylaxis of chemically induced benign and malignant epithelial tumors by vitamin A (retinoic acid). Eur J Cancer 8:689-693

Fujita J, Yoshida O, Yuasa Y, Rhim JS, Hatanaka M, Aaronson SA 1984 Ha-*ras* oncogenes are activated by somatic alterations in urinary tract tumours. Nature (Lond) 309:464-466

Grubbs CJ, Moon RC, Squire RA et al 1977 13-*cis*-Retinoic acid: inhibition of bladder carcinogenesis induced in rats by *N*-butyl-*N*-(4-hydroxybutyl)nitrosamine. Science (Wash DC) 198:743-744

Hicks RM 1980 Multistage carcinogenesis in the urinary bladder. Br Med Bull 36:39-46

Hicks RM 1983 Pathological and biochemical aspects of tumour promotion. Carcinogenesis (Lond) 4:1209-1214

Hicks RM 1984 Carcinogenesis. A multistage process. In: Javadpour N (ed) Bladder cancer. Williams and Wilkins, Baltimore, MD (International perspectives in urology series, vol 12) p 37-49

Hicks RM, Wakefield J StJ 1972 Rapid induction of bladder cancer in rats with *N*-methyl-*N*-nitrosourea. I. Histology. Chem Biol Interact 5:139-152

Hicks RM, Chowaniec J, Wakefield J StJ 1978 Experimental induction of bladder tumours by a two-stage system. In: Slaga TJ et al (eds) Mechanisms of tumour promotion and co-carcinogenesis. Raven Press, New York (Carcinogenesis: a comprehensive survey, vol 2) p 475-489

Hicks RM, Chowaniec J, Turton JA, Massey ED, Harvey A 1982 The effect of dietary retinoids on experimentally induced carcinogenesis. In: Arnott MS et al (eds) Molecular interrelations of nutrition and cancer. Raven Press, New York (MD Anderson symposia in fundamental cancer research, 34th) p 419-447

Hicks RM, Turton JA, Tomlinson CN et al 1985 The effect of three synthetic retinoids on BBN-induced carcinogenesis in the B6D2F1 mouse bladder. Carcinogenesis (Lond), in press

Ito N, Hiasa Y, Tamai A, Okajima E, Kitamura H 1969 Histogenesis of urinary bladder tumours induced by *N*-butyl-*N*-(4-hydroxybutyl)nitrosamine in rats. Gann 60:401-410

Kark JD 1980 The relationship of serum vitamin A and serum cholesterol to the incidence of cancer in Evans County, Georgia. J Chronic Dis 33:311-322

Moon RC, McCormick DL 1982 Inhibition of chemical carcinogenesis by retinoids. J Am Acad Dermatol 6:809-814

Moon RC, McCormick DL, Becci PJ et al 1982 Influence of 15 retinoic acid amides on urinary bladder carcinogenesis in the mouse. Carcinogenesis (Lond) 3:1469-1472

Murasaki G, Miyata Y, Babaya K, Arai M, Fukushima S, Ito N 1980 Inhibitory effect of an aromatic retinoic acid analog on urinary bladder carcinogenesis in rats treated with *N*-butyl-*N*-(4-hydroxybutyl)nitrosamine. Gann 71:333-340

Slaga T, Klein-Szanto A, Fischer S, Weeks C, Nelson K, Major S 1980 Studies of mechanism of action of anti-tumour promoting agents: their specificity in two-stage promotion. Proc Natl Acad Sci USA 77:2251-2254

Spandidos DA, Wilkie NM 1984 The mutant Ha-*ras*-1 gene from T24 human bladder carcinoma cell line induces complete malignant transformation of early passage cells, while normal Ha-*ras*-1 proto-oncogene only induces immortalisation. Nature (Lond) 310:469-475

Sporn MB, Newton DL 1979 Chemoprevention of cancer with retinoids. Fed Proc 38:2528-2534

Sporn MB, Squire RA, Brown CC, Smith JM, Wenk ML, Springer S 1977 13-*cis*-Retinoic acid: inhibition of bladder carcinogenesis in the rat. Science (Wash DC) 195:487-489

Sporn MB, Newton DL, Smith JM, Acton N, Jacobson AE, Brossi A 1979 Retinoids and cancer prevention: the importance of the terminal group to the retinoid molecule in modifying activity and toxicity. In: Griffin AC, Shaw CR (eds) Carcinogens: identification and mechanisms of action. Raven Press, New York (MD Anderson symposia in fundamental cancer research, 31st) p 441-453

Studer UE, Biedermann D, Chollet P et al 1984 Prevention of recurrent superficial bladder tumours by oral etretinate: preliminary results of a randomised, double-blind multicenter trial in Switzerland. J Urol 131:47-49

Thompson HJ, Becci PJ, Grubbs CJ et al 1981 Inhibition of urinary bladder cancer by *N*-(ethyl)-all-*trans*-retinamide and *N*-(2-hydroxyethyl)-all-*trans*-retinamide in rats and mice. Cancer Res 41:933-936

Turton JA, Hicks RM, Gwynne J, Hunt R, Hawkey CM 1985 Retinoid toxicity. In: Retinoids, differentiation and disease. Pitman, London (Ciba Found Symp 113) p 220-246

Wagner G, Habs M, Schmähl D 1983 Inhibition of the promotion phase in two-step carcinogenesis in forestomach epithelium of mice by the aromatic retinoid etretinate. Arzneim Forsch 33:851-852

Wald N, Idle M, Boreham J, Bailey A 1980 Low serum vitamin A and subsequent risk of cancer. Lancet 2:813-815

Willett WC, Polk BF, Underwood BA et al 1984 Relation of serum vitamins A and E and carotenoids to the risk of cancer. New Engl J Med 310:430-434

Yuspa SH 1983 Retinoids and tumor promotion. In: Roe D (ed) Diet, nutrition and cancer: from basic research to policy implications, Alan Liss, New York, p 95-109

DISCUSSION

Papers by Moon et al (p 156–167) and by Hicks et al

Malkovský: What is the mechanism of action of the retinoids in your system, Professor Hicks? Is the action of the carcinogen somehow suppressed directly or is the antiproliferative effect of the retinoids important? Or do retinoids function through the immune system? It seems unlikely to be purely an antiproliferative effect because your tumour incidence curves for retinoid-treated and control animals are essentially parallel.

Hicks: I think the effects of retinoids in the bladder are comparable to their effects in the skin. We have done trials in which rats are killed at various times throughout the experiment and their bladders examined. We give the carcinogen over a period of weeks and by the time we have finished giving it the urothelium has responded with quite significant hyperplasia. During the latent period that hyperplasia persists but does not progress; there is an increase in the thickness of the epithelium but no identifiable markers of preneoplastic modification are produced. We have looked for surface effects, the development of microvilli, basal lamina effects and everything else we could think of, but during the latent period nothing happens. It is only after this period, when exponential growth commences, that microvilli start appearing on the surface of the cells and we see microinvasion through the basal lamina into the mesenchyme. In animals that are fed with placebo this happens at time 't', and in animals that are given the retinoid it happens at time 't + n', about 10 weeks later. There is nothing different about the urothelial pathology. The retinoid does not prevent the hyperplasia developing and it does not make it regress; the hyperplasia simply persists for a longer period of time before tumour growth commences. I think this is comparable to what happens in skin cultures *in vitro*, but I cannot tell you what the mechanisms are. I think the retinoid is delaying the clonal expansion of the premalignant, initiated cells, but how I don't know.

Yuspa: I agree that there are many similarities between the bladder system and skin system. You suggest that retinoids probably act on something analogous to the promotion phase in skin, so why don't you study the mechanism in the two-stage bladder tumorigenesis system which you pioneered, or does this not mimic the human disease?

Hicks: One of the reasons why we have not looked at retinoid effects in the two-stage system is that it takes a long time to do whole-animal experiments, and we need a large number of animals. We need a minimum of 50 animals per experimental group, preferably 100, and a two-year experimental period. We have been trying to mimic the human disease in which we know that the urothelium has been exposed to a carcinogenic dose of something because the patient has developed cancer. We have used, therefore, the full carcinogen-treated system *in vivo*. We are trying to use the two-stage system, but *in vitro* in organ cultures of rat bladder because it should be quicker than doing it in the whole animal.

Yuspa: There is strong evidence that retinoic acid can be a promoter in mouse skin under the right conditions (Hennings et al 1982), and that is a frightening thought, particularly if your bladder model is similar to complete carcinogenesis in the skin. Under conditions of repeated carcinogen administration to mouse skin, retinoic acid definitely enhances tumour induction (Verma et al 1982). So I think there is a danger that retinoids could enhance the expression of malignancy in the bladder, if it is analogous to the skin model. Have you ever made any observations in the bladder system which might suggest that this is a possibility? In Table 1 (p 175) you showed that, in mice treated with 30 mg *N*-butyl-*N*-(4-hydroxybutyl)nitrosamine (BBN), *N*-(4-hydroxyphenyl)retinamide (HPR) actually increased the number of hyperplastic lesions.

Hicks: Yes, but at the same time it reduced the number of carcinomas on a percentage basis and this is the reason that there were more hyperplasias. This is an effect of the retinoid on the time course of the developing lesion; the hyperplasias progress to carcinomas and if they are prevented from developing into carcinomas they persist as hyperplasias. There was no actual increase in the incidence of hyperplasias, merely a slowing of their progression. We have not increased the total number of tumours in any of our experiments with those retinoids that we have used, but all retinoids are not the same. We were concerned about the effect on the bladder of tetrazol-5-ylretinamide, which did appear to increase the number of carcinomas-in-situ. There were also some very aggressive P3 cancers in animals treated with this retinoid. It is possible that this particular compound might increase the risk of developing invasive malignant disease, but that is an impression only, gained from a group of 50 animals, and I would want to repeat the experiment several times before reaching any conclusion.

Moon: Using 100 animals, we saw a decrease in tumour incidence on a short-term basis with that particular compound, but the 13-*cis* derivative had no effect (McCormick et al 1982).

Yuspa: The bladder system worries me more than the mammary system because of the similarities with skin.

Wald: Dr Moon, how do you take account of animals that die through toxicity in your analyses?

Moon: We always run a control group, but at the levels of retinoid that we use there is no toxicity. We have done experiments with life-time administration of retinoid, and mortality is no greater than in animals that receive a placebo.

Wald: So you don't actually need to use actuarial survival methods?

Moon: We use survival tables for intercurrent mortality, but this is really not necessary to get meaningful results.

Wald: Professor Hicks, you mentioned the relative toxicities of HPR, etretinate and 13-*cis*-retinoic acid in humans, but what are the corresponding figures for efficacy and how do you judge it?

Hicks: It depends on the particular system being used. We can only talk about efficacy in rodent models because it has not yet been studied in people. In rodents, HPR is more effective as a chemopreventive agent than 13-*cis*-retinoic acid, and much more effective than etretinate which is not effective at all against bladder cancer unless it is given before the carcinogen.

Moon: In our short-term studies we found that the efficacies of 13-*cis*-retinoic acid and HPR for reducing cancer incidence in the BBN mouse model were about the same, but HPR was less toxic (Moon et al 1982). So we feel that HPR is the best compound because of its reduced toxicity. In a long-term study you may get different results; our study lasted only six months.

Lotan: Is there a difference between the target cells affected by retinoid in the bladder and mammary systems? In mammary epithelial cells you have an inhibition of proliferation of uninitiated cells, which can be seen even before carcinogen treatment, but in the bladder system you see inhibition of the proliferation of initiated cells.

Moon: I think that in the mammary system we get both. We can get inhibition of normal cyclic proliferation in the rat, but this can be overcome by an intense hormonal stimulation such as occurs in pregnancy. In animals treated with *N*-methyl-*N*-nitrosourea (MNU), we can also get an inhibition of proliferation of initiated cells. In models such as the GR mouse, in which intense hormonal stimulation (as in pregnancy) is required to produce tumours, one cannot override the effect by the retinoid. So there are both hormonal and retinoid effects on proliferation, and retinoids can affect both cell types: initiated and normal.

Hartmann: Have you done experiments where you used a cytotoxic drug like mitomycin C or cyclophosphamide together with the retinoid? It could be that with the retinoids you keep one subpopulation of tumour cells in check but not another, which will finally kill the animal.

Hicks: An experiment on these lines, financed by the British Cancer Research Campaign, is in progress. We are using the accepted cytotoxic chemotherapeutic agent thiotepa, which is widely used in clinical practice. We thought that if we could delay the latent period for tumour development with a retinoid and then hit the animals with a cytotoxic chemotherapeutic agent just as the growth of the preneoplastic foci was starting, we might get a synergistic effect. The experiment is not yet complete, but the preliminary results suggest that there is indeed some degree of synergy.

Hartmann: It would be especially useful if you could decrease the amount of cytotoxic drug required.

Malkovský: Do retinoids have an effect on tumour latency in immunocompromised animals?

Hicks: We have not tried immunocompromising animals, so I don't know if it has anything to do with the immune system. MNU is an immunosuppressive agent but I don't think BBN is.

Malkovský: Do retinoids work in your system in immunocompromised animals, Dr Moon?

Moon: I don't know. The only things that we have looked at are HPR and MVE-2 (maleic anhydride-divinyl ether copolymer), which is an immunostimulator. In rats, MVE-2 seems to inhibit carcinogenesis with MNU. If we combine MVE-2 with a retinoid, we get no further suppression of tumorigenesis. This might indicate that the two mechanisms may be the same (McCormick et al 1982).

Malkovský: Have you looked at the effects of retinoids in any of the established human urinary bladder carcinoma cell lines like T24, Professor Hicks?

Hicks: That work is in progress. We have organ cultures of human bladder tissues and also human urothelial explant cultures. We are also working with established cell lines and are currently looking at the interplay of retinoids and growth factors in these *in vitro* systems.

Malkovský: If retinoids just extend the latent period for tumour development they may not be beneficial to the patient at all. The cancer is already there.

Hicks: But the patient may benefit; this is very important. The bladder cancer patient has a urothelium that contains preneoplastic cells at different stages of development. The surgeon removes those that have progressed into visible papillomas and at that point the patient is started on retinoids. Those transformed cells that have already started the exponential phase of growth will continue to grow, but there will also be cells all round the bladder that have been initiated but that have not yet progressed that far, and these should be further restrained by the retinoid. After one or perhaps even two years you should see, as Dick Moon has already shown in rat mammary carcinoma, that the *rate* of recurrence of the tumours is reduced. To begin with probably there would not be much difference between untreated and retinoid-treated groups, but then you would expect to see some benefit from the retinoids as those cells that would have grown into tumours are delayed in their development, possibly by a period of years. This is an advantage for the patient. Usually, the first bladder tumour is the forerunner of a 'lethal cascade'. To begin with you can control the disease by physically removing the tumours, but when it gets to the point when tumours are cropping up over the whole bladder or are causing obstruction, there are few options left. Even after a cystectomy the patient may still die because the tumours have been present long enough to metastasize.

Malkovský: You are trying to hit the cells that are in their latent period, but if you wanted to attack metastatic cancer or a tumour that could not be removed, I wouldn't expect retinoids to be of any use.

Sporn: No. We are not talking about using retinoids for the treatment of invasive tumours; there is no clinical evidence that retinoids affect invasive metastatic disease. It is proposed to use these agents purely for preventive purposes, for people with preneoplastic disease.

Hicks: This is not an alternative therapy; it is an additional way of managing the bladder cancer patient. It does not in any way replace surgery or radiotherapy but we hope it will decrease the *rate* of recurrence. Also, because of the differentiating effects that retinoids have on the urothelium, retinoid treatment may extend the period of time for which there are only well-differentiated small papillary growths that are amenable to surgery.

Wald: If one can achieve a frame shift of five years, that would be an enormous advance in the prevention of bladder cancer. We can compare this with the effects of stopping smoking. You can imagine that smoking is a human experiment: you are exposed to the carcinogens in smoke, but when you stop smoking you remove the carcinogens. For non-smoking British doctors the incidence of lung cancer was found by Doll & Hill to be about five cases per 100000 per year, but for cigarette smokers the figure was about 100 cases per 100000 per year. Incidence increases with time (i.e. age), and the lines for smokers and non-smokers are parallel on a semilog scale (Fig. 1). If you stop

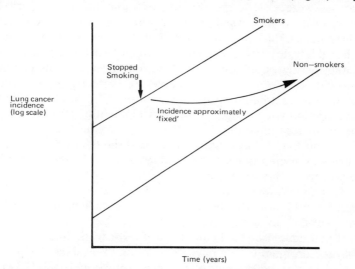

FIG. 1. (*Wald*) Lung cancer incidence in smokers and non-smokers.

smoking you move to the non-smokers' line; you do not actually change the rate. This is analogous to a frame shift, and in human terms it is regarded as prevention. The difference between the lines is so great that you are likely to die of something other than lung cancer if you give up smoking. A few people will die of lung cancer but 10–20 years later; most will die of other causes. I mention this because some people seem to think that one does not achieve anything because the slopes of the survival curves for lung cancer or bladder cancer are not changed. They don't have to be.

Sporn: In addition to Marian Hicks' clinical trial at the Middlesex Hospital, London, a study with HPR is being started by Professor Umberto Veronesi in a breast cancer trial at the Tumour Institute in Milan. Five thousand women who have had a mastectomy, and for whom there is a risk of recurrence in the form of a second primary in the contralateral breast, will be involved. The risk of recurrence is not as high as it is for bladder cancer, which is 70% in a five year period, but it is still about 4% over five years. Dr Moon has shown the beneficial effects of this retinoid on mammary tumorigenesis in rats (Moon et al

1979), so we have a good animal model. He has also found that the combination of surgery plus retinoid gives some added protection (McCormick et al 1983).

Hicks: There is a possible way of managing bladder cancer that we have not explored, but that parallels Dick Moon's approach with mammary cancer. Bertram & Craig (1972) looked at the effects of BBN on C57BL/6 black mice. They produced dose–response curves for male and female mice and found that the responses were exactly the same except that the latent period before bladder cancer developed was longer for female than for male mice. If they castrated the males, the curve shifted towards the female response, and if they gave the females testosterone their curve shifted towards the male response. So it is possible that if you eliminated testosterone from animals you might get synergism with the retinoid effect. People are perfectly prepared to accept castration for the treatment of prostatic disease, and if you are going to die of bladder cancer and you are aged 60 or 70, you might be prepared to consider castration as an additional therapy for prolonging your life.

Koeffler: You found that retinoids extended the latent period for bladder cancer in rats by some weeks. But why does a 10 week latent period in a rat translate into five years in humans?

Hicks: It is an extrapolation based on relative life-span.

Koeffler: I would think that human and murine malignant cells would have similar rates of cell division.

Hicks: If you have tried to do any *in vitro* work with human cells, you will know what the problem is. We have been trying to produce malignant transformation of human cells *in vitro* in organ culture with MNU, which is a direct-acting carcinogen. You can achieve malignant transformation of rat and mouse organ cultures in about the same length of time as it takes *in vivo*, but although we can keep human organ cultures alive for over a year, they look absolutely normal at the end of that time despite receiving high doses of MNU. Richard Peto has put forward the hypothesis that the latency between exposure to carcinogen and development of malignancies is proportional to the life-span of the animal. There are a lot of data to support this, and, since we have been unable to measure the latent period *in vitro* using human tissue, we have simply used the relationship between the life-span of a mouse or rat and that of a human to estimate the latent period.

Sporn: We don't really know how latency in experimental animals translates into latency in humans. You can give a Sprague-Dawley rat intravenous MNU and get palpable breast cancers 6–8 weeks later, but clinicians estimate that the latency for the development of human breast cancer is of the order of 20 years. For cigarette smoking and asbestos exposure, the latent periods are known for humans. They are not of the order of weeks, as in experimental rodent systems, but 20 years or more. The obvious reason for doing clinical trials with retinoids is to see if we *can* get a five-year extension of this latency for bladder cancer. It is important to realize that the extra five years would be tumour-free life, life without morbidity. In the case of chemotherapeutic agents, a five-year increase in survival is the standard hallmark of acceptability, but a lot of that increase may be associated with morbidity.

Koeffler: One treatment for *in situ* bladder cancer is cystectomy. Alternatively, only detectable malignancies can be removed, but statistically we know that many sites of malignancy still exist in the bladder that cannot be detected, and within several years the patient will develop one or several clinical recurrences. I will be pleased but very surprised if retinoids are able to increase the latency period by five years.

Hicks: We can't tell until we try. With human bladder cancer the mean time to detection (which is not necessarily the same thing as latency) following known exposure to 2-naphthylamine is about 20 years. So there is a very long period of time to play with in terms of latent period.

Koeffler: Will you be giving the drug after the first cancer is detected or after exposure to the carcinogen?

Hicks: Obviously after the first cancer; we don't usually know to what carcinogen the patients have been exposed, or when. The patients who would be admitted to our trial are patients who have presented with one superficial cancer. At first presentation 80% of patients have a single papillary tumour which can be removed. After tumour removal, these patients have a pathologically assessed T_0 urothelium; in other words they are 'cured'. But we know that 70% of such patients will develop another tumour within five years, and for this reason they are always followed up routinely at regular intervals. These are the patients we think we can help. A patient with carcinoma-in-situ would not be returned to a T_0 urothelium after treatment, and therefore would not be admitted into this trial. The idea of a chemopreventive agent is to avoid interfering with the clinician's right to treat the patient in the way he or she considers best. We are just hoping to give the patient the chance of a few extra years of symptom-free life before the next episode. This is not an alternative to conventional therapy, but an addition.

REFERENCES

Bertram JS, Craig AW 1972 Specific induction of bladder cancer in mice by butyl-(4-hydroxybutyl)nitrosamine and the effects of hormonal modifications on the sex difference in response. Eur J Cancer 8:587-594

Hennings H, Wenk M, Donahoe R 1982 Retinoic acid promotion of papilloma formation in mouse skin. Cancer Lett 16:1-5

McCormick DL, Becci PJ, Moon RC 1982 Inhibition of mammary and urinary bladder carcinogenesis by a retinoid and a maleic anhydride–divinyl ether copolymer (MVE-2). Carcinogenesis (Lond) 3:1473-1477

McCormick DL, Sowell ZS, Thompson CA, Moon RC 1983 Inhibition by retinoid and ovariectomy of additional primary malignancies in rats following surgical removal of the first mammary cancer. Cancer (Phila) 51:594-599

Moon RC, McCormick DL, Becci PJ et al 1982 Influence of 15 retinoic acid amides on urinary bladder carcinogenesis in the mouse. Carcinogenesis (Lond) 3:1469-1472

Moon RC, Thompson HJ, Becci PJ et al 1979 N-(4-Hydroxyphenyl)retinamide, a new retinoid for prevention of breast cancer in the rat. Cancer Res 39:1339-1346

Verma A, Conrad E, Boutwell R 1982 Differential effects of retinoic acid and 7,8-benzoflavone on the induction of mouse skin tumors by the complete carcinogenesis process and by the initiation–promotion regimen. Cancer Res 42:3519-3525

Effect of retinoids on rheumatoid arthritis, a proliferative and invasive non-malignant disease

CONSTANCE E. BRINCKERHOFF, LYNN A. SHELDON, MARY C. BENOIT, DAVID R. BURGESS* and RONALD L. WILDER†

*Department of Medicine, Dartmouth-Hitchcock Medical Center, Hanover, New Hampshire, *Department of Anatomy and Cell Biology, University of Miami School of Medicine, Miami, Florida, and †Arthritis and Rheumatism Branch, NIH, Bethesda, Maryland, USA*

Abstract. In rheumatoid arthritis synovial tissue proliferates and destroys articular cartilage, bone and tendons. Collagenase is a major mediator of the connective tissue degradation. This enzyme is produced in large quantities by rheumatoid tissue and its synthesis can be inhibited by retinoids. However, knowledge of mechanisms controlling retinoid inhibition of collagenase production and of factors possibly controlling synovial cell proliferation is limited. We found that transforming growth factor β in combination with epidermal growth factor, epidermal growth factor alone and immune interferon increased proliferation of cultured human and rabbit synovial fibroblasts. Only transforming growth factor β caused a piling up of cells into foci resembling those seen in primary cultures of human rheumatoid tissue. All the factors were antagonized by retinoids but not by glucocorticoids or indomethacin. Adding retinoids or glucocorticoids to collagenase-producing cells decreased hybridizable collagenase mRNA by 50% within 24 h. Oral administration of retinoids to rats with experimental arthritis decreased clinical disease without toxicity, and inhibited collagenase synthesis by synovial cells taken from treated animals. Retinoids are both antiproliferative and anti-invasive, and therefore may be potential therapeutic agents in the treatment of rheumatoid arthritis.

1985 Retinoids, differentiation and disease. Pitman, London (Ciba Foundation Symposium 113) p 191–211

Rheumatoid arthritis is a non-malignant proliferative and invasive disease in which synovial tissue—consisting of immune lymphocytes, mono-cytes/macrophages, polymorphonuclear leucocytes and fibroblast-like cells—organizes into a mass that invades and destroys articular cartilage, bone and tendons. The aetiology of rheumatoid arthritis is not understood, but a favoured hypothesis is that an unknown antigen elicits an immune response within the joints of genetically susceptible individuals. This immune response leads to a chronic inflammation of the synovial lining tissue. Associated with and probably caused by the inflammation is the proliferative/destructive component of the disease, which is mediated by the heterogeneous mixture of cell types composing rheumatoid synovium (Harris 1985).

Mononuclear cell factor (MCF), an interleukin 1-like substance, is well-characterized for its ability to stimulate both synovial cell proliferation and connective tissue degradation, the latter occurring because MCF induces the synthesis and secretion of large quantities of collagenase by the synovial cells (Mizel et al 1981). Collagenase is the only enzyme able to initiate breakdown of the interstitial collagens, and the role of this enzyme in the joint destruction accompanying rheumatoid disease is well recognized (Harris 1985).

Retinoids inhibit collagenase production by synovial cells (Brinckerhoff et al 1980, Brinckerhoff & Harris 1981b) and antagonize the proliferative action of growth factors in a number of cell-culture and animal systems (Sporn et al 1984). With the exception of MCF, the growth factors operative in rheumatoid arthritis are not yet defined and the mechanism by which retinoids inhibit collagenase synthesis is unknown. The goals of this study were therefore: (1) to identify other growth factors that may be operative in rheumatoid arthritis and to monitor the ability of retinoids to antagonize their action; (2) to document the mechanism behind retinoid inhibition of collagenase production; and (3) to measure the effect of orally administered retinoids on animal models of arthritis. This approach allows us to study the effects of retinoids on synovial cells at the levels of animal and cell biology, biochemistry and molecular genetics.

Antagonism of growth factors by retinoids

We tested the ability of three growth factors, transforming growth factor β (TGF-β), epidermal growth factor (EGF), and immune or gamma interferon (IFN-γ), to affect the proliferation of synovial cells. TGF-β has been defined operationally by its ability to permit anchorage-independent growth of cells in soft agar, provided that epidermal growth factor (also known as a form of TGF-α) is also present (Roberts et al 1981). TGF-β has been found in many normal tissues including kidney, salivary gland (Roberts et al 1981) and platelets (Assoian et al 1983), and the question arises as to whether it may play a physiological role in a non-malignant proliferative disease such as rheumatoid arthritis. Because TGF-β depends absolutely on EGF for its proliferative effect, we also tested the effect of EGF.

IFN-γ was tested for its mitogenicity for several reasons. IFN-γ is a product of activated immune T cells, cells that are commonly found in inflamed rheumatoid synovium (Harris 1985). In addition, interferons are found in the sera of patients with autoimmune disorders (Harris 1985, Preble et al 1982).

Table 1 demonstrates the proliferative action of TGF-β and EGF on rabbit synovial fibroblasts and compares the ability of *all-trans*-retinoic acid (10^{-6}M), indomethacin (10^{-7}M) and dexamethasone (10^{-7}M) to antagonize this proliferation (Brinckerhoff 1983). Glucocorticoids (such as dexamethasone) and non-steroidal anti-inflammatory drugs (such as indomethacin) are

TABLE 1 Effect of *all-trans*-retinoic acid, dexamethasone and indomethacin on cell proliferation induced by growth factors

Cell treatment for expt I	[³H] Thymidine incorporated (c.p.m./mg cell protein ± SD)		Cell treatment for expt II
	Expt I	Expt II	
Untreated	370 ± 36	390 ± 135	Untreated
EGF only (2 ng/ml)	668 ± 7	1370 ± 270	EGF (5 ng/ml)
TGF-β + EGF	1020 ± 129	—	—
TGF-β + EGF + RT	492 ± 20	570 ± 170	EGF + RT
TGF-β + EGF + DX	1206 ± 34	2315 ± 170	EGF + DX
TGF-β + EGF + INDO	1327 ± 55	3415 ± 36	EGF + INDO

Duplicate cultures of confluent rabbit synovial fibroblasts in 10% fetal calf serum were incubated with *all-trans*-retinoic acid, dexamethasone or indomethacin for 24 h. Medium and drugs were renewed and growth factors were added and the incubation continued for an additional 48 h. The cultures were then pulse-labelled with [³H]thymidine for 4 h and the amount of protein in each culture was measured. In Experiment I cultures were treated with TGF-β (200 ng/ml) supplemented with EGF (2 ng/ml). In Experiment II cultures were treated with EGF (5 ng/ml). RT, *all-trans*-retinoic acid (10^{-6}M); DX, dexamethasone (10^{-7}M); INDO, indomethacin (10^{-7}M). [Reprinted from Brinckerhoff (1983) with permission of the publishers.]

important experimental and therapeutic agents in rheumatoid arthritis and were for this reason included in the experiment. The cultures were treated with drugs for 24 h. Medium and drugs were renewed and growth factors were added; incorporation of [³H]thymidine was measured at 48 h. In agreement with data from other investigators (Roberts & Sporn 1984), our results show that TGF-β in combination with EGF and EGF alone stimulated synovial cell growth and that retinoic acid antagonized this proliferation. In contrast, dexamethasone and indomethacin further enhanced cell proliferation, perhaps by reducing the synthesis of endogenous prostaglandins which suppress growth (Korn et al 1980) or by increasing the number of EGF receptors on the cell (Baker et al 1978).

We also investigated the effect of treating human synovial fibroblasts with IFN-γ and either *all-trans*-retinoic acid (10^{-6}M) or the glucocorticoid prednisolone (10^{-7}M) (Fig. 1). A 72 h treatment with IFN-γ only gave a dose-dependent increase in [³H]thymidine incorporation. As in the experiments shown in Table 1, retinoic acid was able to antagonize the increase, while the steroid hormone enhanced the mitogenicity of the growth factor.

IFN-γ is generally thought to be an antiproliferative agent and in our hands it did, in fact, inhibit growth of HeLa cells (Brinckerhoff & Guyre 1985). Its ability to stimulate fibroblast proliferation suggests that this compound may

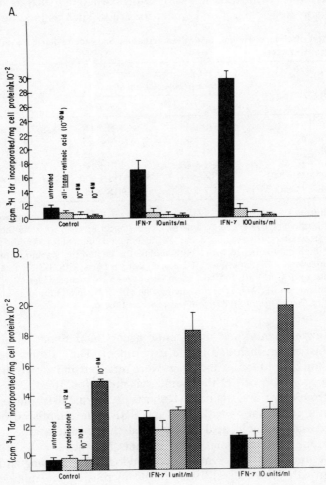

FIG. 1. Antagonism by *all-trans*-retinoic acid and enhancement by prednisolone of immune interferon-induced synovial cell proliferation. Duplicate cultures of confluent human synovial fibroblasts in 10% fetal calf serum were treated for 72 h with immune interferon (IFN-γ) at 1, 10, or 100 anti-viral units/ml in the presence or absence of (A) *all-trans*-retinoic acid (10^{-6}, 10^{-8} or 10^{-10} M) or (B) prednisolone (10^{-8}, 10^{-10} or 10^{-12} M). The cultures were then pulse-labelled for 3 h with [³H]thymidine [³H Tdr] and the amount of radioactivity incorporated and the protein content of the cultures were determined. [Reprinted from Brinckerhoff & Guyre (1985) with permission of the publishers.]

play a role in the proliferative lesion of rheumatoid arthritis and may affect fibroblasts more generally. Since retinoic acid can inhibit the mitogenic effects of IFN-γ, this interferon can be added to the list of polypeptides against which retinoids can act.

We also studied the effect of growth factors on cell morphology and correlated this morphology with the organization of actin filaments within the cell (Fig. 2). Control cultures were typically flat with highly organized bundles of actin filaments (Fig. 2A,B). Treatment with EGF changed the cell shape and pattern of actin only slightly (Fig. 2C,D). In contrast, TGF-β, in combination with EGF, caused a marked piling up of cells and a disruption of actin filaments (Fig. 2E,F). Note that, with respect to both cell morphology and distribution of actin filaments, TGF-β-treated rabbit cells resemble human rheumatoid synovial cells in primary culture (Fig. 2G,H). Note, too, that all-trans-retinoic acid antagonized the morphological transformation that occurred when monolayers of rabbit synovial fibroblasts were treated with TGF-β (Fig. 2I,J). These TGF-β-induced morphological changes are similar to those that were observed when non-neoplastic Rat-1 cells were treated with transforming growth factors (Ozanne et al 1980, Roberts & Sporn 1984). These authors found that the changes induced by transforming growth factors correlated well with the morphology and actin patterns seen in malignant cells.

Despite the fact that TGF-β-treated rabbit synovial cells morphologically resembled rheumatoid synovial cells in primary culture, we do not yet know whether polypeptide mediators like TGF-β have a pathophysiological role in rheumatoid arthritis. It is intriguing to speculate that the presence of transforming growth factors in rheumatoid tissue may have accounted for the ability of this tissue to survive, even for a limited period of time, in nude mice (Brinckerhoff & Harris 1981a), whereas a single-cell suspension of normal synovial fibroblasts disappeared from the injection site within days. When injected subcutaneously into nude mice, dissociated rheumatoid synovial cells organized into a mass remarkably similar to rheumatoid tissue. The mass retained, for at least three to four weeks, its ability to synthesize collagenase and prostaglandin E_2, but, in contrast to malignant cells, it did not go on to form a tumour.

Mechanism of inhibition of collagenase production by retinoids

Until recently, glucocorticoids have been the only agents known to inhibit the synthesis of collagenase and hence, at least potentially, to be able to ameliorate the joint destruction seen in rheumatoid arthritis (Harris 1985). However, the side-effects associated with long-term use of these drugs render

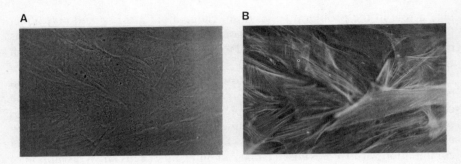

FIG. 2A,B. Control rabbit synovial fibroblasts: (A) phase contrast; (B) phallocidin-stained.

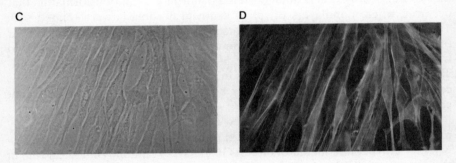

FIG. 2C,D. Rabbit synovial fibroblasts treated with epidermal growth factor: (C) phase contrast; (D) phallocidin-stained.

FIG. 2. Effect of transforming growth factors on synovial cell morphology and actin filaments and antagonism by *all-trans*-retinoic acid. Confluent cultures of rabbit synovial fibroblasts in 10% fetal calf serum were grown on glass cover-slips and treated with transforming growth factors α and β at a final concentration of 200 μg total protein per ml or with epidermal growth factor (a form of transforming growth factor α) at 5 ng/ml for 48 h. Selected cultures were treated with both transforming growth factors and *all-trans*-retinoic acid (10^{-6} M). Also cultured on cover-slips in 10% fetal calf serum for five days were primary human rheumatoid synovial cells (See Brinckerhoff 1983). To fix the cells, cultures were washed twice in phosphate-buffered saline and treated with a 3.7% formaldehyde solution in buffered saline for 10 min. The cover-slips were washed twice in buffered saline and extracted with a solution of acetone at 4 °C for 5 min. After a quick water wash, actin filaments were stained with NBD-phallocidin (Molecular Probes, Plano, Texas) at 60 units stain/ml phosphate-buffered saline for 30 min at room temperature. The cover-slips were washed rapidly twice, mounted cell-side down on a glass slide and the cells visualized by phase-contrast and immunofluorescence microscopy with a 40 × oil-immersion objective.

E

F

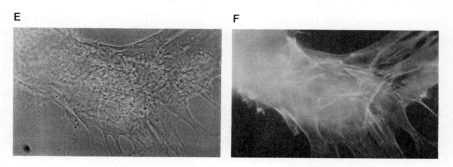

FIG. 2E,F. Rabbit synovial fibroblasts treated with transforming growth factors: (E) phase contrast; (F) phallocidin-stained.

G

H

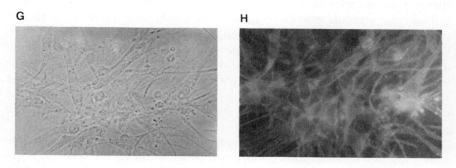

FIG. 2G,H. Primary culture of human rheumatoid synovial cells: (G) phase contrast; (H) phallocidin-stained.

I

J

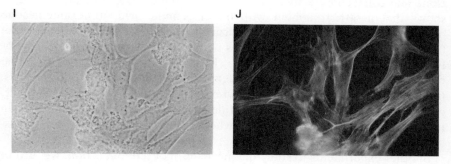

FIG. 2I,J. Rabbit synovial fibroblasts treated with transforming growth factors and *all-trans*-retinoic acid: (I) phase contrast; (J) phallocidin-stained.

them undesirable as therapeutic agents for the treatment of rheumatoid arthritis. The discovery several years ago that retinoids can also prevent collagenase synthesis (Brinckerhoff et al 1980, Brinckerhoff & Harris 1981b) advances these compounds to the top of the list of potential candidates for therapeutic use.

Despite the fact that retinoids are thought to affect transcription (Sporn et al 1984), much like the glucocorticoids, the specific mechanisms of retinoid action are largely unknown. We have used a model system of monolayer cultures of rabbit synovial fibroblasts to study the effect of retinoids (Brinckerhoff et al 1980, Brinckerhoff & Harris 1981b, Brinckerhoff et al 1985, Gross et al 1984). Unless stimulated experimentally with agents such as phorbol myristate acetate (12-O-tetradecanoylphorbol-13-acetate, TPA; 10^{-8} M), these rabbit cells secrete negligible amounts of collagenase. Treatment with an inducer increases collagenase levels to those seen in primary cultures of rheumatoid synovial cells and, conversely, treatment of induced cells with retinoids or steroids prevents synthesis of collagenase protein. In further studies, using a cDNA clone for rabbit synovial cell collagenase, we showed that collagenase mRNA (2.7×10^3 bases) appeared in the cell by 5 h (Gross et al 1984). The increase in collagenase mRNA correlated with an increase in immunoreactive collagenase protein (M_r 57K and 61K) that was detectable in the culture medium by 10 h. Thus the increase in collagenase mRNA in the cell was paralleled by an increase in collagenase protein in the culture medium.

We compared the ability of *all-trans*-retinoic acid (10^{-6} M) and dexamethasone (10^{-7} M) to affect collagenase mRNA levels in rabbit synovial fibroblasts. For these experiments, populations of cells were induced with TPA (10^{-8} M). At time zero, the inducer was removed and cultures were treated with either retinoic acid or dexamethasone. At intervals, the amount of collagenase mRNA in the cell was measured by hybridization to a cDNA clone and collagenase activity in the culture medium was determined. Fig. 3 shows that a 60 h treatment with either retinoic acid or dexamethasone reduced the amount of mRNA hybridizing to the cDNA clone for collagenase (Fig. 3A), and that this decrease was paralleled by a decrease in collagenolytic protein in the culture medium (Fig. 3B). Results obtained with the retinoid and the steroid are nearly identical; this suggests that, like steroids, retinoids act on transcription to decrease collagenase synthesis. However, effects on mRNA half-life cannot be ruled out. Thus, these experiments begin to address the question of how, not whether, retinoids modify gene expression.

In other experiments to study the mechanism of retinoid action, we further extended the parallel between steroid and retinoid behaviour and asked whether retinoids, like steroids, might induce a regulatory protein. In macrophages this regulatory protein is called macrocortin ($M_r \approx 40K$ and 15K). It is an autocrine protein found in resting cells and it regulates

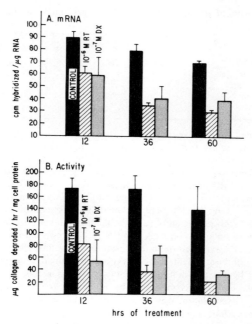

FIG. 3. Effect of *all-trans*-retinoic acid and dexamethasone on collagenase mRNA and collagenase activity. Confluent cultures of rabbit synovial fibroblasts in serum-free medium were stimulated to produce collagenase by 48 h treatment with TPA (10^{-8}M). TPA was then removed, medium was renewed and selected cultures received *all-trans*-retinoic acid (RT, 10^{-6}M) or dexamethasone (DX, 10^{-7}M). At intervals (12, 36 and 60 h), selected cultures were terminated by harvesting the medium and RNA. Medium and drugs were renewed in the remaining cultures at each time point. RNA was harvested by pelleting through guanidine hydrochloride (Chirgwin et al 1979) and collagenase mRNA was measured by dot-blot analysis (Gross et al 1984) with 1 µg and 2 µg of RNA spotted in duplicate onto nitrocellulose filters. The RNA on the filters was hybridized with a cDNA clone for rabbit synovial cell collagenase (Gross et al 1984) radiolabelled, by nick-translation, with ^{32}P. The amount of hybridization was quantified by excising the dots and measuring their radioactivity. Collagenase activity in the medium taken from these cultures was determined in a standard fibril assay with radiolabelled collagen (Brinckerhoff & Harris 1981a).

prostaglandin levels within the cell by regulating phospholipase A_2. After addition of steroids to macrophages, pre-existing macrocortin is released extracellularly, the synthesis of new material is stimulated and the level of prostaglandin synthesis decreased (Blackwell et al 1983).

In our studies of conditioned medium taken from resting cultures of rabbit synovial fibroblasts we found that synovial fibroblasts synthesize and secrete an autoregulatory protein that inhibits collagenase production. Table 2 lists

TABLE 2 Characteristics of an autoregulatory protein that inhibits production of rabbit synovial fibroblast collagenase

M_r	12.5K, 25–50K, 150K
Trypsin sensitivity	+
Heat sensitivity	+
Dithiothreitol sensitivity	+
pI	3.2–3.7
Glycosylation	+

Serum-free conditioned medium was harvested from confluent and resting monolayer cultures of rabbit synovial fibroblasts. The medium was concentrated, dialysed exhaustively and lyophilized. Lyophilized material was reconstituted (to 1.5–2.0 mg protein/ml, a concentration 10 times that of the original medium) and tested for the characteristics listed. To test the treated media for biological activity, the 10× concentrated samples were dialysed vs. leucine-free medium, sterilized by filtration, and then placed on cultures of fibroblasts along with both [^{3}H]leucine and TPA (10^{-8}M). After 30 h at 37 °C, culture medium was assayed for immunoprecipitable [^{3}H]collagenase with monospecific antiserum. M_r determinations were made by gel filtration of concentrated conditioned medium on AcA$_{44}$ and AcA$_{54}$. Trypsin sensitivity was tested by 60 min treatment with 100 μg/ml trypsin at room temperature followed by addition of a four-fold excess of soy-bean trypsin inhibitor. Heat sensitivity was determined by boiling concentrated conditioned medium for 10 min, and dithiothreitol sensitivity was assessed by 30 min treatment with 10 mM-dithiothreitol. The pI was measured by isoelectric focusing with ampholines with a pH range of 2.5–4. Glycosylation was determined by chromatography on Concanavalin A-Sepharose. [See Brinckerhoff et al (1985) for details.]

the properties of this inhibitory protein (Brinckerhoff et al 1985) and Fig. 4 shows first, the ability of this protein to suppress collagenase production and second, the augmentation of this suppressive effect in conditioned medium from retinoic acid-treated cells. In this experiment, conditioned medium was harvested from untreated cells or from cells treated with *all-trans*-retinoic acid (10^{-6}M) or dexamethasone (10^{-7}M). The medium was dialysed exhaustively and lyophilized. It was then reconstituted at a concentration (1.5–2.0 mg protein/ml) that was 10 times (10×) that of the original medium and was placed on cultures of fibroblasts along with both [^{3}H]leucine and an inducer of collagenase. Its ability to suppress collagenase synthesis was measured by immunoprecipitation of [^{3}H]collagenase from the culture medium. Lanes 1 and 2 of Fig. 4 show the immunoprecipitable collagenase synthesized by cells treated with non-conditioned medium (negative control). Lanes 3 and 4 show that increased amounts of collagenase were produced by cells treated with non-conditioned medium and the collagenase inducer TPA (positive control). Lanes 5 and 6 show that, compared with the positive control, cells treated with 10× conditioned medium and TPA synthesized decreased amounts of collagenase. Cells treated with 10× conditioned medium from retinoic acid-treated cells produced even smaller amounts (lanes 7 and 8) but cells treated with 10× conditioned medium from dexamethasone-treated cells (lanes

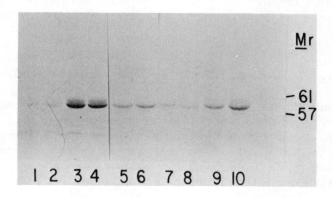

FIG. 4. Ability of conditioned medium from untreated and retinoic acid-treated cells to inhibit collagenase synthesis. Conditioned medium was harvested from confluent cultures of rabbit synovial fibroblasts that had been treated for 48 h with serum-free medium alone, or with serum-free medium containing *all-trans*-retinoic acid (10^{-6} M) or dexamethasone (10^{-7} M). The medium was concentrated, dialysed and lyophilized. Lyophilized material was reconstituted (to 1.5–2.0 mg protein/ml, a concentration 10 times that of the original medium) in leucine-free medium with added [^{3}H]leucine (20 μCi/ml, 6.7 Ci/mmol; New England Nuclear, Boston, MA) and also TPA (10^{-8} M), and was placed on cultures of synovial fibroblasts to assess its ability to inhibit collagenase synthesis. After 30 h at 37 °C, [^{3}H]collagenase was immunoprecipitated from the culture medium (see Brinckerhoff et al 1985). The amount of immunoprecipitated material was visualized by gel electrophoresis and autoradiography.
Lanes 1,2: non-conditioned medium (1×)
      3,4: non-conditioned medium (1×) + TPA
      5,6: conditioned medium (10×) + TPA
      7,8: conditioned medium (10×, from retinoic acid-treated cells) + TPA
      9,10: conditioned medium (10×, from dexamethasone-treated cells) + TPA

9 and 10) did not. It is important to point out that the pattern of proteins synthesized and secreted into the culture medium by cells treated with conditioned media was the same as that of control cultures, indicating that the suppression of collagenase production was not due to a generalized toxic effect on protein synthesis (Brinckerhoff et al 1985).

These data provide preliminary evidence that treatment of synovial cells with retinoids, but not with steroids, enhances production/activity of a protein that inhibits collagenase synthesis, in much the same way as steroids enhance the action of macrocortin by macrophages. In our studies, conditioned medium obtained by treatment of cells with dexamethasone was not effective in inhibiting collagenase. Retinoids and steroids act synergistically to inhibit collagenase synthesis (Brinckerhoff & Harris 1981b), which suggests independent pathways of action; our disparate data on the effects of conditioned medium from retinoid- or steroid-treated cells on collagenase synthesis support this concept of synergy.

Effect of orally administered retinoids on streptococcal cell wall arthritis

Rheumatoid arthritis is an illness peculiar to humans. However, several animal models have been developed to help our understanding of the pathophysiology of the disease. Collagen-induced arthritis and adjuvant arthritis are two commonly used rat models, and we have tested the effect of retinoids on these with differing results. Collagen-induced arthritis, essentially an autoimmune disease, was exacerbated (Trentham & Brinckerhoff 1982) while adjuvant arthritis, an inflammatory disorder with immunological aspects, was suppressed (Brinckerhoff et al 1983).

A third model, streptococcal cell wall arthritis, shares a number of important features with rheumatoid arthritis. It is a biphasic clinical disease with an initial acute inflammatory component that subsides within 7–10 days. This is followed at 10–14 days by a chronic proliferative polyarticular synovitis that, in contrast to synovitis in other models, persists for many months. As with collagen-induced and adjuvant arthritis, streptococcal cell wall arthritis has inflammatory and immune components that involve monocytes/macrophages and immune lymphocytes (Cromartie et al 1977, B. Haraoui et al, unpublished work 1984).

To test the ability of retinoids to modulate streptococcal cell wall arthritis, we induced arthritis in female LEW/N rats by a single i.p. saline injection of streptococcal cell walls. At the time of the injection, animals were started on chow containing either vehicle or the retinoid N-(4-hydroxyphenyl)retinamide at 1 or 2 mmol/kg diet. We used this retinoid because it is less toxic and has a longer half-life than 13-cis-retinoic acid and it is not stored in the liver (Sporn et al 1984). The severity of disease was assessed at intervals by determining the articular indices (the degree of erythema and swelling on all joints distal to the knee or elbow). On day 60, synovium from rats in each group was evaluated histologically and its ability to synthesize collagenase and prostaglandin E_2 was measured.

Fig. 5A shows that all three groups of rats developed typical acute polyarthritis within 24 h of cell wall injection. This acute phase disease subsided within 10–14 days. Animals in the untreated group went on to develop a chronic erosive synovitis manifested both clinically (Fig. 5A) and histologically (Fig. 5B) while animals treated with *N*-(4-hydroxy-phenyl)retinamide showed a dose-dependent suppression of disease (Fig. 5A,C). Furthermore, synovium cultured from rats treated with the retinoid displayed a dose-dependent reduction in collagenase and prostaglandin E_2 synthesis (Table 3). Results of other experiments showed that a 7–10 day pretreatment of animals with the retinoid prevented development of the acute as well as chronic phase of the disease. Further studies demonstrated that the

A

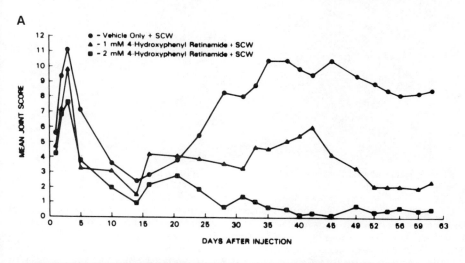

B

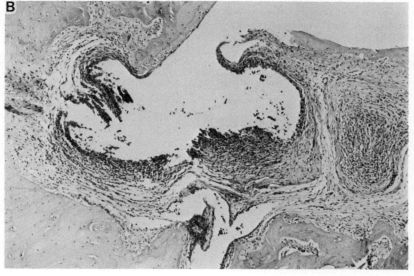

FIG. 5. (*ctd on p. 204*)

FIG. 5. (ctd)

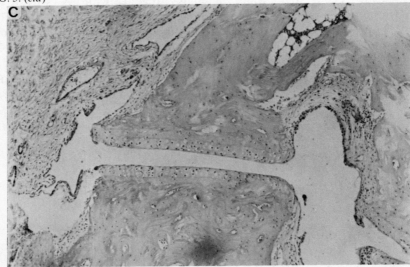

FIG. 5. Effect of oral administration of N-(4-hydroxyphenyl)retinamide on streptococcal cell wall-induced arthritis. On day 0, Group A streptococcal cell wall arthritis was induced in 100 g LEW/N female rats by i.p. injection of streptococcal cell walls (15 μg/g body weight) in saline. On the day of cell wall injection, the animals were placed into three groups (10 animals/group) and fed *ad libitum* with chow containing 0, 1, or 2 mmol N-(4-hydroxyphenyl)retinamide/kg. Articular indices (on a scale of 0–16) were determined at regular intervals. On day 60, synovium from selected rats was processed for histological evaluation (B. Haraoui et al, unpublished work 1984). (A) Clinical course of the disease; SCW, streptococcal cell walls. (B) Histology of synovium from rat treated with streptococcal cell walls but not with retinoid (80× magnification). (C) Histology of synovium from rat treated with streptococcal cell walls and with retinoid (2 mmol/kg chow) (80× magnification).

TABLE 3 Collagenase and prostaglandin E₂ production by synovial cells taken from arthritic rats

Treatment of rats	Collagenase activity[a]	Prostaglandin E_2[b]
Control (no treatment)	<5	<10
SCW	81 ± 15	1357 ± 272
SCW + 1 mmol retinoid/kg diet	68 ± 5	900 ± 117
SCW + 2 mmol retinoid/kg diet	16 ± 2	608 ± 73

Streptococcal cell wall arthritis was induced in female LEW/N rats on day 0. Animals received vehicle or 1 or 2 mmol N-(4-hydroxyphenyl)retinamide/kg diet for 60 days. Rats were killed and synovium was cultured for 48 h in 10% fetal calf serum for production of collagenase or prostaglandin E_2. Collagenase activity was determined in a fibril assay with radiolabelled collagen and prostaglandin E_2 was measured by radioimmunoassay (B. Haraoui et al, unpublished work 1984).
[a] Measured as μg collagen degraded per mg cell protein per hour (±SD).
[b] Measured as ng/mg cell protein (±SD).
SCW, streptococcal cell walls.

effectiveness of the drug in ameliorating established disease depended on how advanced the disease was: animals with severe progressive illness were more recalcitrant than were those with earlier milder stages. These results are intriguing, and future work must address the question of the clinical usefulness of retinoids in the treatment of patients with rheumatoid arthritis.

Conclusions

Our data show that retinoids act on a number of pathways operative in rheumatoid arthritis to modulate both synovial cell behaviour and the outcome of the clinical disease. First, in keeping with a 'hall-mark' of retinoid action, retinoids decreased synovial cell proliferation. They antagonized the mitogenic effects of three growth factors, transforming growth factor β, epidermal growth factor and the T cell product, immune interferon. Although none of these is known for certain to have a physiological role in the proliferative lesion of rheumatoid arthritis, the evidence presented here suggests that this is a logical supposition. The demonstration that retinoids can have antiproliferative effects by antagonizing these factors may help to explain the decrease in the severity of clinical and histological disease in retinoid-treated rats.

The morphological similarity between primary cultures of human rheumatoid synovial cells and the transforming growth factor β-treated rabbit synovial fibroblasts is striking and provides further circumstantial evidence for a physiological role of these growth factors in proliferative but non-malignant disorders such as rheumatoid arthritis. The demonstration that retinoic acid can at least partially reverse the TGF-induced morphology suggests one possible mechanism by which retinoids may influence synovial cell function.

Second, retinoids inhibited collagenase synthesis, thus abrogating an enzyme essential to the destructive component of rheumatoid arthritis. Our initial evidence indicated that retinoids may be acting on transcription to halt synthesis of collagenase mRNA. Further initial data suggest that collagenase production by synovial fibroblasts is inhibited by an autoregulatory protein and that retinoids, but not steroids, may augment production/activity of this protein. Thus, augmentation of this regulatory protein by retinoids may explain retinoid suppression of collagenase synthesis. It will be important in further experiments to confirm this finding with purified protein and to determine whether it can modulate other cell functions, such as cell growth, prostaglandin levels and matrix production.

Third, retinoids decreased the manifestations of the clinical disease in rats. This was shown by decreased articular inflammation and erythema as well as by radiological and histological evaluation. The clinical manifestations result from the migration of polymorphonuclear leucocytes, immune lymphocytes and macrophages/monocytes to the joint. They also result from pathological changes caused by the secretion of mediators such as interleukin 1, prostaglandin E_2 and collagenase, which are already documented, and transforming growth factors and immune interferon, which are implicated. For retinoids to be effective therapeutic agents in the treatment of rheumatoid arthritis they will probably need to act at several sites along the pathogenic pathways operative in this disease:

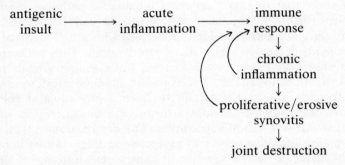

The ability of retinoids to modulate acute inflammation, T and B cell physiology and macrophage/monocyte function in other systems has been noted (Sporn et al 1984). Here we have described their actions, as we presently understand them, on synovial cell function. Elucidating precisely how retinoids influence the function of the other cell types contributing to the pathophysiology of rheumatoid arthritis remains a task for the future, as does assessment of their clinical usefulness in treating patients with rheumatoid disease.

Acknowledgements

We are grateful to Dr Anita Roberts and Dr Michael Sporn, both of NIH, who generously supplied transforming growth factor β and epidermal growth factor. We thank Ms Valerie Hunt for her excellent preparation of the manuscript. This work was supported by USPHS grants CA 32476 and AM-31643, by grants from the National and NH Chapters of the Arthritis Foundation and by a grant from Hoffmann-La Roche Inc.

REFERENCES

Assoian RK, Komoriya A, Meyers CA, Miller DM, Sporn MB 1983 Transforming growth factor β in human platelets. J Biol Chem 258:7155-7160
Baker JB, Barsh GS, Carney DH, Cunningham DD 1978 Dexamethasone modulates binding and

action of epidermal growth factor in serum-free cultures. Proc Natl Acad Sci USA 75:1882-1886

Blackwell GJ, Carnuccio R, DiRosa M et al 1983 Suppression of arachidonate oxidation by glucocorticoid-induced anti-phospholipase peptides. Adv Prostaglandin Thromboxane Leukotriene Res 11:65-71

Brinckerhoff CE 1983 Morphologic and mitogenic responses of rabbit synovial fibroblasts to transforming growth factor β require transforming growth factor α or epidermal growth factor. Arthritis Rheum 26:1370-1379

Brinckerhoff CE, Guyre PM 1985 Recombinant immune interferon increases the proliferation of cultured human synovial fibroblasts. J Immunol, in press

Brinckerhoff CE, Harris ED Jr 1981a Survival of human rheumatoid synovium implanted into nude mice. Am J Pathol 103:411-418

Brinckerhoff CE, Harris ED Jr 1981b Modulation by retinoic acid and corticosteroids of collagenase production by rabbit synovial fibroblasts treated with phorbol myristate acetate or polyethylene glycol. Biochim Biophys Acta 677:424-432

Brinckerhoff CE, McMillan RM, Dayer JM, Harris ED Jr 1980 Inhibition by retinoic acid of collagenase production in rheumatoid synovial cells. New Engl J Med 303:432-436

Brinckerhoff CE, Coffey JW, Sullivan AC 1983 Inflammation and collagenase production in rats with adjuvant arthritis reduced with 13-cis-retinoic acid. Science (Wash DC) 221:756-758

Brinckerhoff CE, Benoit MC, Culp WJ 1985 Autoregulation of collagenase production by a protein synthesized and secreted by synovial fibroblasts: a cellular mechanism for control of collagen degradation. Proc Natl Acad Sci USA, in press

Chirgwin JM, Przybyla AE, MacDonald RJ, Rutter WJ 1979 Isolation of biologically active ribonucleic acid from sources enriched in ribonuclease. Biochemistry 18:5294-5299

Cromartie WJ, Craddock JG, Schwab JH, Anderle SK, Yang C-H 1977 Arthritis in rats after systemic injection of streptococcal cells or cell walls. J Exp Med 146:1585-1602

Gross RH, Sheldon LA, Fletcher CF, Brinckerhoff CE 1984 Isolation of a collagenase cDNA clone and measurement of changing collagenase levels during induction in rabbit synovial fibroblasts. Proc Natl Acad Sci USA 81:1981-1985

Harris ED Jr 1985 Pathogenesis of rheumatoid arthritis. In: Kelley WN et al (eds) Textbook of rheumatology. WB Saunders, Philadelphia, p 879-915

Korn JH, Haluska PV, LeRoy EC 1980 Mononuclear cell modulation of connective tissue function: suppression of fibroblast growth by stimulation of endogenous prostaglandin production. J Clin Invest 65:543-554

Mizel SB, Dayer JM, Krane SM, Mergenhagen SE 1981 Stimulation of rheumatoid synovial cell collagenase and prostaglandin production by partially purified lymphocyte-activating factor (interleukin 1) Proc Natl Acad Sci USA 78:2474-2477

Ozanne B, Fulton RH, Kaplan PL 1980 Kirsten murine sarcoma virus transformed cell lines and a spontaneously transformed rat cell line produce transforming factors. J Cell Physiol 105:163-180

Preble OT, Black RJ, Friedman RM, Klippel JH, Vilcek J 1982 Systemic lupus erythematosis: presence in human serum of an unusual acid-labile interferon. Science (Wash DC) 216:429-431

Roberts AB, Sporn MB 1984 Cellular biology and biochemistry of the retinoids. In: Sporn MB et al (eds) The retinoids. Academic Press, Orlando, vol 2:209-286

Roberts AB, Anzano MA, Lamb LC, Smith JM, Sporn MB 1981 A new class of transforming growth factors potentiated by epidermal growth factor: isolation from non-neoplastic tissues. Proc Natl Acad Sci USA 78:5339-5343

Sporn MB, Roberts AB, Goodman DS (eds) 1984 The retinoids. Academic Press, Orlando

Trentham DE, Brinckerhoff CE 1982 Augmentation of collagen arthritis by synthetic analogues of retinoic acid. J Immunol 129:2668-2672

DISCUSSION

Strickland: Does the antibody that you used for your experiments on collagenase production immunoprecipitate procollagenase?

Brinckerhoff: Yes, it was procollagenase that we measured. The antibody precipitates both procollagenase and active collagenase but the cell secretes the enzyme as procollagenase.

Strickland: Have you looked at proteases other than procollagenase? Dexamethasone has been shown to induce an inhibitor of plasminogen activator in HTC hepatoma cells (Cwikel et al 1984) and there is some evidence that serine proteases are involved in the activation of procollagenase (Werb et al 1977). Could dexamethasone inhibit procollagenase activation, and could this explain the synergistic effects of retinoids and steroids on collagenase synthesis?.

Brinckerhoff: I don't think so because there does not seem to be 'spontaneous' activation of procollagenase in these cultures. When the cells secrete collagenase into the culture medium they secrete only the latent enzyme and it must be activated with trypsin or some other protease added exogenously. There is virtually no active enzyme in these cultures.

Strickland: It will not activate if you just incubate it?

Brinckerhoff: No. You may then ask how joint disruption occurs in rheumatoid tissue if the rheumatoid cells are making only latent collagenase. The answer is that in the rheumatoid joint proteases are present that are capable of activating the latent enzyme.

Eccles: Have you looked at any other collagenases, for example that produced by tumour cells, that might be important for invasion, break-down or cartilage reconstruction?

Brinckerhoff: No.

Koeffler: Does your rabbit collagenase cDNA clone hybridize to human mRNA?

Brinckerhoff: No, we tested that in the first experiment we did. But the antibodies do cross-react. We apparently did not clone the sequences that show homology.

Koeffler: There are a lot of neutrophils and macrophages in synovial fluid in rheumatoid arthritis. These haemopoietic cells synthesize collagenase. How do you know that synovial fibroblasts are the most important cell type and that you are not examining the wrong cell?

Brinckerhoff: When we put the cells in culture, we can show that it is the synovial cells that are making the most collagenase.

Koeffler: Don't macrophages make a lot of collagenase?

Brinckerhoff: I don't want to rule out a role for macrophage collagenase but in terms of quantity it is much less significant than that from the synovial cells.

Hartmann: Is there a difference between collagenase derived from macrophages and that from synovial fibroblasts?

Brinckerhoff: There is a collagenase from macrophages and from polymorphonuclear leucocytes that is a presynthesized and stored enzyme, whereas that from synovial cells is not stored. Once an increase in mRNA levels occurs, the enzyme is synthesized and secreted within an hour.

Yuspa: Is there some link between collagenase production and proliferation?

Brinckerhoff: At this point, the association seems to be indirect. Neither transforming growth factor β nor immune interferon has any effect on collagenase or prostaglandin production. The only compound I know that affects both collagenase and proliferation is interleukin 1 (Dayer et al 1976, Mizel et al 1981), but the two effects may not be linked mechanistically.

Yuspa: Does 12-*O*-tetradecanoylphorbol-13-acetate (TPA) cause the same morphological changes in synovial fibroblasts that the transforming growth factor does?

Brinckerhoff: No. The fibroblasts become 'skinny' but they do not migrate over the dish and pile up.

Yuspa: Is there any effect on collagenase?

Brinckerhoff: Yes. TPA is a potent inducer of rabbit synovial fibroblast collagenase.

Breitman: When you see the large increase in thymidine incorporation in fibroblasts treated with growth factors, what percentage of the cells incorporate the thymidine? Is an increase in cell number observed? That is, is there cell proliferation?

Brinckerhoff: Yes. I haven't looked at what percentage of the cells are incorporating thymidine, but I have done cell counts and measured the protein content. In all cases an increase in thymidine incorporation correlates with an increase in cell number and with an increase in the protein content of the cultures.

Dennert: You found that treating animals with retinoid suppresses the first peak of the streptococcal cell wall-induced disease. Do you think this is due to an effect of the retinoid on the immune system?

Brinckerhoff: I would think it's due to suppression of the inflammatory response.

Dennert: So it would be an immune response.

Brinckerhoff: Yes, involving polymorphonuclear lymphocytes. But I don't know the mechanism.

Dennert: How does rheumatoid arthritis progress in a thymus-deficient animal?

Brinckerhoff: Dr Ronald Wilder and his colleagues at NIH have addressed this question (Wilder & Allen 1984). They found both thymic and athymic rats developed the acute inflammatory phase of streptococcal cell wall-induced arthritis. However, only those animals with a thymus went on to develop chronic/erosive/proliferative disease, which strongly suggests that the immune system and cell-mediated immune responses play a role in driving the chronic phase of the disease.

Wald: Are there any clinical trials in the pipeline?

Brinckerhoff: Not that I know of.

Peck: Fritsch et al (1984) have used Ro 13-6298 [(*E*)-4-[2-(5,6,7,8-tetrahydro-5,5,8,8-tetramethyl-2-naphthalenyl)-1-propenyl]benzoic acid ethyl ester] in humans for psoriatic arthritis and rheumatoid arthritis with success. I don't know whether or not Fritsch is planning a study with *N*-(4-hydroxyphenyl)retinamide (HPR). It is interesting that in another disease in which there is an excessive production of collagenase, recessive epidermolysis bullosa dystrophica, retinoids, particularly 13-*cis*-retinoic acid, also inhibit collagenase activity (Bauer 1984, Bauer et al 1977, 1983).

Brinckerhoff: In this case skin fibroblasts rather than synovial fibroblasts are making the enzyme.

Sporn: Is the collagenase involved in the pathogenesis of the disease?

Peck: Yes. Blister formation occurs within the superficial dermis so the excessive collagenase production by the skin is considered to be a primary part of the disease.

Sporn: Is there a useful clinical response to retinoids in that disease?

Peck: Clinical studies are currently in progress, but thus far the therapeutic effect has not been dramatic (E.A. Bauer, unpublished observations). One hopes to prevent new blisters from forming as well as enhance the healing of old lesions. However, even if retinoids only partially reduced the incidence of new blisters, they would be of value in this severe, debilitating disease.

Strickland: Is it possible that retinoids could be used as a treatment for gout?

Brinckerhoff: Yes. By the time there is a proliferative erosive synovitis, whether in psoriatic arthritis, gout or haemophilia when there is bleeding into joints, the pathology is similar. There are increased levels of procollagenase and presumably of collagenase mRNA with all these disorders, so I think that if retinoids ever do become a clinically appropriate treatment for proliferative synovitis, gout will be on the list.

Hicks: How many different retinoids have you tried?

Brinckerhoff: In collaboration with Dr David Trentham at Harvard, Dr Ronald Wilder at NIH and Drs John Coffey and Ann Sullivan at Hoffmann-La Roche we have tried retinoids in three different animal models of arthritis. In adjuvant arthritis (Brinckerhoff et al 1983) and streptococcal cell wall arthritis (B. Haraoui et al, unpublished work) we see a nice suppressive effect of retinoids on the disease, but in collagen-induced arthritis (Trentham & Brinkerhoff 1982) retinoids exacerbate the disease. With collagen-induced arthritis we have used 13-*cis*-retinoic acid and HPR; with the adjuvant arthritis we have used 13-*cis*-retinoic acid but not HPR; with the streptococcal cell wall arthritis we have used only HPR. When we use the two drugs in the same model we get consistent results.

Sporn: Do you have any idea why in two experimental models arthritis is alleviated by retinoids whereas in the third it is made worse? What is the difference in the biology of the disease in the different models?

Brinckerhoff: Collagen-induced arthritis is basically an autoimmune model, but although Dr Trentham looked at various immunological parameters he found no differences between the untreated and the retinoid-treated arthritic rats (Trentham & Brinckerhoff 1982).

Sporn: Is collagen-induced arthritis considered to have less relevance to human rheumatoid arthritis?

Brinckerhoff: There is an immune component to human rheumatoid arthritis, but it is not the only component.

REFERENCES

Bauer EA, Gedde-Dahl T, Eisen AZ 1977 The role of human skin collagenase in epidermolysis bullosa. J Invest Dermatol 68:119-124

Bauer EA, Seltzer JL, Eisen AZ 1983 Retinoic acid inhibition of collagenase and gelatinase expression in human skin fibroblast cultures. Evidence for a dual mechanism. J Invest Dermatol 81:162-169

Bauer EA 1984 Inhibition of collagenase expression in normal and recessive dystrophic epidermolysis bullosa fibroblast cultures by retinoic acid. Dermatologica 169:234

Brinckerhoff CE, Coffey JW, Sullivan AC 1983 Inflammation and collagenase production in rats with adjuvant arthritis reduced with 13-*cis*-retinoic acid. Science (Wash DC) 221:756-758

Cwikel BJ, Barouski-Miller PA, Coleman PL, Gelehrter TD 1984 Dexamethasone induction of an inhibitor of plasminogen activator in HTC hepatoma cells. J Biol Chem 259:6847-6851

Dayer J-M, Krane SM, Russell RGG, Robinson DR 1976 Production of collagenase and prostaglandins by isolated adherent rheumatoid synovial cells. Proc Natl Acad Sci USA 73:945-949

Fritsch P, Rauschmeier W, Neuhofer J 1984 Response of psoriatic arthropathy to arotinoid (Ro 13-6298): a pilot study. In: Cunliffe WJ, Miller AJ (eds) Retinoid therapy. MTP Press, Lancaster, p 329-333

Mizel SB, Dayer JM, Krane SM, Mergenhagen SE 1981 Stimulation of rheumatoid synovial cell collagenase and prostaglandin production by partially purified lymphocyte-activating factor (interleukin 1). Proc Natl Acad Sci USA 78:2474-2477

Trentham DE, Brinckerhoff CE 1982 Augmentation of collagen arthritis by synthetic analogues of retinoic acid. J Immunol 129:2668-2672

Werb Z, Mainardi CL, Vater CA, Harris ED 1977 Endogenous activation of latent collagenase by rheumatoid synovial cells: evidence for a role of plasminogen activator. N Engl J Med 296:1017-1023

Wilder RL, Allen JB 1984 The role of thymus-dependent lymphocytes in the pathogenesis of Group A streptococcal cell wall-induced arthritis. Arthritis Rheum 27:S36

General discusssion II

Retinoids and glycoprotein synthesis

Lotan: I became interested in retinoids because of the reports that they could modulate glycoprotein glycosylation (De Luca 1977). In my initial experiments on the effects of retinoids on glycoprotein synthesis I found that both retinoic acid and retinyl acetate inhibited the growth of melanoma cells, so for a while I was distracted and studied the antitumour action of retinoids. But now I would like to return to the original question of what happens to glycoprotein synthesis in retinoid-treated melanoma cells.

Glycoprotein synthesis starts in the rough endoplasmic reticulum. Several sugars, delivered as oligosaccharide chains by dolichol phosphate intermediates, are attached covalently onto the growing polypeptide chain. There is then some restructuring involving removal of the glucose and some of the mannose residues in the Golgi apparatus and addition of peripheral sugars such as mannose, *N*-acetylglucosamine, galactose and sialic acid by the glycosyl transferases, which work in tandem. When the carbohydrate side-chains are completed, vesicles from the Golgi move through the cell, fuse with the membrane and expose the sugar-bearing glycoproteins to the exterior. One can study the synthesis of glycoproteins by providing cells with radioactive precursors such as glucosamine or mannose, and look at the exposed sugars by oxidizing sialic acid with periodate and then reducing with sodium borotritiide. One can also label galactose residues by removing the sialic acid with neuraminidase, oxidizing the galactose residues with galactose oxidase and then reducing with borotritiide. We have done this with melanoma cells, solubilized the labelled cells and analysed them by gel electrophoresis and autoradiography. In cells treated with 10 µM-retinoic acid (or even with only 0.1 µM) a glycoprotein with an apparent M_r of 160 000 (gp160) seemed to incorporate more glucosamine and mannose than in control cells. There was more sialic acid on the cell surface gp160 and, after removal of the sialic acid, more galactose residues were labelled on gp160 in treated cells than in control cells (Fig. 1). Since there was no increase in the incorporation of amino acids into gp160, we concluded that retinoids increase the glycosylation of a pre-existing protein rather than induce the synthesis of a new protein.

Kreisel et al (1980) have presented data suggesting that the oligosaccharide side-chain of a plasma membrane glycoprotein from rat liver normally undergoes degradation and resynthesis several times before the polypeptide is completely degraded. We think that retinoids might stimulate the rebuilding or

212

restructuring of such oligosaccharides. The enzymes involved in these reactions are glycosyl transferases. Sialyl transferase, for example, uses a terminal galactose residue as an acceptor and CMP-sialic acid as a donor. The activity of sialyl transferase can be measured by supplying the enzyme with CMP-[^{14}C]sialic

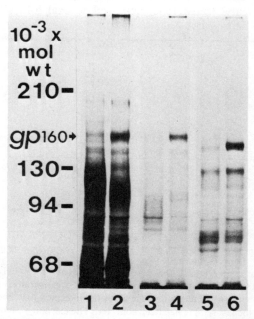

FIG. 1. (*Lotan*) Effect of retinoic acid on cellular and cell-surface glycoproteins. S91 melanoma clone C-2 cells were grown for six days in the absence (lanes 1, 3 and 5) or presence (lanes 2, 4 and 6) of 10 μM-retinoic acid and radiolabelled either by incubation with [^{3}H]glucosamine (lanes 1 and 2) or by oxidation of intact cells with NaIO$_4$ (lanes 3 and 4) or with galactose oxidase (after neuraminidase treatment) (lanes 5 and 6) followed by reduction with NaB^{3}H$_4$. The cells were solubilized with detergents and the labelled molecules were separated by polyacrylamide gel electrophoresis and analysed by fluorography.

acid and with a glycoprotein that has terminal galactose residues (e.g. desialylated fetuin). We can then measure incorporation of sialic acid into the acceptor. In cultures treated with >0.1 μM-retinoic acid we found increased incorporation of sialic acid into the exogenous glycoprotein acceptor, which was added in excess. This effect on sialyl transferase activity was evident 24 hours after retinoic acid was added to the cultures and peaked at about 72 hours (Deutsch & Lotan 1983). Even without the exogenous acceptor, when we just let the cells incorporate sialic acid into whatever glycoprotein they had, we still saw increased glycosylation (Fig. 2).

The question arose of whether these changes are related to the growth-inhibitory effects of retinoids. We have recently isolated melanoma cell mutants that are resistant to the growth-inhibitory effect of retinoic acid (Lotan et al 1983) and we have used them to probe whether or not the changes in glycosylation are related to growth inhibition. In sub-clones of the sensitive cells the level of sialyl transferase increased several fold after retinoic acid

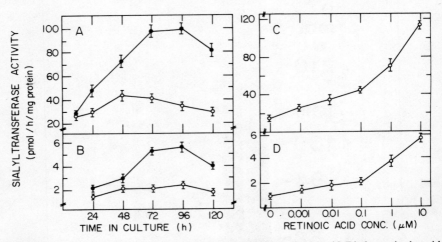

FIG. 2. (*Lotan*) Time course (A,B) and dose–response relationship (C,D) for retinoic acid-induced stimulation of sialyl transferase activity in S91 clone C-2 melanoma cells. (A,B) Cells were grown without (○) or with (●) 10 µM-retinoic acid for the indicated time and then detached and frozen at −70 °C until the last sample had been collected. The enzymic activity was assayed either with the exogenous acceptor asialofetuin (A) or with endogenous acceptors (B). (C,D) Cells were grown in the absence of retinoic acid or in the presence of the indicated concentrations of retinoic acid for five days and then solubilized and assayed for sialyl transferase activity either with asialofetuin (C) or with endogenous acceptors (D).

treatment, whereas in resistant mutant clones the level of sialyl transferase was not affected by retinoids. Corresponding effects were observed on sialic acid labelling on the cell surface: in the sensitive cells, retinoic acid caused an increased sialylation of gp160, but in the resistant cells there was no increase in the glycosylation (Lotan et al 1984). Similar increases in sialyl transferase were also observed in a variety of other cell lines, with the exception of a human chondrosarcoma, and corresponding increases in sialic acid labelling were found on the cell surface. However, the sialylated components were not the same in every cell type. In the human osteosarcoma line there was an increase in a 190 kDa component; in the human chondrosarcoma there was an increase in this component but also in two others (Meromsky & Lotan 1984); in B16 melanoma there was an increase in a 170 kDa glycoprotein (Lotan et al 1982).

Our results demonstrate that in a variety of cells retinoic acid modifies membrane glycoprotein sialylation and this effect is correlated with an increase in sialyl transferase. We have also found increases in galactosyl transferase and fucosyl transferase. Others have reported retinoid-induced increases in sialyl transferase, for example in HL-60 cells (Durham et al 1983) and in human KB carcinoma cells (Moskal et al 1980). There is therefore good evidence that retinoids modify glycoprotein synthesis by inducing an enzyme rather than by working as a cofactor. We have looked hard for the formation of a doubly labelled intermediate of retinoic acid with mannose using high performance liquid chromatography but have been unable to demonstrate any such intermediate in cells in which we see a change in membrane glycoprotein synthesis (A. Clifford et al, unpublished work). The implications of these modifications of the cell surface have yet to be investigated, but membrane components are known to act as receptors for hormones and growth factors and changes in the negative charge of glycoproteins could affect the affinity for some of these factors and could block or increase their binding.

Sporn: Has anybody looked at the effect of retinoic acid on the functional properties of the receptor for any growth factor? Is there any possibility that the affinity of a growth factor receptor for its ligand is changed by the degree or nature of its glycosylation?

Lotan: The effects of retinoids have not been studied, but it has been shown that the epidermal growth factor (EGF) receptor that is produced by tunicamycin-treated cells is much less effective than normal at binding EGF. So it is possible that changes in glycosylation could alter the affinity or turnover of receptors. When we treated B16 melanoma cells with retinoic acid to increase glycosylation of gp160, and then treated the cells with tunicamycin, we found that the half-life of the modified glycoprotein was much longer than that of glycoprotein in control cells (Lotan et al 1982). If we treat control cells for 24 hours with tunicamycin we 'shave off' all the membrane glycoproteins so we don't see them, whereas in retinoic acid-treated cells we can still see all the glycoproteins even one week after we remove the retinoid. This means that the more glycosylated and more sialylated glycoproteins survive longer and are more resistant to proteolysis. This could obviously affect their function.

Jetten: Carpenter (1984) has looked at the effects of several factors on the binding of EGF to its receptor. Tunicamycin inhibits N-linked glycosylation and under these conditions the EGF receptor does not reach the cell surface. Swainsonine inhibits the conversion of high mannose-type carbohydrates to complex-type carbohydrates, but in the presence of this inhibitor EGF can still bind to its receptor and generate biological responses. So changes in the carbohydrate structure of the EGF receptor may or may not have a dramatic effect.

Yuspa: We have looked at the influence of retinoids on glycoprotein synthesis (Adamo et al 1979) and have found that changes occur fairly rapidly. In your studies, Dr Lotan, the sialyl transferase induction is really a delayed response; measurable changes are delayed by 24 hours and the effect does not peak until 48–72 hours. What are the kinetics of your cell-surface labelling changes and what sort of kinetics do you see in your metabolic labelling studies?

Lotan: In the metabolic studies it takes about two days for incorporation of labelled glucosamine, so we cannot get very high labelling in a short period of time. With cell surface labelling, it takes at least 12 hours before we can see a twofold increase in gp160 labelling. We would therefore not see rapidly occurring changes in glycoproteins. Sialyl transferase is found in only very small quantities and it is quite difficult to measure changes in its activity unless they are very marked. Before measuring the enzyme activity we usually solubilize the cells with Triton but we have also done studies where we have not added any detergent, but have just homogenized the cells. After subcellular fractionation on a sucrose gradient, a fraction that corresponds to the Golgi was isolated. In this fraction we observed an increase of up to 15-fold in enzyme activity after about 12 hours of treatment. Effects on cell proliferation, in terms of changes in the cell cycle, are observed by 24 hours. So it is possible that first of all retinoic acid induces the sialyl transferase, then there are changes in the cell surface and only subsequently are there changes in growth. This relationship to growth is also supported by the fact that if we inhibit growth by other means, e.g. with 12-*O*-tetradecanoylphorbol-13-acetate, we do not see any changes in sialyl transferase or the cell-surface glycoproteins.

Vitamin A in growth and differentiation

Lotan: If you give vitamin A to vitamin A-deficient animals they eat and gain weight. But what exactly is the function of vitamin A in promoting growth *in vivo*? What type of cell is affected? What can be defined as a 'physiological' effect of vitamin A?

Pitt: One view is that normal growth depends upon normal differentiation in a wide variety of tissues. In the absence of vitamin A normal differentiation cannot take place, and the normal pattern of development comes to a halt; hence growth stops.

Sporn: But when an animal on a vitamin A-deficient diet reaches a growth plateau, which particular cells are affected? All epithelial cells or mesenchyme throughout the body? One of the striking things about epithelial lesions in vitamin A deficiency is that they are spotty and punctate, so in the trachea, bladder etc. you certainly haven't knocked out all cell function. So what causes growth to stop?

Pitt: It's difficult to say. When rats, for example, become vitamin A deficient they cease to grow, or lose weight, and they stop eating. It is hard to elucidate whether the initial effect on growth precedes or follows the changes in appetite, but loss of appetite will rapidly affect growth and many other biological mechanisms. Nockels et al (1984) have reported that hypothyroidism is a very early sign of vitamin A deficiency in chicks and may account for some of the observed deficiency signs.

Sporn: Why do the animals stop eating? And when you give them a dose of retinoic acid, why do they start eating again?

Pitt: Nobody really knows. Loss of appetite is usually one of the first things to happen in vitamin A deficiency in experimental animals. But vitamin A deficiency stops growth in the chick embryo and that cannot be explained by its stopping eating.

Lotan: The definition of vitamin A-like compounds is that they cause regaining of weight after vitamin A deficiency, but I don't think that is a very good criterion because the gaining of weight may not be a direct effect.

Pitt: The traditional nutritionist's viewpoint is that this is what can be measured in the living animal, and in vitamin A-deficient rats weight changes are among the earliest signs observed. A paradox is that while in many tissues cell multiplication is reduced in vitamin A deficiency, as might be expected, in some tissues vitamin A deficiency has been reported to cause an *increase* in cell division.

Sporn: Yes. Wolbach & Howe (1925) stated very definitely that retinoid-deficient tissues have enhanced growth activity. Many retinoid-deficient epithelia are hyperplastic. In terms of the bifunctional action of retinoids it is interesting that Malcolm Maden (this volume) has found that the growth of extra limbs in axolotls is preceded by some growth suppression.

Sherman: Enhanced tissue growth activity in retinoid-deficient animals, despite inhibition of whole animal growth, fits well with the observation that differentiation is inhibited, because in most systems you suppress growth when you induce differentiation. Retinoids are powerful inducers of differentiation in several systems. The question is: do they suppress growth by inducing differentiation or vice versa?

Sporn: We don't know what the relationship is between differentiation and proliferation or whether differentiation and proliferation can be dissociated from one another.

Hicks: It's an oversimplification to say that a lack of vitamin A prevents differentiation. A vitamin A-deficient animal does not retain all its cells in an undifferentiated state and therefore stop growing; there is differentiation but it is different from normal. A normal level of vitamin A maintains normal differentiation. If you give either too much or too little you alter the differentiation. In vitamin A-deficient animals, tissues that are normally mucous membranes become highly differentiated and keratinized. It is not quite as simple as stopping differentiation; many cells do differentiate but take the wrong pathway. For example, vitamin A-deficient tracheal cultures will keratinize, but if you give them vitamin A they stop keratinizing.

Sporn: You may get metaplastic effects in epithelia but not necessarily in mesenchyme.

Hicks: Yes, but the point is that differentiation is not stopped, just disturbed.

Pitt: It is difficult to get at mechanisms in orthodox vitamin A-deficient animals. Many observed effects may be secondary. These animals are not eating; they have keratinized epithelia more susceptible to infection; and their immune systems may not be working as well as normal. None of these things holds for the vitamin A-deficient chick embryo, which, in this respect, is a better preparation to study.

Sporn: Retinoid-deficient embryos are sitting in a pool of yolk and albumin, so they are not missing nutrients, but they die. Some aspect of the cell programme for proliferation or differentiation must be inhibited.

Pitt: Growth always stops by Hamburger & Hamilton stage 13.

Sporn: We have no idea why it stops at that point. The elements that are most significantly affected in the early embryo deficient in vitamin A are not epithelial but mesenchymal. In contrast, in adult vitamin A-deficient animals, epithelial abnormalities are most prominent. At the very beginning of development, mesenchymal rather than epithelial elements seem to require the retinoids.

Pitt: What is surprising about vitamin A in evolutionary terms is that many organisms do not need vitamin A except for vision. Lower animals get on fine without any. If one keeps insects on a vitamin A-deficient diet, they go blind, but are otherwise all right; the rest of the body seems normal. But higher animals must have vitamin A to stay alive. One can imagine they might seize on retinoids for one specific vital function, but in recent years we have had reports of the apparent participation of retinoids in a variety of roles in cellular physiology. It does not make great evolutionary sense for them to be used by higher animals for a number of different functions.

REFERENCES

Adamo S, DeLuca LM, Silverman-Jones CS, Yuspa SH 1979 Mode of action of retinol: involvement in glycosylation reactions of cultured mouse epidermal cells. J Biol Chem 254:3279-3287

Carpenter G 1984 The receptor for epidermal growth factor: aspects of its metabolism, structure and function. Proc Am Assoc Cancer Res 25:413-414

De Luca LM 1977 The direct involvement of vitamin A in glycosyl transfer reactions in mammalian membranes. Vitam Horm 35:1-57

Deutsch V, Lotan R 1983 Stimulation of sialyltransferase activity of melanoma cells by retinoic acid. Exp Cell Res 149:237-245

Durham JP, Ruppert M, Fontana JA 1983 Glycosyltransferase activities and the differentiation of human promyelocytic (HL-60) cells by retinoic acid and a phorbol ester. Biochem Biophys Res Commun 110:348-355

Kreisel W, Volk BA, Buchsel R, Reutter W 1980 Different half-lives of the carbohydrate and protein moieties of a 110,000 dalton glycoprotein isolated from plasma membranes of rat liver. Proc Natl Acad Sci USA 77:1828-1831

Lotan R, Irimura T, Nicolson GL 1982 Inhibition of experimental pulmonary metastases by suppression or enhancement of the glycosylation of melanoma cell membrane glycoproteins. In: Galeotti T et al (eds) Membranes in tumor growth. Elsevier Biomedical Press, Amsterdam

Lotan R, Stolarsky T, Lotan D 1983 Isolation and analysis of melanoma cell mutants resistant to the antiproliferative action of retinoic acid. Cancer Res 43:2863-2875

Lotan R, Lotan D, Meromsky L 1984 Correlation of retinoic acid-enhanced sialyltransferase activity and glycosylation of specific cell surface sialoglycoproteins with growth inhibition in a murine melanoma cell system. Cancer Res 44:5805-5812

Maden M 1985 Retinoids and the control of pattern in regenerating limbs. In: Retinoids, differentiation and disease. Pitman, London (Ciba Found Symp 113) p 132-155

Meromsky L, Lotan R 1984 Modulation by retinoic acid of cellular, surface-exposed, and secreted glycoconjugates in cultured human sarcoma cells. J Natl Cancer Inst 72:203-215

Moskal JR, Lockney MW, Marvel GC et al 1980 Regulation of glycoconjugate metabolism in normal and transformed cells. In: Sweeley CC (ed) Cell surface glycolipids. ACS (Am Chem Soc) Symp Ser 128:241-260

Nockels CF, Ewing DL, Phetteplace H, Rittaco KA, Mero KN 1984 Hypothyroidism: an early sign of vitamin A deficiency in chickens. J Nutr 114:1733-1736

Wolbach SB, Howe PR 1925 Tissue changes following deprivation of fat soluble A vitamin. J Exp Med 42:753-757

Retinoid toxicity

J. A. TURTON*, R. M. HICKS*, J. GWYNNE*, R. HUNT† and C. M. HAWKEY‡

*School of Pathology, Middlesex Hospital Medical School, Riding House Street, London W1P 7LD, †Transplantation Biology Section, Clinical Research Centre, Watford Road, Harrow, Middlesex HA1 3UJ, and ‡Institute of Zoology, Zoological Society of London, Regent's Park, London NW1 4RY, UK

Abstract. The long-term effects of N-ethylretinamide (NER) on the haematology of the rat, and the dose-related effects of retinoids on lymphoid organs of the mouse and rat were investigated. Retinoid-induced long-bone changes were used to develop a method for quantifying skeletal effects. This technique was used to investigate the activity of five retinamides in inducing long-bone changes in the rat. The ability of non-steroidal anti-inflammatory compounds (NSAICs) to prevent retinoid-induced skeletal effects was examined, and preliminary investigations made into the mechanisms of retinoid-induced long-bone remodelling.

NER-fed rats had reduced red blood cell counts and fibrinogen values. Retinoids caused dose-related proliferation of the spleen and lymph nodes in the mouse and to a lesser extent in the rat. They induced dose-related reductions in femoral diaphysis and medullary cavity diameters in both rats and mice. Aspirin prevented NER-induced changes of rat long bones, but subsequent studies indicated this effect may be closely dependent on the dose level of both the retinoid and NSAIC administered. Retinoids induce rapid long-bone remodelling in the rat which tends to revert on feeding a control diet, but remodelling processes are different in the young growing rat and the mature animal.

1985 Retinoids, differentiation and disease. Pitman, London (Ciba Foundation Symposium 113) p 220–251

The preclinical and clinical toxicology of retinoids was reviewed recently by Kamm et al (1984); previous reviews include those of Moore (1967), Teelman (1981), Windhorst & Nigra (1982), Cunningham & Ehmann (1983) and DiGiovanna & Peck (1983). With the increasing clinical use of retinoids, their adverse side-effects are becoming more apparent, but rational assessment of their toxicity (DiGiovanna & Peck 1983) will assist rather than detract from the full exploitation of their therapeutic potential. However, although the activity of retinoids in the therapy of many disease conditions is now well established, the more wide-spread clinical use of these drugs may be limited by their adverse reactions.

It is informative to consider the adverse reactions of retinoids in the context of hypervitaminosis A and this approach has been adopted by Windhorst &

Nigra (1982), DiGiovanna & Peck (1983) and Kamm et al (1984). The side-effects of retinoids in humans are clearly similar to those documented for vitamin A (Table 1). A comparative approach is also constructive when studying retinoid and vitamin A toxicity in laboratory animals and it emerges that almost all the side-effects of vitamin A and retinoids in laboratory animals also occur in humans (see Kamm et al 1984).

Within the overall pattern of beneficial and adverse effects there are, however, dissimilarities between the activities of particular retinoids in animals and humans. Retinoid efficacy varies: the spectrum of disorders suitable for treatment with 13-*cis*-retinoic acid (13CRA) is different from that for etretinate (Cunningham & Ehmann 1983); Moon & Itri (1984) have detailed the activity or inactivity of numerous retinoids in preventing specific neoplastic conditions in experimental animals; Willhite & Shealy (1984) demonstrated 13CRA to be embryotoxic in the hamster whereas *all-trans-N*-ethylretinamide (NER) was non-toxic, and Stinson et al (1980) showed that NER caused atrophy of the testicular germinal epithelium of hamsters whereas 13CRA did not. As well as such qualitative differences between retinoids, there are also quantitative differences. For example, Sani & Meeks (1983) in a study of the toxicity of retinoids in mice showed *all-trans*-retinoic acid to be more toxic than *all-trans-N*-(2-hydroxyethyl)-retinamide > *all-trans-N*-(4-hydroxyphenyl)retinamide > NER. Similarly in humans, Kamm et al (1984) described quantitatively different side-effects in response to four orally administered retinoids. These qualitative and quantitative differences in the effects of retinoids have been emphasized recently by Bollag, who concluded that almost every retinoid possessed a different spectrum of properties, both therapeutic and side-effects. These qualitative and quantitative variations between the activity and toxicity of different retinoids are encouraging, and imply that the development of retinoids with greater selective action and therapeutic margins should be feasible.

This paper reports some effects of retinoids on the haematology, lymphoid organs and skeletal parameters of rodents, with emphasis on variations in the activity of retinoids and species-related differences in their effects in the rat and mouse.

Methods and abbreviations

Female F344/N rats and B6D2F1 mice were fed a ground diet (normal diet) to which retinoids were added as dietary supplements. Crystalline retinoids were dissolved in ethanol–trioctanoin with antioxidants, and blended into the normal diet; 13-*cis*-retinoic acid (13CRA), vitamin A (retinyl) acetate (VAA) and beta-carotene (BC) were added to the normal diet as beadlet preparations.

TABLE 1 Symptoms described in 200 reports of cases of hypervitaminosis A[a]

Symptoms	No. of times cited in reports
Skin changes:	
Mouth or lip fissures, chapping	41
Pale, dry or scaly skin	34
Pruritis, itching	29
Haemorrhages, petechiae (bleeding gingivae, membranes, nose, skin)	25
Rash, erythema or scaly rash	22
Desquamation, exfoliation, eruption	16
Pigmentation	12
Brittle or soft nails	7
Changes associated with skull and brain:	
Headache	56
Bulging fontanelle	51
Elevated CSF pressure, cranial hypertension, pseudotumour cerebri	33
Cranial hyperostosis, increased size of head	18
EEG anomalies	5
Changes in blood and serum:	
Elevation of serum vitamin A	50
Elevation of serum alkaline phosphatase	18
Hypercalcaemia	13
Anaemia	10
Prolonged prothrombin time	7
Elevated serum transaminase	5
Elevated serum lipids	4
Bone symptoms:	
Tenderness in extremities, aching, swelling	31
Joint pains	25
Long bone thickening, widening, hyperostosis	18
Nausea, vomiting	68
Visual changes:	
Diplopia, distorted or blurred vision	27
Papilloedema	24
Exophthalmos	10
Conjunctivitis	5
Fatigue, malaise, lethargy, somnolence, weakness, asthenia	64
Hair loss:	
Alopecia	32
Hair sparse and/or coarse	18
Changes in liver and spleen:	
Hepatomegaly, palpable or tender liver	32
Splenomegaly	11
Irritability	37
Loss of equilibrium (vertigo, dizziness, giddiness, ataxia, walking problems)	34

TABLE 1 (*ctd*)

Symptoms	No. of times cited in reports
Oedema (of face, eyelids, abdomen, legs)	20
Weight loss	18
Diarrhoea	11
Fever	11
Insomnia	11
Urination (altered patterns)	9
Nervous complaints	8
Menstrual changes	6
Thirst, polydipsia	4

[a]Published with modifications from: Bauernfeind JC 1980 The safe use of vitamin A. The Nutrition Foundation, Washington (with the permission of the author and publisher). CSF, cerebrospinal fluid; EEG, electroencephalogram.

Animals were fed control (placebo) diets containing ethanol–trioctanoin and antioxidants without retinoid, or placebo beadlets of 13CRA or BC with no active ingredient (see Hicks et al 1982, Festing et al 1984). Concentrations of retinoids and BC are given in millimoles per kg of diet (mmol/kg diet). Retinoids are abbreviated as follows: N-ethylretinamide, NER; N-(2-hydroxyethyl)retinamide, OHNER; N-butylretinamide, NBR; N-(4-hydroxyphenyl)retinamide, 4HPR; N-tetrazol-5-ylretinamide, TZR; 13-*cis*-N-ethylretinamide, 13cisNER; etretinate, ETR. The non-steroidal anti-inflammatory compounds (NSAICs) aspirin (ASP) and flurbiprofen (FRO) (Boots) were mixed into the placebo diet as dry crystals or powder; indomethacin (INDO) was added to the drinking water as an ethanolic solution containing 10 mg/ml; *dl*-alpha-tocopherol (TOC) was added to the normal diet in ethanol–trioctanoin solution. Weights of the femur, humerus, spleen, thymus, liver and lymph nodes (pooled axillary, brachial and inguinal nodes) are expressed in g/kg body weight. Bone parameters were measured with engineer's callipers. Long bones were cut with a jeweller's saw or decalcified and cut with a razor blade; measurements were taken by means of a binocular microscope with a micrometer eye piece. Where statistical analyses are used, groups fed dietary supplements are compared with the relevant control (placebo diet) group.

The effects of NER on rat blood

The haematological findings for rats fed 2 mmol NER/kg diet for one year, and 1 mmol or 2 mmol NER/kg diet for two years, are given in Table 2. A

TABLE 2A Haematological findings (means, standard deviations and observed ranges*) for female F344 rats fed NER for one year (49 weeks)

Treatment group	1	2	3
Dietary NER supplement (mg NER/kg diet)	Normal diet	Placebo diet	NER (654)
No. of rats autopsied†	14 (10)	15 (9)	14 (8)
Haemoglobin (g/dl)	15.09 ± 0.43 (14.5–16.0)	15.29 ± 0.68 (14.2–16.5)	14.11[bbb] ± 0.67 (13.5–15.5)
Red blood cell count ($\times 10^{12}$/litre)	7.67 ± 0.42 (6.89–8.69)	7.84 ± 0.40 (7.19–8.47)	7.61 ± 0.49 (6.78–8.41)
Packed cell volume (litre/litre)	0.466 ± 0.012 (0.450–0.485)	0.476 ± 0.016 (0.455–0.505)	0.443[bbb] ± 0.013 (0.420–0.460)
Mean cell volume (fl)	60.95 ± 3.60 (55.2–68.2)	60.90 ± 3.72 (54.3–67.6)	58.41 ± 3.72 (53.8–65.0)
Mean cell haemoglobin (pg)	19.73 ± 1.24 (17.5–21.8)	19.54 ± 1.45 (17.0–22.6)	18.61 ± 1.41 (16.7–21.1)
Mean cell haemoglobin concentration (g/dl)	32.35 ± 0.62 (31.3–33.3)	32.11 ± 1.08 (30.2–34.7)	31.85 ± 0.86 (30.7–33.7)
Reticulocytes (% red blood cells)	1.60 ± 0.85 (0.5–3.3)	1.81 ± 1.07 (0.5–3.3)	2.31 ± 1.46 (0.3–4.3)
White blood cell count ($\times 10^{9}$/litre)	2.16 ± 1.01 (1.0–4.1)	2.97[a] ± 1.02 (1.8–5.6)	3.06 ± 0.98 (1.2–4.9)
Platelets ($\times 10^{9}$/litre)	460 ± 52 (395–520)	421 ± 66 (292–510)	483[b] ± 78 (311–616)
Fibrinogen (g/litre)	2.19 ± 0.22 (1.79–2.50)	2.27 ± 0.23 (1.91–2.65)	2.01[bb] ± 0.29 (1.51–2.44)
Erythrocyte sedimentation rate‡ (mm in 1 h)	0	0	0
(mm in 24 h)	0	0	0§

TABLE 2A (*ctd*)

Treatment group	1	2	3
Dietary NER supplement (mg NER/kg diet)	Normal diet	Placebo diet	NER (654)
Neutrophils (×10⁹/litre)	0.708 ± 0.229 (0.374–1.036)	0.740 ± 0.205 (0.480–1.094)	0.986 ± 0.463 (0.384–1.660)
Lymphocytes (×10⁹/litre)	1.590 ± 0.838 (0.660–2.952)	2.301 ± 0.881 (1.408–3.660)	1.765 ± 0.745 (0.768–3.234)
Monocytes** (×10⁹/litre)	0.059 ± 0.064 (0–0.186) [7]	0.078 ± 0.083 (0–0.224) [6]	0.709 ± 0.105 (0–0.300) [5]
Eosinophils (×10⁹/litre)	0.042 ± 0.049 (0–0.164) [7]	0.048 ± 0.067 (0–0.192) [4]	0.024 ± 0.042 (0–0.098) [3]
Basophils (×10⁹/litre)	— [0]	0.004 ± 0.011 (0–0.034) [1]	— [0]

* Maximum and minimum values in parentheses.

† Numbers in parentheses are numbers used for the differential white blood cell counts; for technical reasons, these counts were only obtained for 27 of the 43 rats killed.

‡ Wintrobe method.

§ One rat in group 3 had an erythrocyte sedimentation rate of 1 mm in 24 h; this result has been excluded from the analysis.

** Square brackets [ ] show number of rats in each group in which cells of this type were observed.

Differences between the groups were analysed by Student's *t* test. Levels of significance: 'a' for groups 1/2; 'b' for groups 2/3. $^a P < 0.05$, $^{aa} P < 0.02$, $^{aaa} P < 0.001$ etc.

TABLE 2B Haematological findings (means, standard deviations and observed ranges*) for female F344 rats fed NER for two years (109 weeks)

Treatment group	4	5	6	7
Dietary NER supplement (mg NER/kg diet)	Normal diet	Placebo diet	NER (327)	NER (654)
No. of rats autopsied	13	13	14	16
Haemoglobin (g/dl)	14.25 ± 1.45 (11.7–17.8)	13.92 ± 1.39 (11.3–15.7)	13.64 ± 1.05 (10.7–15.1)	13.28 ± 1.59 (9.8–14.9)
Red blood cell count ($\times 10^{12}$/litre)	7.46 ± 0.84 (5.72–8.98)	7.48 ± 0.85 (5.91–8.71)	7.29 ± 0.79 (5.33–8.06)	6.97 ± 0.92 (4.61–8.26)
Packed cell volume (litre/litre)	0.449 ± 0.050 (0.355–0.575)	0.443 ± 0.043 (0.365–0.490)	0.433 ± 0.038 (0.320–0.470)	0.421 ± 0.047 (0.310–0.470)
Mean cell volume (fl)	60.29 ± 2.38 (57.5–64.0)	59.41 ± 3.42 (53.1–66.6)	59.59 ± 2.67 (56.6–66.1)	60.76 ± 3.81 (54.6–69.4)
Mean cell haemoglobin (pg)	19.16 ± 0.98 (17.5–20.9)	18.65 ± 0.89 (17.1–20.6)	18.79 ± 1.25 (17.0–21.7)	19.15 ± 1.33 (17.3–21.5)
Mean cell haemoglobin concentration (g/dl)	31.69 ± 0.75 (30.4–33.0)	31.44 ± 0.75 (30.6–33.1)	31.53 ± 1.09 (30.0–33.4)	31.59 ± 0.98 (29.9–33.9)
Reticulocytes (% red blood cells)	5.25 ± 3.71 (0.2–12.7)	7.78 ± 5.10 (1.5–17.6)	7.06 ± 3.08 (3.1–15.6)	6.06 ± 3.53 (0.4–13.5)
White blood cell count ($\times 10^{9}$/litre)	4.46 ± 1.61 (1.7–7.0)	5.85[c] ± 1.65 (2.5–8.5)	5.38 ± 1.95 (2.2–8.5)	5.60 ± 2.52 (2.2–12.8)
Platelets ($\times 10^{9}$/litre)	535 ± 118 (260–724)	580 ± 185 (402–1010)	619 ± 125 (403–794)	660 ± 100 (526–892)
Fibrinogen (g/litre)	3.35 ± 1.11 (1.88–5.53)	4.21 ± 2.41 (2.36–9.91)	2.68[d] ± 0.70 (1.72–4.16)	3.85[ff] ± 1.43 (2.08–6.89)
Erythrocyte sedimentation rate† (mm in 1h)	0	0	0	0

TABLE 2B (ctd)

Treatment group	4	5	6	7
Dietary NER supplement (mg NER/kg diet)	Normal diet	Placebo diet	NER (327)	NER (654)
Neutrophils ($\times 10^9$/litre)	1.758 ± 0.822 (0.350–3.021)	2.876^{cc} ± 1.315 (0.675–5.355)	2.424 ± 1.186 (0.806–5.040)	2.417 ± 0.946 (0.726–4.125)
Lymphocytes ($\times 10^9$/litre)	2.671 ± 1.339 (0.782–5.950)	2.898 ± 0.844 (1.722–4.464)	2.898 ± 1.192 (1.364–6.120)	3.140 ± 1.972 (1.265–9.728)
Monocytes‡ ($\times 10^9$/litre)	— [0]	0.050^c ± 0.074 (0–0.255) [6]	0.005^d ± 0.014 (0–0.045) [2]	0.004^c ± 0.015 (0–0.060) [1]
Eosinophils ($\times 10^9$/litre)	0.033 ± 0.034 (0–0.100) [7]	0.022 ± 0.033 (0–0.094) [5]	0.053 ± 0.090 (0–0.320) [6]	0.040 ± 0.072 (0–0.256) [7]
Basophils ($\times 10^9$/litre)	— [0]	— [0]	— [0]	— [0]

* Maximum and minimum values in parentheses.

† Wintrobe method.

‡ Square brackets [ ] show number of rats in each group in which cells of this type were observed.

Differences between the groups were analysed by Student's t test. Levels of significance: 'c' for groups 4/5; 'd' for groups 4/5; 'e' for groups 5/7; 'f' for groups 6/7. $^cP < 0.05$, $^{cc}P < 0.02$, $^{ccc}P < 0.001$ etc.

Post mortem, one rat in each of groups 4, 6 and 7 was found to have leukaemia. These animals have been excluded from this table.

principal components analysis of some of these data has been reported recently (Festing et al 1984). At one year, in rats fed NER, there were significant reductions in haemoglobin, packed cell volumes and fibrinogen values; platelet counts were slightly increased. At two years, significant reductions in fibrinogen values were again seen in NER-fed rats. Monocyte counts in rats fed NER were reduced but this is of questionable significance (Festing et al 1984). The interpretation of the two-year data was complicated by age-related changes in blood values, reflected by large variations in many parameters including haemoglobin, packed cell volumes and fibrinogen. Published reports on the effect of retinoids on blood are reviewed by Kamm et al (1984). A reduction in red cell values by retinoids has been reported by several workers, but no satisfactory account for these changes has been proposed. We have recently demonstrated that the effects of long-term treatment with 13CRA and 4HPR on rat red cell values are similar to those shown in Table 2, and that coincidentally there is significant extramedullary haemopoeisis in both spleen and liver. The relevance of these effects of retinoids on red blood cell parameters in rodents in relation to the possible clinical significance in humans has not been considered elsewhere in depth.

The effects of retinoids on the lymphoid organs of the mouse and rat

VAA, retinyl palmitate and retinoic acid stimulate skin graft rejection in mice (e.g. Medawar & Hunt 1981), and in animals given these supplements there is hypertrophy of lymphoid tissues (Dennert 1984). The effects of several retinoids on the proliferation of lymphoid tissues in our mice and rats are shown in Table 3, Fig. 1 and Fig. 2. In the first study in mice (Table 3), VAA and 13CRA were each fed at three dose levels. Body weight was unaffected except with VAA fed at the highest level. In general VAA and 13CRA induced a dose-related increase in lymph node and spleen weight, and the effect of VAA tended to be greater than that of the same dose of 13CRA. Thymus weight increased in some VAA-fed and 13CRA-fed groups but this was not consistently dose related. In a second trial (Fig. 1), VAA was fed at three dose levels and compared with NER, OHNER and BC. VAA again caused a dose-related increase in lymph node (Fig. 1B) and spleen (Fig. 1C) weight which was greater than that elicited by the two retinamides. Effects on the thymus (Fig. 1D) were difficult to interpret. BC did not affect spleen or thymus weight but increased that of the lymph nodes.

For comparison 13CRA, VAA and BC were also fed at three dose levels to rats (Fig. 2). In contrast to the mouse, the responses were not uniform, consistent or statistically significant when dose related. Lower levels of 13CRA reduced the spleen weights (Fig. 2B) but lower levels of VAA slightly

TABLE 3 Body and organ weights and femur measurements[a] of female B6D2F1 mice fed diets containing VAA or 13CRA

	Placebo diet	VAA Low	Mid	High	13CRA Low	Mid	High
No. of mice	21	21	22	22	21	20	21
Body weight (g)	28.0	29.2	29.4	25.0***	30.0	28.6	27.1
	(2.7)	(2.0)	(2.8)	(2.0)	(2.8)	(2.5)	(2.4)
Lymph node wt[b]	0.878	1.232***	1.390***	4.637***	0.959*	1.206***	1.191***
(g/kg body wt)	(0.130)	(0.231)	(0.283)	(3.138)	(0.142)	(0.244)	(0.219)
Spleen weight	3.758	3.681	3.860	6.848**	3.376	4.168*	4.403**
(g/kg body wt)	(0.638)	(0.691)	(0.718)	(3.627)	(0.582)	(0.691)	(0.681)
Thymus weight	2.141	2.386*	2.502**	2.379	2.131	2.429**	2.545**
(g/kg body wt)	(0.264)	(0.632)	(0.542)	(0.499)	(0.406)	(0.365)	(0.501)
Liver weight	57.84	58.22	59.84	63.40**	55.42	58.96	61.83*
(g/kg body wt)	(5.85)	(4.75)	(5.27)	(6.37)	(5.16)	(4.18)	(6.08)
Femur weight	3.597	2.937***	2.950***	2.749***	2.948***	2.844***	2.963***
(g/kg body wt)	(0.543)	(0.332)	(0.478)	(0.371)	(0.402)	(0.295)	(0.405)
Femur length (cm)	1.574	1.589	1.589	1.560	1.602**	1.602*	1.602*
	(0.035)	(0.056)	(0.031)	(0.043)	(0.028)	(0.046)	(0.050)
Femur diameter (mm)	1.58	1.51***	1.44***	1.27***	1.46***	1.41***	1.30***
	(0.06)	(0.06)	(0.08)	(0.12)	(0.06)	(0.06)	(0.08)

[a] Mean values; SD in parentheses.
[b] Pooled axillary, brachial and inguinal nodes.
Mice were 3–4 weeks old (15.1 g) at the beginning of the experiment and were fed diets for 12.5 weeks. VAA and 13CRA were fed at 0.19, 0.38 or 0.75 mmol per kg diet (low, mid and high dose levels respectively). Differences between each VAA and 13CRA group and the control (placebo-fed) group were analysed by Student's t test: $*P < 0.05$, $**P < 0.01$, $***P < 0.001$.

increased spleen weight. Higher levels of 13CRA increased lymph node weight (Fig. 2C), whereas VAA was inactive. Thymic responses (Fig. 2D), as in the mouse, were difficult to interpret and showed no clear overall patterns of change. Recently the effects of three retinoids on the lymphoid organs of the rat have been re-examined (Table 4A) and no general patterns of change in the spleen, lymph nodes or thymus were detected. It is concluded that in the mouse, but not in the rat, retinoids cause a dose-related hypertrophy of the spleen and lymph nodes.

Further examination of retinoid-treated mouse lymphomedullary organs in collaboration with Dr D. Katz (Middlesex Hospital Medical School) has shown that VAA and 13CRA cause an expansion of the paracortical region in the spleen with almost no reactive follicle formation. In mice fed VAA or 13CRA for 4, 7 or 12 weeks (Table 5), density gradient analysis has enabled an increase in the accessory cell population to be identified. With VAA, by 12

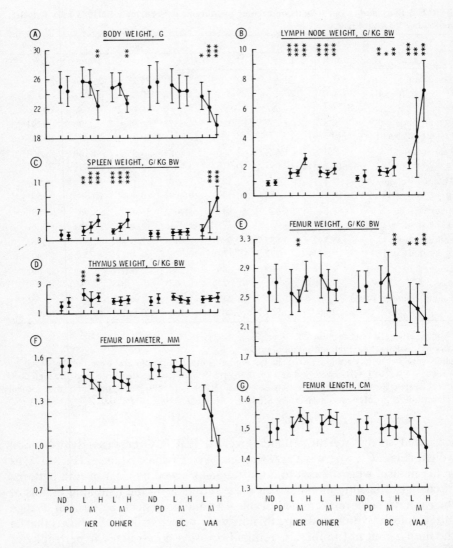

FIG. 1. Mean (± SD) of body and organ weights, and femoral measurements of mice fed low (L), mid (M) or high (H) levels of NER, OHNER, BC or VAA. The experiment was in two parts (NER plus OHNER, and BC plus VAA) each with a control normal diet (ND) and placebo diet (PD). Low, mid and high dose levels of NER and OHNER were 0.5, 1.0 and 1.5 mmol/kg diet; for BC, 0.2, 1.0 and 5.0 mmol/kg diet; and for VAA, 0.25, 0.5 and 0.75 mmol/kg diet. Mice were 16 g at the beginning of the experiment and autopsies were 13 weeks later; there were 17–20 mice in each group. *P<0.05; **P<0.01; ***P<0.001. Statistical analyses were not carried out on femoral diameters or lengths.

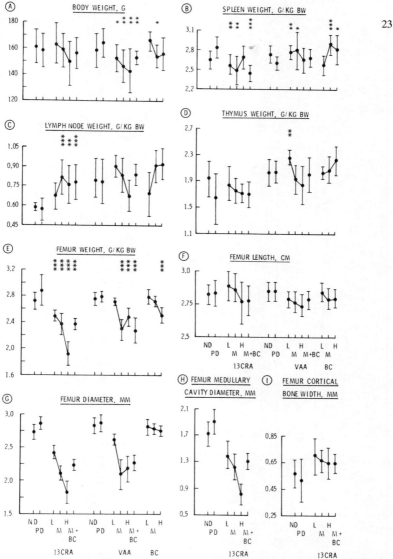

FIG. 2. Mean (± SD) of body and organ weights, and femoral measurements of rats fed low (L), mid (M) or high (H) levels of 13CRA, VAA or BC. The experiment was in two parts (13CRA, and VAA plus BC) each with a control normal diet (ND) and placebo diet (PD). Low, mid and high dose levels of 13CRA were 0.4, 0.8 and 1.6 mmol/kg diet; for VAA, 0.5, 1.0 and 2.0 mmol/kg diet; for BC, 0.2, 1.0 and 5.0 mmol/kg diet. In addition, 1.0 mmol BC was fed with 0.8 mmol 13CRA and 1.0 mmol VAA/kg diet. Rats weighed 60 g at the beginning of the experiment and were autopsied seven weeks later; there were 7–12 rats in each group. P values as in Fig. 1. Statistical analyses were not carried out on femoral measurements except weight. Femora were sectioned and measured microscopically only from animals fed 13CRA and the two relevant control groups.

TABLE 4A Body and organ weights[a] of rats fed a placebo diet or diet containing the retinamides NER, TZR or 4HPR[b]

	Week 0[c]	Week 6				Week 15	
	PD	PD	NER	TZR	4HPR	PD	NER
Body weight (g)	58.1	145.4	148.1	150.1	139.3	188.3	188.2
	(14.9)	(6.7)	(7.9)	(6.5)	(8.5)	(9.3)	(10.9)
Spleen weight[d]	3.43	2.25	2.31	2.16	2.27	2.04	1.84***
	(0.46)	(0.15)	(0.23)	(0.19)	(0.10)	(0.12)	(0.08)
Lymph node weight[d]	0.29	0.23	0.25	0.18*	0.18	0.16	0.14
	(0.08)	(0.06)	(0.07)	(0.03)	(0.06)	(0.03)	(0.05)
Thymus weight[d]	3.50	1.89	1.78	1.78	2.00	1.16	1.05*
	(0.23)	(0.26)	(0.20)	(0.26)	(0.21)	(0.10)	(0.11)
Liver weight[d]	47.12	45.13	49.97*	44.56	49.41*	41.82	44.71
	(5.03)	(5.10)	(4.00)	(2.92)	(3.75)	(4.46)	(3.62)

[a] Mean values; SD in parentheses. Ten rats per group, except in the placebo diet (PD) group at 15 weeks with 11 rats.
[b] Fed at 2.5 mmol/kg diet for six weeks (NER, TZR, 4HPR) or 15 weeks (NER).
[c] Beginning of experiment.
[d] Expressed as g/kg body weight.
Each group fed a retinamide diet was compared with the appropriate PD group by Student's t test: $*P < 0.05$, $***P < 0.001$.

TABLE 4B Body weight, femoral and numeral measurements[a] of rats fed a placebo diet (PD) or diet containing the retinamides NER, TZR or 4HPR with or without ASP for six weeks[b]

	Week 0[c]	Week 6							
	PD	PD	PD + ASP	NER	NER + ASP	TZR	TZR + ASP	4HPR	4HPR + ASP
Body weight (g)	58.1 (14.9)	145.4 (6.7)	144.5 (9.1)	148.1 (7.9)	134.9*** (4.8)	150.1 (6.5)	144.3 (7.0)	139.3 (8.5)	135.7 (12.0)
Femur weight[d]	3.00 (0.14)	2.65 (0.15)	2.41** (0.15)	2.36††† (0.18)	2.79*** (0.13)	2.29 (0.17)	2.23 (0.14)	2.28 (0.05)	2.28 (0.13)
Femur diameter (mm)[e]	1.87 (0.19)	2.36 (0.08)	2.24** (0.09)	2.23††† (0.07)	2.39** (0.13)	2.03 (0.07)	2.05 (0.07)	1.98 (0.09)	2.01 (0.08)
Femur medullary cavity diameter (mm)[f]	1.50 (0.16)	1.43 (0.08)	1.16*** (0.14)	1.20††† (0.14)	1.50*** (0.11)	1.17 (0.08)	1.16 (0.06)	1.07 (0.07)	1.12 (0.10)
Femur cortical bone width (mm)[f]	0.27 (0.04)	0.53 (0.04)	0.63** (0.09)	0.56 (0.06)	0.64* (0.09)	0.60 (0.05)	0.58 (0.06)	0.58 (0.06)	0.57- (0.03)
Humerus weight[d]	1.52 (0.16)	1.09 (0.06)	0.94*** (0.05)	1.01† (0.08)	1.16*** (0.08)	0.93 (0.06)	0.95 (0.07)	0.92 (0.06)	0.92 (0.03)
Humerus diameter (mm)[e]	1.59 (0.10)	1.54 (0.04)	1.46*** (0.04)	1.46† (0.08)	1.63*** (0.07)	1.37 (0.10)	1.31 (0.07)	1.29 (0.07)	1.33 (0.09)
Humerus medullary cavity diameter (mm)[f]	1.23 (0.21)	0.67 (0.16)	0.58 (0.11)	0.52† (0.11)	0.82*** (0.12)	0.45 (0.13)	0.43 (0.10)	0.44 (0.09)	0.48 (0.07)
Humerus cortical bone width (mm)[f]	0.23 (0.07)	0.52 (0.05)	0.47* (0.04)	0.52 (0.10)	0.47 (0.06)	0.47 (0.06)	0.48 (0.04)	0.47 (0.05)	0.46 (0.03)

[a] Mean values; SD in parentheses. Ten rats per group.
[b] Retinamides were fed at 2.5 mmol/kg diet and ASP at 0.15% of the diet.
[c] Beginning of experiment.
[d] Expressed as g/kg body weight.
[e] Calliper measurement (these are made with the bones in different orientations to the microscopic measurements).
[f] Microscopic measurement.
Each group of rats fed a diet + ASP was compared by Student's t test with the appropriate group without dietary ASP: *$P < 0.05$, **$P < 0.01$, ***$P < 0.001$. Rats fed NER diet were compared with those fed PD: †$P < 0.05$, †††$P < 0.001$.

weeks there is up to a 10-fold increase in low density non-adherent cells in the mixed cell pool; with 13CRA the effect is comparable, but less marked. Both compounds increase the potential for accessory cell–T cell interaction, and increase the number of cells expressing the markers linked to this interaction (Drzymala et al 1984).

TABLE 5 Numbers of low density accessory cells[a] from the spleens of female B6D2F1 mice fed diets containing 13CRA or VAA

| Dietary supplement | Cell count $(\times 10^{-6})$ | | |
	4 weeks	7 weeks	12 weeks
Placebo diet	0.2	0.3	1.6
13CRA	0.3	0.8	4.2
VAA	0.5	1.2	19.4

[a]Mean of results from three experiments.
Diets were fed for 4, 7 or 12 weeks. 13CRA and VAA were fed at 1.8 and 1.0 mmol per kg diet respectively. In each group 8–10 mice were killed at each time point. For further information see Drzymala et al (1984).

The effects of retinoids on the long bones of the rat and mouse

The effect of retinoids on rodent bones has been investigated: firstly, changes in long-bone parameters have been used to monitor retinoid activity in both the rat and mouse; secondly, the modulation by NSAICs of retinoid-induced bone changes has been investigated; thirdly, the effect of age on retinoid-induced long-bone remodelling in the rat has been studied.

Development of a short-term test to quantify retinoid-induced long-bone changes in the rat and mouse

In initial studies (Table 3) measurements of femoral weight, diaphysis (shaft) diameter, and length were made in young mice fed retinoids for 12.5 weeks. With VAA and 13CRA fed at three dose levels, femoral weight emerged as a sensitive indicator of retinoid activity, with weight reductions evident at the lowest dose levels tested. Femur length was not reduced, suggesting that the functioning of the epiphyseal plate was not adversely affected. Diaphysis diameter was reduced in a dose-related fashion. Other experiments with humeri showed similar changes to those obtained with the femur (Table 3). The scapula, a flat bone, did not exhibit any obvious retinoid-induced changes in weight or overall measurement, and we did not see evidence of the fenestration reported by other workers to be associated with hypervitaminosis A

(Barnicot & Datta 1972). Sesamoid, short or irregular bones from mice or rats have not been examined.

In a second study with young mice (Fig. 1), VAA was compared with BC, NER and OHNER. BC was without effect except to reduce femur weight at the highest level tested (Fig. 1E) and the reasons for this reduction are not clear. Neither NER nor OHNER reduced femoral weight in a dose-related fashion, but with VAA significant reductions occurred. Diaphysis diameter (Fig. 1F) and length (Fig. 1G) were not analysed statistically; since such measurements are proportional to body weight, relative, not absolute, values would be more meaningful. Femur length in NER-fed and OHNER-fed mice was not reduced, even at the highest dose levels which reduced body weight (Fig. 1A), and in some retinoid-fed groups femur length actually appeared to be increased. VAA induced a dose-related decrease in femur length, but with the body weight reductions in these groups, these bone lengths were the same as in normal mice of the same body weight. Femoral diameters were consistently reduced in a dose-related fashion in groups fed NER, OHNER and VAA.

Bones from the rat are larger and thus easier to work with than mouse bones and have been used for most of the work reported here. Fig. 2 shows the effect of three dose levels of 13CRA, VAA and BC on the femora of young (growing) rats, and the effect of co-administering BC with 13CRA and VAA. BC hardly affected the long-bone parameters. The length was not significantly reduced by 13CRA (Fig. 2F), and the slight reductions with VAA reflect body weight decreases (Fig. 2A); as in the mouse, the function of the epiphyseal plate does not appear to be adversely affected by 13CRA or VAA. Dose-related reductions in both femoral weight and femoral diameter were evident in 13CRA-fed and VAA-fed rats (Fig. 2E,G).

Femora of young rats fed 13CRA were further examined by transversely sectioning the diaphysis to measure cortical bone thickness, medullary cavity and diaphysis diameter. Concomitant with the reduction in overall femoral diameter there was a reduction in medullary cavity diameter, but cortical bone thickness was slightly increased (Fig. 2H,I). These sectioned femora from young rats fed 13CRA demonstrated that dose-dependent periosteal resorption and endosteal deposition occur in these animals to produce a remodelled bone with increased cortical thickness. Sections at different positions along the shaft confirmed that this remodelling occurred relatively uniformly along the length of the diaphysis, and that a single transverse section from the mid-diaphysis region was representative.

Retinoids induced similar changes in another long bone, the humerus (Table 4B), but the overall dimensions and weight of the scapula showed no variation in retinoid-treated rats and no fenestration. The mandible also showed no retinoid-induced effects. Scanning electron microscopy showed

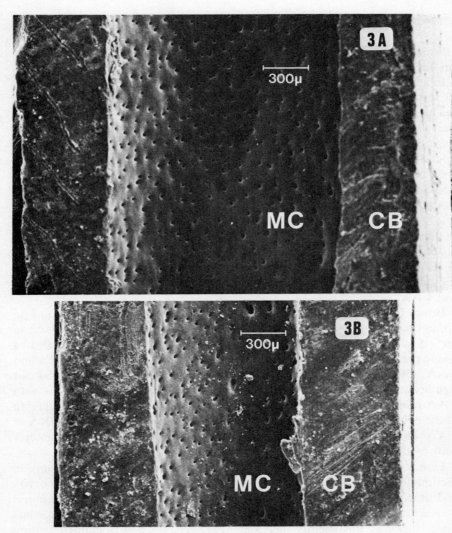

FIG. 3. Scanning electron micrographs of sectioned rat femoral diaphyses. (A,B) Longitudinally sectioned sodium hypochlorite-treated femora from rats fed placebo diet (A) or 2.5 mmol NER/kg diet (B) for 15 weeks. Rats weighed 58 g at the beginning of the experiment. The mean diaphysis diameter, medullary cavity diameter and cortical bone thickness of 10 rats in the placebo diet group were 2.852, 1.492 and 0.680 mm, and of 10 rats in the NER-fed group, 2.356, 0.946 and 0.705 mm respectively. (C,D) Transversely sectioned diaphyses of rats fed placebo diet (C) or 1.6 mmol 13CRA/kg diet (D) for 78 weeks. Rats weighed 84 g at the beginning of the experiment. The mean diaphysis diameter, medullary cavity diameter and cortical bone thickness of eight rats in the placebo diet group were 2.758, 1.596 and 0.581 mm, and of 10 rats in the 13CRA-fed group 2.208, 1.132 and 0.538 mm respectively. Fig. 3D shows the most extreme reduction of diaphysis and medullary cavity diameter in the 10 rats examined. CB, cortical bone; MC, medullary cavity.

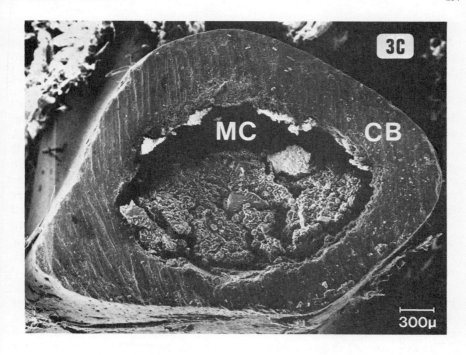

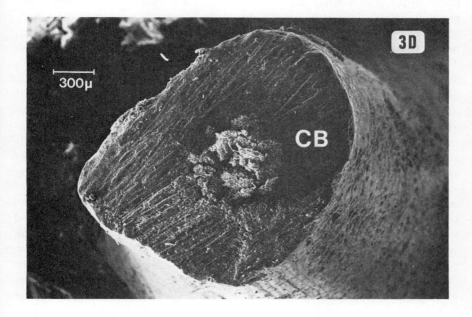

that the femur from young NER-fed animals (Fig. 3B) had decreased diaphysis and medullary cavity diameters and increased cortical bone thickness by comparison with the control (Fig. 3A). The femur from rats fed 1.6 mmol 13CRA/kg diet for 78 weeks also had a significantly reduced medullary cavity (Fig. 3D) by comparison with controls (Fig. 3C).

Dose-related reductions in the body weight of rodents fed retinoids have been used by many authors as a quantitative index of retinoid activity (e.g. Hicks et al 1982). In humans, weight loss is a feature of hypervitaminosis A (Table 1) but has not been reported as a side-effect of retinoid therapy. It is not clear why retinoids cause weight loss in laboratory animals and, although easy to assess, body weight loss in rodents may be a relatively insensitive index of retinoid activity. This is illustrated in Table 3 where, of the three dose levels of VAA and 13CRA fed, only the highest dose of VAA caused significant body weight loss, whereas lymph nodes, and femur weight and diameter were all affected by the lowest dose of both compounds. Bone measurements may thus be a more sensitive index of retinoid activity, but this is counterbalanced by the fact that they are only available *post mortem*.

The decreases in red blood cell values (Table 2), and associated extra-medullary haemopoeisis, also reported in other rodent trials of retinoids, may well be associated with reductions in the medullary cavity volumes in retinoid-fed animals.

The objective of these initial experiments was to provide a base-line for a short-term test to compare the activity of several retinamides fed at equimolar concentrations. As part of a much larger study, weanling rats were fed three retinamides (NER, TZR, 4HPR) at 2.5 mmol/kg diet for six weeks (Fig. 4A), but at this dose they did not produce rapidly a large enough femoral response. The retinamide dose level was therefore increased to 2.875 mmol/kg diet. Weanling rats were fed five retinamides at 2.875 mmol/kg diet for six weeks, also as part of a more extensive trial (Fig. 4C). In addition, rats were fed NER at 5.75 mmol/kg diet, and a mixture of 2.875 mmol NER and 2.875 mmol 4HPR/kg diet (Fig. 4B). Of the five retinamides fed at equimolar concentrations, NBR produced the greatest response, followed by TZR, NER, 4HPR and OHNER; NER at 5.75 mmol/kg diet and the NER plus 4HPR diet were more potent than NBR. This experiment indicated that medullary cavity measurements (endosteal deposition) are a more sensitive indicator of 'retinoid toxicity' than measurements of diaphysis diameter (periosteal resorption). From these results a dose level of retinamide can be calculated which should produce a particular effect on the femur. For example, a 35% reduction in medullary cavity diameter (65% of the control value) would be produced by feeding 3.07 mmol OHNER, 2.91 mmol 4HPR, 2.80 mmol NER, 2.65 mmol TZR, or 2.16 mmol NBR/kg diet to weanling rats for six weeks; a 25% reduction in shaft diameter (75% of control value) would be

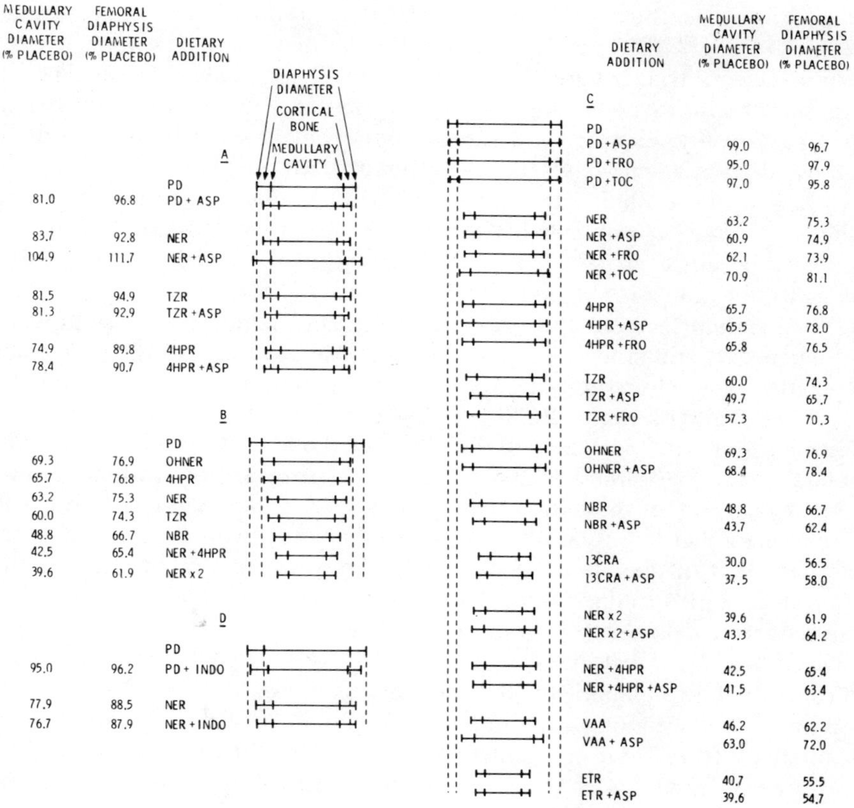

MEDULLARY CAVITY DIAMETER (% PLACEBO)	FEMORAL DIAPHYSIS DIAMETER (% PLACEBO)	DIETARY ADDITION
		A
		PD
81.0	96.8	PD + ASP
83.7	92.8	NER
104.9	111.7	NER + ASP
81.5	94.9	TZR
81.3	92.9	TZR + ASP
74.9	89.8	4HPR
78.4	90.7	4HPR + ASP
		B
		PD
69.3	76.9	OHNER
65.7	76.8	4HPR
63.2	75.3	NER
60.0	74.3	TZR
48.8	66.7	NBR
42.5	65.4	NER + 4HPR
39.6	61.9	NER x 2
		D
		PD
95.0	96.2	PD + INDO
77.9	88.5	NER
76.7	87.9	NER + INDO

Diagram labels: DIAPHYSIS DIAMETER / CORTICAL BONE / MEDULLARY CAVITY

DIETARY ADDITION	MEDULLARY CAVITY DIAMETER (% PLACEBO)	FEMORAL DIAPHYSIS DIAMETER (% PLACEBO)
C		
PD		
PD + ASP	99.0	96.7
PD + FRO	95.0	97.9
PD + TOC	97.0	95.8
NER	63.2	75.3
NER + ASP	60.9	74.9
NER + FRO	62.1	73.9
NER + TOC	70.9	81.1
4HPR	65.7	76.8
4HPR + ASP	65.5	78.0
4HPR + FRO	65.8	76.5
TZR	60.0	74.3
TZR + ASP	49.7	65.7
TZR + FRO	57.3	70.3
OHNER	69.3	76.9
OHNER + ASP	68.4	78.4
NBR	48.8	66.7
NBR + ASP	43.7	62.4
13CRA	30.0	56.5
13CRA + ASP	37.5	58.0
NER x 2	39.6	61.9
NER x 2 + ASP	43.3	64.2
NER + 4HPR	42.5	65.4
NER + 4HPR + ASP	41.5	63.4
VAA	46.2	62.2
VAA + ASP	63.0	72.0
ETR	40.7	55.5
ETR + ASP	39.6	54.7

FIG. 4. Results of three experiments presented as diagrams of sectioned femora showing the diaphysis and medullary cavity diameter, and cortical bone thickness (cf. Fig. 3A,B); diaphysis and medullary cavity diameters are also expressed as percentages of the control (placebo diet, PD) values. In each experiment the structure of the control femur is projected as a broken line to illustrate relative bone deposition and resorption in other groups. All trials lasted six weeks . (A) Experiment 1: NER, TZR and 4HPR were fed at 2.5 mmol/kg diet, and ASP at 0.15% of the diet; there were 10 rats per group averaging 58 g at the beginning of the experiment. (B,C) Experiment 2: five retinamides (OHNER, 4HPR, NER, TZR, NBR) were fed at 2.875 mmol, 13CRA at 1.6 mmol, ETR at 0.155 mmol and VAA at 2.0 mmol/kg diet; NER was also fed at 5.75 mmol/kg diet (2.875 × 2), and at 2.875 mmol/kg diet plus 4HPR at 2.875 mmol/kg diet; ASP was fed at 0.3%, FRO at 0.01% and TOC at 0.71% of the diet. There were 8–10 rats per group averaging 64 g at the beginning of the experiment. Results of Fig. 4B are taken from 4C. (D) Experiment 3: NER was fed at 2.875 mmol/kg diet and INDO added to the drinking water to give 10.9 mg/l. There were five or six rats per group which averaged 153 g at the beginning of the experiment. Measurements of diaphysis and medullary cavity diameter and cortical bone thickness of rats fed placebo diet in the three experiments were: (A) 2.310, 1.430, 0.440 mm; (B,C) 2.761, 1.865, 0.448 mm; (D) 2.953, 1.711, 0.621 mm respectively.

produced by feeding 2.95 mmol OHNER, 2.94 mmol 4HPR, 2.88 mmol NER, 2.85 mmol TZR, or 2.56 mmol NBR/kg diet.

In a recent trial where 2.0 mmol 13cisNER/kg diet was fed, not to weanlings but to rats weighing 186 g (17 weeks old), no changes in femoral or medullary cavity diameter and relative bone weight were found in animals autopsied after 20 and 40 weeks, even though body weight was reduced. This contrasts with the finding that 2.0 mmol NER/kg diet fed to rats of similar weight produced highly significant femoral effects which were identified at one year (Hicks et al 1982). 13cisNER may thus be unusual among retinamides in not inducing skeletal changes.

When similar techniques were used to measure femoral changes in weanling mice fed retinamides for 12 weeks, periosteal resorption and endosteal deposition were observed but the overall effects were less marked than in the rat. Mice fed 0.42 mmol 4HPR/kg diet had a medullary cavity 94.8% of controls, and with 3.0 mmol 4HPR/kg diet, 86.5% of controls; shaft diameters were 96.4% and 90.3% of controls respectively; with 0.42 mmol NBR/kg diet the medullary cavity was 97.6% of controls, and with 3.0 mmol/kg diet, 66.2% of controls; shaft diameters were 97.1% and 79.7% respectively. Collected findings from several mouse experiments indicate that to produce a 10% reduction in diaphysis diameter the following concentrations of retinoids (in mmol/kg diet) would be effective: 13CRA, 0.37; VAA, 0.39; NBR, 1.32; NER, 1.44; 4HPR, 1.81; OHNER, 1.88. This sequence of efficacy for the retinamides is the same as in the rat (Fig. 4B).

When using weanling rats and mice it should be remembered that the weanling rat (e.g. Table 4A) grows from 58 g to 145 g in six weeks and to 188 g in 15 weeks, increases of 150.2% and 224.1% respectively. By contrast, the 15.1 g weanling mouse (Table 3) grows to 28.0 g in 12 weeks, which is only an 85.4% increase. Thus, any effects on the growing skeleton will be relatively greater in the rat than in the mouse and hence the young rat may be more sensitive to retinoids than the young mouse.

Modification of retinoid-induced long-bone changes in the rat by NSAICs

The steroid pregnenolone-16α-carbonitrile prevents vitamin A-induced bone lesions in the rat (Tuchweber et al 1976), and retinoic acid-induced bone fractures in mice can be reduced by the co-administration of NSAICs (Hixson & Harrison 1981). We have examined such drug interactions using bone remodelling as a sensitive index of retinoid effects.

Weanling rats were fed 2.5 mmol NER, TZR and 4HPR/kg diet, with dietary ASP at 0.15% for six weeks (Fig. 4A). A 'retinoid-like' effect was produced in rats fed ASP alone, which reduced diaphysis diameter and medullary cavity and affected other bone parameters (Table 4B). This remodelling by ASP parallels that reported by Sudmann (1975) for indomethacin, which causes bone remodelling in rabbit ear chambers. However, in rats fed ASP in conjunction with NER, the effect of NER was reversed; medullary cavity and diaphysis diameters increased (Fig. 4A) as did femoral and humeral weight and humeral diameters (Table 4B). These results thus confirmed the previous reports of prevention of retinoid-induced skeletal changes by NSAICs. However, ASP did not reverse the retinoid-induced effects in rats fed TZR or 4HPR (Fig. 4A, Table 4B), which implies that its ability to modify retinoid action depends upon the particular retinoid used.

To investigate the interactions of ASP and retinoids further, five retinamides were fed at 2.875 mmol/kg diet, 13CRA at 1.6 mmol, ETR at 0.155 mmol and VAA at 2.0 mmol/kg diet; ASP was fed at 0.3% of the diet in these experiments. Dietary FRO (0.01%) was also tested against three retinamides. Vitamin E has been reported to prevent hypervitaminosis A, and so dietary TOC (0.71%) was tested against NER. As the lesions induced by Hixson & Harrison (1981) were severe and included bone fractures, NER was also fed at the increased level of 5.75 mmol/kg diet, and a diet containing NER at 2.875 mmol plus 4HPR at 2.875 mmol/kg diet was also used. The results are summarized in Fig. 4C. Retinamide-induced changes were not affected by the co-administration of FRO. We found initially that ASP prevented NER-induced changes (Fig. 4A), but this effect was not reproduced by the higher dose levels of NER (2.875 mmol/kg diet) and ASP (0.3%). With other retinoids also the results were negative in that ASP did not reverse retinoid-induced effects, although it slightly reduced VAA-induced changes, and TOC decreased the NER-induced effects. Surprisingly, ASP enhanced the TZR-induced skeletal changes in this second trial. The interaction of NSAICs with retinoids thus appears to be dependent upon the dose levels of both interacting compounds. A trial is planned to study the interaction of NER plus ASP with each compound fed at high and low dose levels.

Since INDO prevented retinoic acid-induced skeletal effects (Hixson & Harrison 1981), it seemed possible that this NSAIC might also prevent NER-induced changes. Alone, like ASP and FRO, INDO had a slight 'retinoid-like' effect on the femur but it did not prevent the NER-induced bone changes (Fig. 4D). Relatively low doses of INDO, averaging about 0.82 mg/kg body weight per day, were used here.

Retinoids are reported to be anti-inflammatory and thus the similarity of action on the skeleton of retinoids, ASP, FRO and INDO is not surprising. The ability of NSAICs to prevent retinoid-(NER)-induced effects (Fig. 4A) is more puzzling; it is not clear whether the ASP inhibits the NER activity or vice versa.

Age-related effects on retinoid-induced long-bone remodelling in the rat

Preliminary investigations of the mechanisms involved in retinoid-induced long-bone remodelling have been made in collaboration with Dr L. Ream (Wright State University, USA) and Dr E. Katchburian (London Hospital Medical College). In the normal weanling rat, femoral remodelling occurs on both periosteal and endosteal surfaces, but little bone is deposited endosteally; increase in cortical bone thickness with age is due to periosteal bone deposition. In weanling rats fed 2.5 mmol NER/kg diet for 15 weeks, remodelling also occurs on both surfaces, but in contrast to the controls, there is significant bone deposition on the endosteal surface. Considerable deposition also occurs on the periosteal surface, but less than in controls. Retinoids induce these changes rapidly, but if young NER-fed rats are then transferred to a control diet, there is rapid reversal of the NER-induced changes; bone is deposited periosteally and resorbed endosteally. In the 15-week trial with 2.5 mmol NER/kg diet, body weight was not reduced and neither femur length nor hindlimb length was affected. Movement (spontaneous activity), measured with an open-field behavioural apparatus, was not affected in NER-fed animals.

Bone shows significant age-related changes, and results from young rats cannot be extrapolated to the ageing animal. In rats fed control diets for nine months, and then when adult 2.0 mmol 4HPR/kg diet for six months, periosteal bone resorption and endosteal deposition in the long bones were induced by the retinoid, but more slowly than in young animals. In rats fed 4HPR for nine months and then control diet for six months, the retinoid-induced changes began to reverse by periosteal deposition and endosteal resorption, but again at a slower rate than in young rats.

The long-term (100 weeks) effects of 1.2 mmol 13CRA/kg diet on remodelling the rat femur were investigated (Fig. 5). In normal 170 g rats (fed placebo diet) remodelling occurred on both the endosteal and periosteal surfaces, but there was little variation in the endosteal (medullary cavity) dimensions during the 100-week period. Cortical bone thickness increased with age by periosteal deposition. In 13CRA-fed rats there was a rapid response, and by 13 weeks significant periosteal resorption and endosteal deposition had occurred. The gross anatomy of this remodelled bone then remained approximately constant for the remaining 87 weeks (Fig. 5). 13CRA thus initiated a remodelling process in the first 13 weeks, which changed the bone dimensions; this could be described as an 'initial change' remodelling process. In subsequent remodelling, still under 13CRA control, the 'initial' remodelled state was retained for the remainder of the experiment.

The ability of retinoids to remodel rodent long bones is thus affected by: (1)

the retinoid under investigation; (2) the dose administered, which in these experiments may depend upon the palatibility of the retinoid; (3) the animal species; (4) the age of the animal on first exposure to retinoids; (5) the duration of administration; (6) the route of administration; and possibly (7) the vehicle in which the retinoid is given.

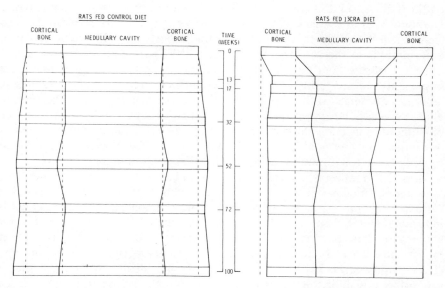

FIG. 5. Diagrams of a series of transverse sections of the femoral diaphyses of rats fed control placebo diet or diet containing 1.2 mmol 13CRA/kg for 100 weeks. Animals weighed 170 g (15 weeks old) at the beginning of the experiment (week 0), and groups of 5–9 rats were killed at 0, 13, 17, 32, 52, 72 and 100 weeks. The structure of the bone at week 0 is projected as broken lines to week 100 to illustrate the relative amounts of bone deposited or removed from the periosteal or endosteal surfaces.

Conclusions

Of the various retinoid-induced changes reported in this paper, those relating to bone changes are probably the most relevant to the clinical use of these compounds. There is increasing concern about retinoid-induced skeletal changes, seen as side-effects of retinoid therapy in humans (Tamayo & Ruíz-Maldonado 1981, Milstone et al 1982, Windhorst & Nigra 1982, DiGiovanna & Peck 1983, Pittsley & Yoder 1983, Gerber et al 1984, Kamm et al 1984). The relationship between the retinoid-induced changes in rodents and in humans has not yet been established. Long-bone effects in humans have been reported both in hypervitaminosis A (Gerber et al 1954, Barnicot

& Datta 1972, DiGiovanna & Peck 1983) and as a result of retinoid therapy (Tamayo & Ruíz-Maldonado 1981, Milstone et al 1982). Similarly, long-bone thinning is a consistent feature of hypervitaminosis A in laboratory animals (Moore 1967, Leelaprute et al 1973, Cho et al 1975), and in animals treated with retinoids (Teelman 1981, Kurtz et al 1984). Furthermore, the overall patterns of skeletal change in laboratory animals and humans with hyper-vitaminosis A are markedly similar and comparable to retinoid-induced changes in humans. In this context, the reports of ETR-induced spinal deformations in the rat (Teelman 1981) and ETR-induced exostoses at the insertion of the foot tendon in the rabbit (Mahrle & Berger 1982) are important and suggest that retinoids do indeed induce similar changes in laboratory animals to those seen in clinical hypervitaminosis A. The remod-elling of long bones in rats and mice described here may be used as an index of retinoid-induced skeletal change in these species and may be regarded as one indicator of retinoid toxicity. We suggest that these changes may prove to be prognostic for retinoid-induced skeletal lesions in humans.

More information is needed on retinoid-induced skeletal changes not only in rodents, but also in those species which, like humans, have haversian systems (e.g. rabbit, dog). Additionally, the relationship between retinoid-induced skeletal changes and the interaction with NSAICs requires further exploration especially as NSAICs are regularly prescribed for the alleviation of retinoid-induced bone and joint pains in patients receiving retinoid therapy (DiGiovanna & Peck 1983, Gerber et al 1984). Finally, we suggest that a promising possibility arises from this work, namely the potential use of retinoids in orthopaedic practice to stimulate bone remodelling in the treatment of certain bone diseases.

Acknowledgements

This work was funded in part with Federal funds from the American Department of Health and Human Services under contract numbers NO1 CP 75938 and NO1 CP05602-56. The contents of this publication do not necessarily reflect the views or policies of the Department of Health and Human Services, nor does the mention of trade names, commercial products or organizations imply endorsement of the US Government. We thank Dr J. Bauernfeind and the International Vitamin A Consultative Group for permission to publish Table 1, R. Wright and K. Nandra for producing Fig. 3, and the Illustration and Photography Units, Middlesex Hospital Medical School for their assistance. We also acknowledge the help of other colleagues who have assisted in many ways over seven years, including N. Comben, E. Chrysostomou, M. Drzymala, M. Festing, M. Hart, J. Harvey, E. Hixson, I. Hunneyball, E. Katchburian, D. Katz, P. Medawar, R. Moon, K. Nandra, L. Palmer, M. Pedrick, D. Pike, G. Pitt, M. Putnam, K. Rainsford, L. Ream, N. Rogers, Y. Shealy, C. Smith, M. Sporn, C. Tomlinson, R. Tudor and R. Wright.

REFERENCES

Barnicot NA, Datta SP 1972 Vitamin A and bone. In: Bourne GH (ed) The biochemistry and physiology of bone. Vol 2. Physiology and pathology. Academic Press, New York, p 197-229

Bauernfeind JC 1980 The safe use of vitamin A. The Nutrition Foundation, Washington

Cho DY, Frey RA, Guffy MM, Leipold HW 1975 Hypervitaminosis A in the dog. Am J Vet Res 36:1597-1603

Cunningham WJ, Ehmann CW 1983 Clinical aspects of the retinoids. Semin Dermatol 2:145-160

Dennert G 1984 Retinoids and the immune system: immunostimulation by vitamin A. In: Sporn MB et al (eds) The retinoids. Academic Press, Orlando, vol 2:373-390

DiGiovanna JJ, Peck GL 1983 Oral synthetic retinoid treatment in children. Pediatr Dermatol 1:77-88

Drzymala M, Katz DR, Turton JA, Hicks RM 1984 Retinoid effects on murine immune responses. In: Klaus GGB (ed) Morphological aspects of the immune system. Plenum Press, New York (Proc 8th Int Conf Lymphatic tissues and germinal centres in immune reactions, Cambridge, 1984), in press

Festing MFW, Hawkey CM, Hart MG, Turton JA, Gwynne J, Hicks RM 1984 Principal components analysis of haematological data from F344 rats with bladder cancer fed N-(ethyl)-all-*trans*-retinamide. Food Chem Toxicol 22:559-572

Gerber A, Raab AP, Sobel AE 1954 Vitamin A poisoning in adults. Am J Med 16:729-745

Gerber LH, Helfgott RK, Gross EG, Hicks JE, Ellenberg SS, Peck GL 1984 Vertebral abnormalities associated with synthetic retinoid use. J Am Acad Dermatol 10:817-823

Hicks RM, Chowaniec J, Turton JA, Massey ED, Harvey A 1982 The effect of dietary retinoids on experimentally induced carcinogenesis in the rat bladder. In: Arnott MS et al (eds) Molecular interrelations of nutrition and cancer. Raven Press, New York, p 419-447

Hixson EJ, Harrison SD 1981 Effect of nonsteroidal anti-inflammatory drugs on sublethal retinoic acid toxicity in Swiss mice. Biochem Pharmacol 30:1714-1716

Kamm JJ, Ashenfelter KO, Ehmann CW 1984 Preclinical and clinical toxicology of selected retinoids. In: Sporn MB et al (eds) The retinoids. Academic Press, Orlando, vol 2:287-326

Kurtz PJ, Emmerling DC, Donofrio DJ 1984 Subchronic toxicity of all-*trans*-retinoic acid and retinylidene dimedone in Sprague-Dawley rats. Toxicology 30:115-124

Leelaprute V, Boonpucknavig V, Bhamarapravati N, Weerapradist W 1973 Hypervitaminosis A in rats. Arch Pathol 96:5-9

Mahrle G, Berger H 1982 DMBA-induced tumours and their prevention by aromatic retinoid (Ro 10-9359). Arch Dermatol Res 272:37-47

Medawar PB, Hunt R 1981 Anti-cancer action of retinoids. Immunology 42:349-353

Milstone LM, McGuire J, Ablow RC 1982 Premature epiphyseal closure in a child receiving oral 13-*cis*-retinoic acid. J Am Acad Dermatol 7:663-666

Moon RC, Itri LM 1984 Retinoids and cancer. In: Sporn MB et al (eds) The retinoids. Academic Press, Orlando, vol 2:327-371

Moore T 1967 Pharmacology and toxicology of vitamin A. In: Sebrell WH, Harris RS (eds) The vitamins, chemistry, physiology, pathology, methods, 2nd edn. Academic Press, New York, vol 1:280-294

Pittsley RA, Yoder FW 1983 Retinoid hyperostosis, skeletal toxicity associated with long-term administration of 13-*cis*-retinoic acid for refractory ichthyosis. N Engl J Med 308:1012-1014

Sani BP, Meeks RG 1983 Subacute toxicity of all-*trans*- and 13-*cis*-isomers of N-ethyl retinamide, N-2-hydroxyethyl retinamide, and N-4-hydroxyphenyl retinamide. Toxicol Appl Pharmacol 70:228-235

Stinson SF, Reznik-Schüller H, Reznik G, Donahoe R 1980 Atrophy induced in the tubules of the testes of Syrian hamsters by two retinoids. Toxicology 17:343-353

Sudmann E 1975 Effect of indomethacin on bone remodelling in rabbit ear chambers. Acta Orthop Scand Suppl (160), p 91-115

Tamayo L, Ruíz-Maldonado R 1981 Long-term follow-up of 30 children under oral retinoid Ro 10-9359. In: Orfanos CE et al (eds) Retinoids, advances in basic research and therapy. Springer-Verlag, Berlin, p 287-294

Teelman K 1981 Experimental toxicology of the aromatic retinoid Ro 10-9359 (etretinate). In: Orfanos CE et al (eds) Retinoids, advances in basic research and therapy. Springer-Verlag, Berlin, p 41-47

Tuchweber B, Garg BD, Salas M 1976 Microsomal enzyme inducers and hypervitaminosis A in rats. Arch Pathol Lab Med 100:100-105

Willhite CC, Shealy YF 1984 Amelioration of embryotoxicity by structural modifications of the terminal group of cancer chemopreventive retinoids. J Natl Cancer Inst 72:689-695

Windhorst DB, Nigra T 1982 General clinical toxicology of oral retinoids. J Am Acad Dermatol 6:675-682

DISCUSSION

Sporn: Dr Peck, how concerned are clinicians about these hyperostoses that Jon Turton has described in rats? Do you see them with the short-term use of 13-*cis*-retinoic acid (isotretinoin) in acne or its long-term use for keratinizing disorders?

Peck: In a retrospective analysis of different groups of patients with disorders of keratinization, Cunningham found an 8–28% incidence of vertebral osteophyte formation in patients receiving isotretinoin daily at 2 mg/kg body weight for two years or longer (W. Cunningham, Retinoid Symposium, Geneva, 20–24 Sep 1984). In a prospective study on patients with disorders of keratinization receiving isotretinoin at this dose, 75% developed small vertebral osteophytes in 12 months (Ellis et al 1984a). In some patients very early osteophyte formation could be detected at the sixth month of treatment. In cystic acne, seven out of 40 patients who were treated daily with isotretinoin at 1–2 mg/kg body weight developed small osteophytes (W. Cunningham, unpublished observations). There are no reports of osteophyte formation with doses below 1 mg/kg per day.

Koeffler: How old were these patients?

Peck: These effects have been seen in teenagers. You see osteophyte formation in the absence of retinoids in a certain percentage of the population over 60; it is a normal ageing change. But the patients on retinoids in whom osteophytes were observed were in their twenties and thirties, so there does seem to be an acceleration of the ageing process.

Moon: This always occurs with long-term administration of the drug. Is there any evidence of problems with short-term treatment in acne patients?

Peck: I don't know precisely how long the acne patients cited by Cunningham were treated. However, the abnormal X rays were taken 12 months after the beginning of therapy. Another effect of the long-term use of isotretinoin is premature closure of the epiphyses in the long bones. Milstone et al (1982) have reported closure of the epiphysis of the right tibia in a child with epidermolytic hyperkeratosis who had been on isotretinoin for 4.5 years at a daily dose of about 3.5–4.5 mg/kg. After an additional two years there was also closure of the proximal tibial epiphysis in the other leg and remodelling of the bone of the tibia and the femur in a way similar to that described by Dr Turton (McGuire et al 1984). The normal columnar shape of these bones was changed to give a more 'pagoda-shaped' structure with narrowed shafts, which would fit with Dr Turton's observations of periosteal resorption. However, I'm not sure how this process relates to skeletal hyperostosis, where the initial change is calcification of ligaments like the anterior spinal ligament and the iliolumbar ligament. How does this process compare with endosteal deposition?

Turton: I don't know. Our initial objective was to develop an index of retinoid activity to confirm that retinoids were not being inactivated in the diet. We looked at changes in several organs (and blood) and found the long bones were a good indicator (Hicks et al 1982). The sensitivity of the technique was increased by sectioning the bones (Turton et al 1983, 1984) allowing, for example, retinamides to be assessed comparatively. The degree of retinoid-induced change depends on the retinoid dose level, the period the retinoid is fed, and the age of the animal. Feeding retinamides at 2.5 to 2.875 mmol/kg diet for six weeks to weanling rats produces significant results and this model has, for example, enabled us to follow up the work of Hixson & Harrison (1981). But really we have only just begun to define our basic parameters. The next phase of study would involve examining the mechanisms involved, and comparing effects and their reversibility in young and old animals. We are working towards this in collaboration with histologists who work on calcified tissues. However, with that as a perspective, we do not know as yet how the observed periosteal resorption and endosteal deposition relate to other bone lesions that may occur in rodents, nor to the calcification of tendons and ligaments and the formation of hyperostoses in humans (Pittsley & Yoder 1983, McGuire et al 1984, Ellis et al 1984b). Clearly we should study our rodents for calcification and hyperostoses, but we may have to use other species. Etretinate causes the formation of exostoses in the rabbit (Mahrle & Berger 1982) and the pattern of changes seen in humans in response to isotretinoin is widely described in the cat and dog in hypervitaminosis A. These larger species also have the advantage of being easier to X-ray.

Sporn: It is anticipated that people at high risk for bladder cancer or breast cancer are going to be put onto N-(4-hydroxyphenyl)retinamide (HPR) at doses of 300–600 mg for a 70 kg person. The doses that you are giving to rats are orders of magnitude greater than that. If you fed animals even very simple dietary nutrients like sodium chloride or sucrose, which are generally recognized as safe, at that sort of dose amplification, I think you could get into severe trouble. If you gave HPR or something like 13-*cis*-ethylretinamide at much more modest levels, would you see changes similar to those that you saw with the high doses?

Turton: The relative retinoid consumption in our rodents would certainly be higher than 300–600 mg HPR per day in humans (4.3 to 8.6 mg/kg body weight). In our long-term trials to assess the activity of 13-*cis*-retinoic acid (13CRA) or HPR against carcinogen-induced bladder cancer in the rat we would feed these compounds routinely at 0.8 and 2.0 mmol/kg diet respectively; in trials with etretinate the retinoid would be fed at about 0.16 mmol/kg diet. A 30-week-old female F344 rat weighs 200 g and eats 11 g of diet a day thus consuming 13.2 mg 13CRA/kg body weight per day, or 43.0 mg HPR/kg body weight per day, or 3.0 mg etretinate/kg body weight per day. It is difficult to extrapolate in a meaningful way from rodent dose rates to those used therapeutically in humans, or those to be used in humans. However, the level for etretinate in the rat is 3.0 times higher than the level used in humans; for 13CRA, 13.2 times higher; for HPR, 6.7 times higher. Therefore, knowing a dose level of retinoid which may show activity in a rodent system does not necessarily enable one to calculate a dose level for use in humans.

We first noticed effects on the long bones in a two-year bladder cancer trial with *N*-ethylretinamide (NER) in rats fed at the usual level of 2.0 mmol/kg diet (Hicks et al 1982). At one year, the control femoral diameter was 3.02 mm, and 2.68 mm in NER-treated rats (11.3% diaphysis reduction). Two other long-term trials were described in the same paper with 0.8 mmol 13CRA and 2.0 mmol NER per kg diet: at 77 weeks in the 13CRA trial control femurs were 3.14 mm and femurs in 13CRA-treated rats were 2.73 mm (13.1% reduction); in the NER trial the results were 3.08 and 2.77 mm (10.1% reduction). In a recent trial feeding 2.0 mmol HPR/kg diet the results at 15 months were 2.95 and 2.60 mm (11.9% reduction). Therefore these reductions are comparable to those in trials in which retinamides were fed at 2.5 mmol/kg diet to weanling rats for six weeks, but we increased the dose level from 2.0 to 2.5 mmol/kg diet to produce a significant effect in a shorter time period. Also, Hixson & Harrison (1981) counted *the number of long-bone fractures per survivor* in trials with non-steroidal anti-inflammatory compounds and retinoic acid in mice, and as Jane Hixson (personal communication) had said that their results were dependent upon the dose levels of retinoic acid and non-steroidal anti-inflammatory compounds used, we decided to increase the basic level of retinoid in our second trial to 2.875 mmol/kg diet.

Hicks: With our doses of HPR, the animals are ingesting daily about 70 mg/kg body weight. If we feed 300 mg per day to people, they will be ingesting only 4 mg/kg body weight. Would you expect to see the bone effects in rats with HPR at 4 mg/kg body weight rather than 70 mg/kg?

Turton: We have even fed a diet to weanling rats containing a mixture of 2.875 mmol 4HPR/kg and 2.875 mmol NER/kg. Again, we had reason to believe that the activity of non-steroidal anti-inflammatory compounds in preventing retinoid-induced changes was dependent upon the dose of retinoid administered, and we wished to test higher retinoid levels as Hixson & Harrison (1981) had done. Your question is difficult because it deals with the extrapolation of data from rodents to humans and vice versa. If we take a situation in humans where HPR may be used to treat bladder cancer, perhaps a 60-year-old patient would be treated for eight years at 4.3 mg/kg body weight per day. This might be equivalent to feeding a 103-week-old female F344 rat for 10 weeks with a 0.24 mmol HPR/kg diet. I don't think we would see any bone changes in this situation. However, in our bladder cancer trials in rats we would normally test HPR for antitumour activity at 2.0 mmol/kg diet.

Sporn: Another point is that you are doing your experiments with weanling rats, which have actively growing bone, but it is generally agreed that, in cancer trials, under no circumstances should any young, growing person be given HPR.

Turton: Many of our experiments have been carried out with weanling, four- to five-week-old rats, mainly because these are the cheapest and easiest to obtain. The model was designed to test retinoids on growing bone, as perhaps this is when calcified tissue is most sensitive to retinoid action. I think the model has provided some interesting results. For example, in a 15-week trial with 2.5 mmol NER/kg diet in collaboration with Dr Larry Ream, control animals and NER-fed animals both grew from 58 g to 188 g. There were no significant effects of NER on body length, tail length, hind-limb length, humerus or femur length, or locomotor activity. However, both humeri and femora were significantly affected in diameter and weight in NER-treated animals. This indicates the sensitivity of bone measurements, as body weight loss is generally considered as an index of retinoid activity. This trial also suggests that the functioning of the epiphyseal plates was not affected. I do not know the details of the proposed clinical trials with HPR against breast and bladder cancer, but bladder cancer is generally a disease of late-middle and old age. It is for this reason that we are now turning our attention to study retinoid effects in older animals, but here one has to maintain them for a year or 18 months before administering retinoid. But to return to your point about retinoid use in young people, Dr Peck has mentioned the use of isotretinoin in treating acne in teenagers, and there are reports of the use of etretinate in children.

Moon: If you start retinoid treatment when the animal is old, do you get the same effects?

Turton: We have only preliminary indications. The results suggest that the long bones of older animals are less sensitive to retinoids; however, the same type of remodelling may occur but at a slower rate than in the young animal. It would also appear that the reversal of retinoid-induced long-bone changes in older animals proceeds at a slower rate. In comparing retinoid activity in young and old rats, one again has to consider the dose levels of retinoid fed to ensure that the same relative doses are given. The level must be increased considerably in older animals to obtain the same consumption as in the weanling animal. For example, if you feed 1.0 mmol retinyl acetate/kg diet to a 58 g weanling F344 rat for six weeks, you must feed 2.06 mmol/kg diet to 290 g 15-month-old animals; both will then receive 35 mg retinyl acetate/kg body weight per day. 13CRA at 0.8 mmol/kg diet fed to a weanling rat is equal to 1.65 mmol 13CRA/kg diet fed to a 290 g rat; both will then consume 25 mg 13CRA/kg body weight per day.

Moon: So in the proposed cancer prevention studies, bone changes may not be a problem in the older patients.

Pitt: We have done experiments similar to Dr Turton's but on adult rats, and using *all-trans*-retinoic acid at high doses. We got what might be called the 'expected' result of hypervitaminosis A. Jon Turton's results are rather different in that the thickness of cortical bone increases. In our adult rats given the higher doses of *all-trans*-retinoic acid we saw a narrower bone diameter, a smaller medullary cavity and a thinning of cortical bone. We found that non-steroidal anti-inflammatory drugs had a marginal effect on these retinoid-induced changes, which was not statistically significant (Harker et al 1984). Similarly C.H. Cashin of Roche Products, Welwyn (unpublished work) found indomethacin (0.5 mg/kg body weight per day) to have no protective effect against the bone fragility [in terms of tibial bone breaking strain (Cashin & Lewis 1984)] induced by etretinate (30 mg/kg body weight per day for 15 days). Anecdotally, rats given etretinate plus indomethacin looked more healthy than those given etretinate alone.

Turton: That is an interesting comment as patients on retinoid therapy do experience bone and joint pains and I believe that non-steroidal anti-inflammatory compounds are prescribed to alleviate the condition (G.L. Peck, personal communication). In a recent trial, we fed etretinate or retinyl acetate to rats and produced clinical signs of hypervitaminosis A (loss of condition, body weight loss, limb paralysis). However, animals fed aspirin with each compound appeared to be in better clinical condition for several days before, and even after, long-bone fractures occurred. Perhaps aspirin was also helping to alleviate the bone and joint pains in these animals.

Tsambaos: Do retinoids have any effects on serum calcium or phosphatases?

Turton: We have not studied this yet, but hope to do so in the near future. It would be relevant to monitor bone changes and possibly relate them to plasma calcium and phosphatase levels. Another related study would be to look at bone changes over a period of weeks or months and examine the relationship between reduction in medullary cavity volume, red blood cell changes and extra-medullary haemopoiesis in the liver and spleen, which we know occurs in rats given 13CRA or HPR. Once we have this information we could possibly conclude whether red cell changes in retinoid-treated rodents have any clinical haematological implications for the use of retinoids in humans.

Peck: In a few instances, elevated calcium concentrations have been observed in humans given isotretinoin (Cassidy et al 1982, Valentic et al 1983). It is a very rare type of toxicity and I don't think it has been seen with etretinate.

Sporn: Do Dr Turton's observations on bone changes have any bearing on what you see in the regenerating axolotl limb, Dr Maden?

Maden: There is no bone present in those limbs; they are all cartilage. With retinoids we see a breakdown of the cartilage rather than remodelling of bone, so the effects are not necessarily equivalent.

REFERENCES

Cashin CH, Lewis EJ 1984 Evaluation of hypervitaminosis A in the rat by measurement of tibial bone breaking strain. J Pharmacol Methods 11:91-95

Cassidy J, Lippman M, Lacroix A, Peck G 1982 Phase II trial of 13-*cis*-retinoic acid in metastatic breast cancer. Eur J Cancer Clin Oncol 18:925-928

Ellis CN, Madison KC, Pennes DR, Martel W, Voorhees JJ 1984a Isotretinoin therapy is associated with early skeletal radiographic changes. J Am Acad Dermatol 10:1024-1029

Ellis CN, Gilbert M, Madison KC, Pennes D, Martel W, Voorhees JJ 1984b Skeletal radiographic changes during retinoid therapy. Dermatologica 169:252

Harker AJ, Barnett P, Lowe JS, Pitt GAJ 1984 Failure of indomethacin and aspirin to protect against retinoid toxicity. Dermatologica 169:226

Hicks RM, Chowaniec J, Turton JA, Massey ED, Harvey A 1982 The effect of dietary retinoids on experimentally induced carcinogenesis in the rat bladder. In: Arnott MS et al (eds) Molecular interrelations of nutrition and cancer. Raven Press, New York

Hixson EJ, Harrison SD 1981 Effect of nonsteroidal anti-inflammatory drugs on sublethal retinoic acid toxicity in Swiss mice. Biochem Pharmacol 30:1714-1716

Mahrle G, Berger H 1982 DMBA-induced tumours and their prevention by aromatic retinoid (Ro 10-9359). Arch Dermatol Res 272:37-47

McGuire J, Milstone LM, Lawson JP 1984 Isotretinoin therapy and bone changes. Dermatologica 169:252

Milstone LM, McGuire J, Ablow RC 1982 Premature epiphyseal closure in a child receiving oral 13-*cis*-retinoic acid. J Am Acad Dermatol 7:663-666

Pittsley RA, Yoder FW 1983 Retinoid hyperostosis. Skeletal toxicity associated with long-term administration of 13-*cis*-retinoic acid for refractory ichthyosis. New Engl J Med 308:1012-1014

Turton JA, Hicks RM, Gwynne J, Hunt R, Palmer L, Medawar PB 1983 Skeletal development as an index of retinoid toxicity. Proc Nutr Soc 42:12A

Turton JA, Hicks RM, Gwynne J, Katz D 1984 Skeletal damage as an early sign of retinoid toxicity in the rat. Proc Nutr Soc 43:3A

Valentic JP, Elias AN, Weinstein GD 1983 Hypercalcemia associated with oral isotretinoin in the treatment of severe acne. JAMA (J Am Med Assoc) 250:1899-1900

The effect of retinoids on haemopoiesis— clinical and laboratory studies

H. PHILLIP KOEFFLER and THOMAS T. AMATRUDA III

Division of Hematology–Oncology, Department of Medicine, UCLA School of Medicine, Los Angeles, California 90024, USA

Abstract. Retinoids affect the growth and differentiation of haemopoietic cells. Individuals deficient in retinoids become anaemic; replacement therapy with retinoids corrects the anaemia. Retinoids enhance the clonal proliferation of erythroid and myeloid precursors in soft-gel culture; *all-trans*-retinoic acid and 13-*cis*-retinoic acid are the most potent. Retinoids also induce the differentiation of HL-60 promyelocytes to functional granulocytes and can induce cells from other relatively mature, myeloid cell lines to undergo partial differentiation. Cells from less mature myeloid leukaemic lines are often resistant to induction of differentiation by retinoids. Like cells from established lines, relatively mature leukaemic cells (promyelocytes, myelomonoblasts) harvested from patients can undergo differentiation *in vitro* in the presence of retinoids. A few reports suggests that a minority of patients with myeloid leukaemia or preleukaemia who receive 13-*cis*-retinoic acid will have improvement in their haemopoiesis. Further studies are required to understand the mechanism of action of retinoids on the growth and differentiation of haemopoietic cells and to explore more fully the therapeutic potential of retinoids.

1985 Retinoids, differentiation and disease. Pitman, London (Ciba Foundation Symposium 113) p 252–267

The importance of retinoids for normal haemopoiesis has been known for nearly 60 years. In contrast, only within the last 5–10 years have models been developed that allow us to explore closely the interaction of haemopoietic cells with retinoids. This paper summarizes the salient features of that interaction and provides glimpses into how that knowledge has and will be applied clinically.

Retinoids and normal haemopoiesis

Normal haemopoietic function is a result of the proliferation of haemopoietic stem cells and their orderly differentiation into mature erythrocytes, granulocytes, macrophages and platelets (Fig. 1). Individual pluripotential stem cells become committed to myeloid, erythroid or megakaryocyte development, leave the proliferative pool, and terminally differentiate into mature, functional cells. The process of haemopoietic proliferation and differentiation is modulated by local and humoral factors including retinoids.

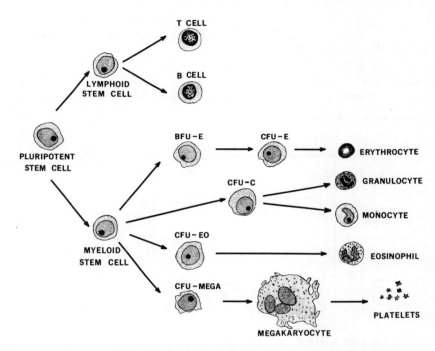

FIG. 1. Scheme of normal haemopoietic differentiation. BFU-E, erythrocyte burst-forming unit; CFU-E, erythrocyte colony-forming unit; CFU-C, colony-forming unit in culture; CFU-EO eosinophil colony-forming unit; CFU-MEGA, megakaryocyte colony-forming unit.

Erythropoiesis

Retinoids affect erythropoiesis in people and in experimental animals (Table 1). Children and adults with retinoid deficiency often become anaemic (Wolbach & Howe 1925, Mohanram et al 1977, Hodges et al 1978) (Table 1). Replacement therapy results in a rapid return to normal of red cell numbers.

In vitro studies (Douer & Koeffler 1982a) suggest that retinoids stimulate the growth of erythroid progenitor cells. Colonies of erythroid cells form in soft-gel culture when circulating blood cells are plated with erythropoietin and a growth factor known as burst potentiating activity. Each colony

TABLE 1 Effects of retinoids on normal haemopoiesis

	Animal	Effects
In vivo		
	Human	Vitamin A deficiency associated with anaemia; vitamin A replacement corrects anaemia.
		Therapeutic use of retinoids in non-haematological disorders does not affect proliferation but may decrease functional activity of myeloid cells.
	Rat	Vitamin A deficiency causes anaemia.
In vitro		
	Human	Retinoids stimulate clonal proliferation of erythroid and myeloid progenitor cells but may decrease several functions of myeloid cells.

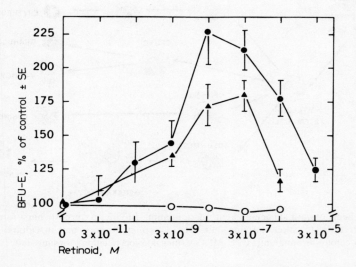

FIG. 2. Effect of retinoids on the formation of normal blood erythroid colonies. BFU-E, erythrocyte burst-forming unit; ●, *all-trans*-retinoic acid; ▲, 13-*cis*-retinoic acid; ○, retinol.

represents the progeny of a single stem cell known as a burst-forming unit-erythroid (BFU-E). Addition of either *all-trans*- or 13-*cis*-retinoic acid at a concentration of 3×10^{-8} M to 3×10^{-7} M to these cultures causes a doubling in proliferation of BFU-E into erythroid colonies (Fig. 2), but retinol is without effect. The study suggests that retinoic acid either directly enhances the proliferation of BFU-E or has an effect on other circulating blood cells which in turn stimulate BFU-E.

Myelopoiesis and myeloid function

Deprivation or excess intake of retinoids is without clear effect on granulocyte or monocyte concentrations in the blood. In contrast, retinoids stimulate the proliferation of myeloid cells *in vitro*. Retinoic acid enhances the clonal proliferation of committed myeloid stem cells (CFU-GM) to form colonies of granulocytes and/or macrophages in soft agar (Douer & Koeffler 1982b). Fig. 3 shows the effects of retinoids on mononuclear cells from normal human

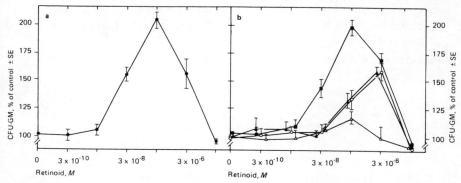

FIG. 3. Effect of (a) *all-trans*-retinoic acid and (b) other retinoids on the formation of colonies by committed myeloid stem cells (CFU-GM) obtained from normal human bone marrow. Cells were plated in soft agar in the presence of colony-stimulating factor. Colony formation is expressed as a percentage of the number of colonies that formed in control dishes not incubated with retinoid. Symbols: ●, *all-trans*-retinoic acid; ■, 13-*cis*-retinoic acid; ▲, retinyl acetate; ○, retinal; △, retinol.

bone marrow plated in soft agar in the presence of colony-stimulating activity (CSA), a growth factor necessary for myeloid colony formation *in vitro*. Addition of retinoic acid (3×10^{-8} M to 3×10^{-6} M) to the cultures significantly increases the number of CFU-GM that form colonies, with a maximal effect at 3×10^{-7} M. Retinal and retinyl acetate are less effective, with a maximal effect at 3×10^{-6} M, while retinol has no stimulating effect. At concentrations of 3×10^{-5} M, all these retinoids inhibit the formation of CFU-GM colonies. Retinoic acid has no direct CSA activity nor does it stimulate CSA production by the cultured marrow cells. These results suggest that retinoids may increase the sensitivity of committed granulocyte–macrophage progenitor cells to stimulation by the growth factor CSA.

Retinoic acid derivatives have an effect on several white cell functions. The chemotactic motility of neutrophils decreased both in psoriatic patients treated with etretinate (Ro 10-9359, 1 mg/kg per day for 15 days) and in patients with cystic acne treated with isotretinoin (Ro 10-3740) (Pigatto et al 1983a, 1983b). Incubation of peripheral blood leucocytes from normal individuals with retinoic acid *in vitro* caused a marked decrease in both superoxide generation and lysozyme concentrations within the cells (Camisa et al 1982, Kensler & Trush 1981).

Retinoids and myeloid cell differentiation

Permanent cell lines that can be induced to differentiate *in vitro* facilitate the study of haemopoietic cell differentiation. The effects and mechanisms of action of pharmacological and physiological inducers of differentiation can be assessed with these homogeneous populations of cells. Table 2 lists several myeloid cell lines.

TABLE 2 Effects of retinoic acid on human myeloid cell lines

Cell line	Stage of differentiation	all-trans-Retinoic acid Effect	ED_{50} (M)	Effect of other inducers of differentiation
KG-1a	Early myeloblast	Clonal proliferation inhibited	2×10^{-9}	No differentiation
KG-1	Myeloblast	Clonal proliferation inhibited	2×10^{-9}	Macrophage differentiation
HL-60 blast	Myeloblast	No differentiation	—	No differentiation
HL-60	Promyelocyte	Granulocyte differentiation	1×10^{-7}	Granulocyte or macrophage differentiation
		Clonal proliferation inhibited	2×10^{-8}	
HL-60 subclones C12 & C13	Promyelocyte	Granulocyte differentiation	—	No differentiation
ML-1, ML-3	Myelomonoblast	Slight differentiation	—	Macrophage differentiation
U-937	Monoblast	Macrophage differentiation	—	Macrophage differentiation
K562	Early blast and/or erythroblast	No effect	—	Enhancement of fetal globin expression

ED_{50}, concentration that produces an effect in 50% of the cells.

HL-60 cells

HL-60 cells morphologically, cytochemically, and functionally resemble normal promyelocytes. The cells can be induced to differentiate terminally *in vitro* either to granulocytes or to macrophages by exposure to a number of agents (Koeffler 1983). Breitman and colleagues (Breitman et al 1980) found that *all-trans*-retinoic acid induces the differentiation of HL-60 promyelocytes to granulocytes in a dose-dependent manner at concentrations between 10^{-6} M and 10^{-9} M (Table 2, Fig. 4). More than 90% of the cells differentiate

(a)

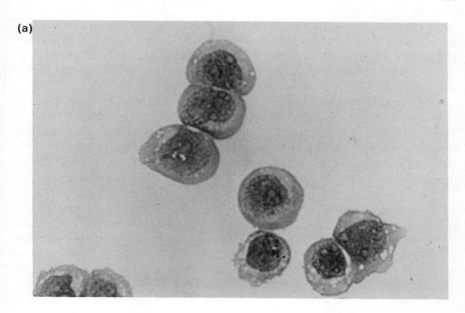

(b)

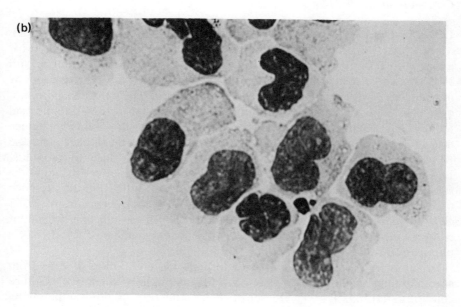

FIG. 4. Morphological differentiation of HL-60 cells induced by retinoic acid. (a) HL-60 promyelocytes; (b) HL-60 granulocytes [after exposure to *all-trans*-retinoic acid (10^{-8} M) for five days].

when HL-60 cells are cultured in the presence of 10^{-6} M-*all-trans*-retinoic acid for six days. Differentiation is maximal after 72–96 h exposure to 10^{-6} M-retinoic acid, but after as little as 3 h there is an effect on differentiation. Thirteen-*cis*-retinoic acid is equally effective but retinol, retinal and retinyl acetate induce differentiation only at concentrations of 10^{-3} M to 10^{-6} M.

Retinoic acid inhibits the growth of HL-60 cells at the same concentrations that induce differentiation (Douer & Koeffler 1982c). Growth inhibition in liquid culture is evident after 48 h incubation and irreversible after exposure to 10^{-6} M-retinoic acid for four days. The cloning efficiency of HL-60 cells in soft agar is decreased by retinoic acid (10^{-9} M to 10^{-6} M) and 50% inhibition of colony formation occurs at 2.5×10^{-8} M (Table 2) (Douer & Koeffler 1982c).

Two subclones of HL-60, known as C12 and C13, are composed of cells that do not differentiate when cultured with most known inducers of HL-60. These cells can, however, differentiate in the presence of retinoids (Ferrero et al 1983a), which suggests that retinoids may trigger the differentiation of myeloid cells by a mechanism different from that of other compounds.

Other human myeloid lines

Other myeloid cell lines show variable responses to retinoids (Table 2). Monoblasts from the U-937 line differentiate toward mature monocytes if exposed to retinoic acid (10^{-8} M to 10^{-6} M) (Olsson 1982). Treatment with 10^{-6} M-retinoic acid causes differentiation in 80% of cells as assessed by morphology, cytochemistry and functional activity. The ML-3 early mono-cytic cells show partial differentiation after exposure to 10^{-6} M-retinoic acid for five days, with about 35% of the cells developing the ability to reduce nitroblue tetrazolium (Ferrero et al 1983b). The KG-1 myeloblasts and the KG-1a early myeloblasts do not differentiate on exposure to retinoids. However, the cells are sensitive to the growth-inhibiting effect of retinoids, showing a 50% inhibition of colony formation when cultured in the presence of 2.4×10^{-9} M-retinoic acid (Douer & Koeffler 1982c). Exposure of KG-1 cells to *all-trans*-retinoic acid (10^{-7} M) for 24–48 h is sufficient to inhibit cloning efficiency by 80%. As with HL-60 cells, the viability of KG-1 cells is unimpaired by retinoid treatment.

In summary, retinoic acid induces terminal differentiation of cells from several human myeloid cell lines *in vitro*. In general, lines of relatively mature cells (e.g. HL-60, U-937) are sensitive to induction of differentiation by retinoids. Usually, cells from these retinoic acid-sensitive lines can also be induced to differentiate by a large variety of other compounds. Studies with leukaemic cell lines show that terminal differentiation induced by retinoids is invariably associated with cessation of growth; however, cessation of growth is not always associated with differentiation (e.g. the clonal proliferation of KG-1 myeloblasts is inhibited by low concentrations of retinoic acid, but retinoic acid is unable to induce differentiation of these cells).

Structure–function relationship for retinoids

The *all-trans* and 13-*cis* forms of retinoic acid are equally effective inducers of HL-60 differentiation, with 50% of the cells undergoing differentiation in the presence of 10^{-7} M of either compound (i.e. ED_{50} 10^{-7} M) (Breitman et al 1980). Retinal and retinyl acetate are less potent, and retinol (10^{-6} M) induces differentiation of less than 20% of HL-60 cells. Etretinate and pyridyl retinoic acid are relatively ineffective as inducers of HL-60 differentiation (Honma et al 1980). Retinoidal benzoic acid derivatives are potent inducers (Strickland et al 1983). The 3-methyl and 3-methoxy derivatives of (*E*)-4-[2-(5,6,7,8-tetrahydro-5,5,8,8-tetramethyl-2-naphthalenyl)-l-propenyl]-benzoic acid are significantly more active than *all-trans*-retinoic acid, with ED_{50} values of 2×10^{-8} M. Other studies (Douer & Koeffler 1982c) show that the effectiveness of retinoids in inhibiting clonal growth of KG-1 parallels their potency in inducing differentiation of HL-60 cells (Douer & Koeffler 1982c). Taken together, these studies show that retinoic acid and its derivatives are much more effective than retinol in inducing differentiation. Consistent with these observations is the finding of Strickland et al (1983) that the terminal carboxyl group (which is present in retinoic acid) is essential for the potency of synthetic retinoids as inducers of differentiation.

Mechanism of action of retinoids on haemopoietic cells

The induction of differentiation and the inhibition of leukaemic growth by retinoic acid are probably more than non-specific toxic effects (Douer & Koeffler 1982c): (1) concentrations of retinoic acid that inhibit clonal growth of leukaemia cells enhance clonal growth of normal human myeloid stem cells; (2) inhibition of growth of leukaemia cells occurs at very low concentrations of retinoic acid; (3) the growth of K562 and M-1 myeloid leukaemia cells is not affected by high concentrations of retinoic acid (10^{-5} M).

A variety of compounds induce the differentiation of HL-60 promyelocytes. Studies with clones of HL-60 that are selectively resistant to specific compounds suggest that differentiation of HL-60 cells may be triggered by several different mechanisms. However, the sequence of developmental events after triggering is very similar with each inducer, suggesting that there is a common pathway for events occurring after the induction of differentiation. The mechanism by which any compound triggers the differentiation of HL-60 cells is unknown.

The recent discovery of cytoplasmic retinoic acid-binding protein (CRABP) in various tissues has led to the hypothesis that the biological effect of retinoic acid might be mediated through this receptor. Differentiation may be mediated through CRABP in normal epithelial tissue and embryonal carcinoma cells. Takenaga et al (1981) have suggested that HL-60 cells contain CRABP, although their ligand displacement studies do not completely support the finding. We were unable to detect CRABP in HL-60, KG-1 or leukaemic blasts from patients (Douer & Koeffler 1982c), but it is possible that the haemopoietic cells possess very low concentrations of CRABP, which our assay is not sensitive enough to detect. However, other mechanisms to explain the biological activities of retinoic acid should be considered. For example, retinoic acid may trigger differentiation or inhibit growth of myeloid leukaemic cells by alteration of the cell membrane.

Retinoids change the structure of the cell membrane of HL-60 cells (Cooper et al 1981). Sterol, phospholipid and phosphatidylcholine synthesis are inhibited, the incorporation of lipids into membranes is decreased, and transmethylation of phosphatidylethanolamine ceases within 5 h of exposing HL-60 cells to retinoic acid. The changes precede by more than 30 h any changes in DNA synthesis or cell growth. Differentiation induced by retinoic acid is associated with decreased membrane fluidity and with an increased ratio of cholesterol to phospholipid. The change in lipid synthesis could relate both to the initiation of myeloid differentiation and to the decrease in myeloid growth after exposure to retinoids.

Activation of a protein kinase is a potential mechanism by which retinoids could trigger the rapid physiological changes which characterize myeloid differentiation. Protein kinase C activation is a key event in the 12-O-tetradecanoylphorbol-13-acetate-induced differentiation of HL-60 cells to macrophages (Feuerstein & Cooper 1984, Weinstein 1983). A similar, but unidentified, protein kinase might be activated by retinoids.

Retinoids and differentiation of leukaemic cells from patients

The phenotypic abnormality in acute myelogenous leukaemia is the inability of the leukaemic cells to differentiate into functional, non-dividing end cells. Theoretically, the identification and utilization of compounds that induce leukaemic cells to differentiate may provide a novel approach to the therapy of leukaemia and other neoplasia.

In vitro studies. Myelogenous leukaemic cells harvested from patients have been cultured with retinoids in attempts to influence their growth and differentiation. Cells from the bone marrow and peripheral blood of 21 patients with acute myelogenous leukaemia were cultured *in vitro* for six days with 10^{-6} M-retinoic acid (Breitman et al 1981). Cells from two patients differentiated into granulocytes when cultured with retinoic acid; both these patients had acute promyelocytic leukaemia. Cells from other patients with acute myelogenous leukaemia, acute myelomonocytic leukaemia and chronic myelogenous leukaemia in blast crisis showed no differentiation when treated with retinoic acid. Therefore, in this study, only leukaemic cells at the promyelocytic stage of maturation (which is the same stage of development as HL-60) differentiated *in vitro* in the presence of retinoic acid. Another study (Honma et al 1983) noted that retinoic acid was effective *in vitro* in inducing granulocytic differentiation of cells from three of four patients with acute promyelocytic leukaemia and from one patient with relatively mature leukaemic myeloblasts. In addition, retinoic acid induced acute myelomonocytic leukaemia cells from one patient to differentiate towards monocytes. In another study, retinoic acid inhibited the clonal growth in soft agar of blast cells from five of seven leukaemic patients, with 50% inhibition of clonal growth at doses of 5×10^{-9} M to 3×10^{-7} M (Douer & Koeffler 1982c). Retinoic acid did not, however, induce differentiation of the leukaemic blast cells.

In vivo studies. Thirteen-*cis*-retinoic acid has been administered to patients with haemopoietic abnormalities. A patient with acute promyelocytic leukaemia had a partial response to 100 mg/m^2 13-*cis*-retinoic acid (Flynn et al 1983). Parallel *in vitro* studies with the bone marrow cells from this patient showed that retinoic acid induced partial maturation of the cells. Approximately 80% of the cells exposed to 13-*cis*-retinoic acid developed the capacity for superoxide production and phagocytosis. These cells were not morphologically normal, showing persistent nuclear immaturity. The case report suggests that retinoids at concentrations that could be used clinically may induce the maturation of myeloid leukaemic cells *in vivo*.

Retinoids have been tested in a phase I clinical trial for the treatment of 17 patients with preleukaemia (Gold et al 1983). Patients in the clinical trial received 20–125 mg/m^2 per day of 13-*cis*-retinoic acid. Hepatic toxicity, reversible on withdrawal of the drug, developed at doses of 125 mg/m^2 per day. Other toxicities were less severe and consisted of dryness and erythema of skin, cheilosis and occasional lethargy. Fifteen of the 17 patients were evaluable. Five patients responded with improvement in granulocytopenia, thrombocytopenia and/or anaemia. Three of these patients also had a decrease in the number of myeloblasts in their bone marrow. Two of the responding patients developed recurrent infections despite increases in white blood cell counts, suggesting that their circulating white cells may not have been functionally normal.

In summary, *in vitro* studies demonstrate that myeloid leukaemic blast cells harvested from patients can occasionally be induced to differentiate to mature cells or can be inhibited in their clonal proliferation by retinoids. Induction of differentiation by retinoids is confined to blast cells that are moderately well differentiated (mature myeloblasts and myelomonoblasts, or promyelocytes). The results of administering 13-*cis*-retinoic acid to a patient with promyelocytic leukaemia and to several patients with preleukaemia suggest that retinoids may occasionally induce myeloid leukaemic cell maturation *in vivo* and lead to improvement in haematological parameters; however, in no case has retinoic acid unequivocally produced a long-term remission. The toxicity of $100 \, mg/m^2$ or less of retinoic acid was limited predominantly to moderate cutaneous disorders.

Retinoids have three effects *in vitro* that could influence the haemopoietic function of patients who receive these drugs. Retinoids appreciably increase the proliferation of normal haemopoietic cells committed to myeloid or erythroid development (CFU-GM or BFU-E). In pharmacological concentrations, retinoids can both inhibit the proliferation and induce the differentiation of cells from several leukaemic cell lines. Finally, retinoids can inhibit the growth of leukaemic cells without any effect on differentiation.

Acknowledgements

Supported in part by National Institutes of Health grants Ca 26038, Ca 3273 and Ca 33936, the Bruce Fowler Memorial Fund and the Jonsson Comprehensive Cancer Center. H.P.K. has a Career Development Award from NIH. T.T.A. is a Fellow of the Leukemia Society of America, Inc. We would also like to thank R. Simon, K. Demos, H. Merriman and S. Negussie for their assistance in the preparation of this paper.

REFERENCES

Breitman TR, Selonick SE, Collins SJ 1980 Induction of differentiation of the human promyelocytic leukemia cell line (HL-60) by retinoic acid. Proc Natl Acad Sci USA 77:2936-2940

Breitman TR, Collins SJ, Keene BR 1981 Terminal differentiation of human promyelocytic leukemic cells in primary culture in response to retinoic acid. Blood 57:1000-1004

Camisa C, Eisenstat B, Ragaz A, Weissmann G 1982 The effects of retinoids on neutrophil functions *in vitro*. J Am Acad Dermatol 6:620-629

Cooper RA, Ip SHC, Cassileth PA, Kuo AL 1981 Inhibition of sterol and phospholipid synthesis in HL-60 promyelocytic leukemia cells by inducers of myeloid differentiation. Cancer Res 41:1847-1852

Douer D, Koeffler HP 1982a Retinoic acid enhances growth of human early erythroid progenitor cells in vitro. J Clin Invest 69:1039-1041

Douer D, Koeffler HP 1982b Retinoic acid enhances colony-stimulating factor-induced clonal growth of normal human myeloid progenitor cells *in vitro*. Exp Cell Res 138:193-198

Douer D, Koeffler HP 1982c Retinoic acid. Inhibition of the clonal growth of human myeloid leukemia cells. J Clin Invest 69:277-283

Ferrero D, Gallo E, Lanfrancone L, Tarella C 1983a Functional and phenotypic characterization of two HL60 clones resistant to dimethylsulfoxide. Exp Cell Res 147:111-118

Ferrero D, Pessano S, Pagliardi GL, Rovera G 1983b Induction of differentiation of human myeloid leukemias: surface changes probed with monoclonal antibodies. Blood 61:171-179

Feuerstein M, Cooper HL 1984 Rapid phosphorylation–dephosphorylation of specific proteins induced by phorbol ester in HL-60 cells. J Biol Chem 259:2782-2788

Flynn PJ, Miller WJ, Weisdorf DJ, Arthur DC, Brunning R, Branda RF 1983 Retinoic acid treatment of acute promyelocytic leukemia: in vitro and in vivo observations. Blood 62:1211-1217

Gold EJ, Mertelsmann RH, Itri LM et al 1983 Phase I clinical trial of 13-*cis*-retinoic acid in myelodysplastic syndromes. Cancer Treat Rep 67:981-986

Hodges RE, Sauberlich HE, Canhan JE et al 1978 Hematopoietic studies in vitamin A deficiency. Am J Clin Nutr 31:876-882

Honma Y, Takenaga K, Kasukabe T, Hozumi M 1980 Induction of differentiation of cultured human promyelocytic leukemia cells by retinoids. Biochem Biophys Res Commun 95:507-512

Honma Y, Fujita Y, Kasukabe T et al 1983 Induction of differentiation of human acute non-lymphocyte leukemia cells in primary culture by inducers of differentiation of human myeloid leukemia cell line HL-60. Eur J Cancer Clin Oncol 19:251-261

Kensler TW, Trush MA 1981 Inhibition of phorbol ester-stimulated chemiluminescence in human polymorphonuclear leukocytes by retinoic acid and 5,6-epoxyretinoic acid. Cancer Res 41:216-222

Koeffler HP 1983 Induction of differentiation of human acute myelogenous leukemia cells: therapeutic implications. Blood 62:709-721

Mohanram M, Kuikarni KA, Reddy V 1978 Hematological studies in vitamin A deficient children. Int J Vitam Nutr Res 47:389-393

Olsson IL, Breitman TR 1982 Induction of differentiation of the human histiocytic lymphoma cell line U-937 by retinoic acid and cyclic adenosine 3′:5′-monophosphate-inducing agents. Cancer Res 42:3924-3927

Pigatto PD, Fioroni A, Riva F et al 1983a Effects of isotretinoin on the neutrophil chemotaxis in cystic acne. Dermatologica (Basel) 167:16-18

Pigatto PD, Riva F, Altomare GF, Brugo AM, Morandotti A, Finzi AF 1983b Effect of etretinate on chemotaxis of neutrophils from patients with pustular and vulgar psoriasis. J Invest Dermatol 81:418-419

Strickland S, Breitman TR, Frickel F, Nürrenbach A, Hädicke E, Sporn MB 1983 Structure–activity relationships of a new series of retinoidal benzoic acid derivatives as measured by induction of differentiation of murine F9 teratocarcinoma cells and human HL-60 promyelocytic leukemia cells. Cancer Res 43:5268-5272

Takenaga K, Honma Y, Hozumi M 1981 Cellular retinoid-binding proteins in cultured human and mouse myeloid leukemia cells. Cancer Lett 13:1-6

Weinstein IB 1983 Protein kinase, phospholipid and control of growth. Nature (Lond) 302:750

Wolbach SB, Howe PR 1925 Tissue changes following deprivation of fat soluble A vitamin. J Exp Med 42:753-777

DISCUSSION

Wald: You say that in the phase I clinical trial on patients with preleukaemia (Gold et al 1983) 25–30% showed improvement. What would you expect without retinoid treatment?

Koeffler: Little or no improvement. We are also conducting a double-blind trial of 13-*cis*-retinoic acid versus placebo in patients with preleukaemia. Patients on placebo either get worse or stay about the same; rarely does an improvement in their haemopoiesis occur.

Hartmann: When one looks at the somewhat disappointing results of the clinical study (Gold et al 1983), one wonders whether preleukaemia is really the right condition to be treated with retinoids. It is a malignant, not premalignant condition; in the bone marrow up to 30% of the nucleated cells are blasts. It might be better to use retinoids for conditions like chronic pancytopenias that arise as a consequence of long-term treatment with cytotoxic drugs. One could try to stimulate the normal stem cells *in vivo* as you have demonstrated in your *in vitro* experiments.

Koeffler: I suspect that in the conditions you are describing severe stem cell damage exists, as after exposure to busulphan. Retinoids may enhance cell reactivity to growth factors, but if there is marked damage of the cell population, retinoids would probably not be very effective. If the cell population is only minimally damaged, retinoids might be helpful.

Hartmann: But there may still be a population of cells that function normally in these patients. These people can carry on at a constant level of pancytopenia, and in most of them cell counts will eventually go up. If you could stimulate remaining cells you might alleviate the pancytopenia.

Breitman: There is experimental evidence indicating that retinoids could be clinically useful in the treatment of patients with acute promyelocytic leukaemia. Each of the 10 samples of cells from patients with acute promyelocytic leukaemia that we have studied has differentiated in primary culture in response to retinoic acid (Breitman et al 1981, 1983). These findings have been confirmed, but not in published reports, by many other investigators and I estimate that *in toto* the cells from at least 50 patients with acute promyelocytic leukaemia have responded to retinoic acid *in vitro*. It is of some interest that retinoic acid-induced differentiation of cells from two patients with acute promyelocytic leukaemia was potentiated by a lymphokine that we call DIA (differentiation-inducing activity) (Breitman et al 1983). DIA is produced by mitogen-stimulated normal mononuclear cells (Olsson et al 1981) and has been purified from the conditioned medium of the human T-lymphocytic leukaemia cell line, HUT-102 (Olsson et al 1984). Combinations of a physiological concentration of retinoic acid (10 nM) (DeRuyter et al 1979) and DIA synergistically induce differentiation of HL-60 cells to cells with many of the functional and morphological properties of monocytes (Breitman et al 1983). Recombinant interferons (gamma and alpha) have effects on HL-60 that are similar to those of DIA (H. Hemmi et al, unpublished work). These results indicate that combinations of retinoic acid and interferon may be clinically useful in the treatment of patients with some leukaemias.

Sporn: How many patients with promyelocytic leukaemia have been treated clinically with retinoids?

Koeffler: One or two cases have been reported.

Breitman: Flynn et al (1983) gave 13-*cis*-retinoic acid orally to a patient with acute promyelocytic leukaemia who was refractory to chemotherapy. Promyelocytes obtained from this patient both before and during treatment differentiated in primary culture in response to retinoic acid. During the 13-day treatment period with 13-*cis*-retinoic acid there was an increase in maturing cells in peripheral blood and in bone-marrow mature elements. However, an evaluation of the utility of 13-*cis*-retinoic acid in this case study is complicated by the fact that the patient died on day 13 from a disseminated *Candida* infection. A clearer response of a patient with acute promyelocytic leukaemia to 13-*cis*-retinoic acid is reported by Nilsson (1984). This patient was in relapse and was refractory to chemotherapy. From October 1982 to the present time (October 1984), 13-*cis*-retinoic acid (1 mg/kg per day; 40 mg twice daily) has been administered orally. Over the initial 16 weeks of treatment the proportion of promyelocytes in the marrow decreased gradually to a normal level. In October 1984 the patient is still well and the treatment [13-*cis*-retinoic acid daily and five-day courses of cytosine arabinoside (ara-C) and thioguanine monthly] has remained unchanged (I.L. Olsson, personal communication October 1984).

Sporn: Have any patients been treated clinically with *all-trans*-retinoic acid?

Koeffler: No.

Sporn: Is there an acceptable conventional cytotoxic chemotherapy available for people with promyelocytic leukaemia? What is the prognosis for these patients?

Koeffler: As a group their prognosis is slightly better than for patients with many other subclasses of acute myelogenous leukaemia. About 70% of these patients will go into remission with ara-C and daunorubicin. Remissions will probably last for 16 months or so, but 20–30% of the patients will have prolonged remissions.

Sporn: So treating these people with a retinoid would involve taking them off some other, more conventional therapy which has a reasonable chance of success.

Breitman: I am not aware of any experimental or clinical evidence that argues against combining retinoid treatment with conventional chemotherapy. On the contrary, there is both experimental and clinical evidence to support a combination treatment. Retinoic acid induces differentiation of HL-60 even in the presence of growth-inhibiting concentrations of either ara-C or hydroxyurea (Ferrero et al 1982), and in Nilsson's clinical study (Nilsson 1984) the patient was also receiving ara-C and thioguanine. Thioguanine is both a cytotoxic agent and an inducer of differentiation of HL-60 cells (Collins et al 1980).

Sporn: Actinomycin D and methotrexate have also been reported to be inducers of differentiation, but I don't understand how they work.

Breitman: The concentrations of actinomycin D and methotrexate that induce differentiation of HL-60 are also very cytotoxic. We have observed an induction of differentiation of HL-60 with the calcium ionophore A23187 but, as with actinomycin D and methotrexate, the dose–response curves for differentiation and for cytotoxicity are almost superimposable (T.R. Breitman & B.R. Keene, unpublished work 1982).

Dr Koeffler, in your cloning assays with myeloid cell lines, how do you distinguish between inhibition of growth and induction of differentiation? If you do not get a colony it may be because you have induced differentiation. There is not necessarily an inhibition of growth.

Koeffler: I do not try to distinguish between them. If we are looking for compounds that might be of use in clinical medicine, an assay for clonal growth may be more useful than an assay for differentiation. A stem cell might not form a colony in soft agar containing a test substance either because the test material inhibits the growth of the stem cell or because the test material induces the stem cell to differentiate into a non-dividing mature cell. The KG-1 myeloblasts do not differentiate in the presence of retinoids in liquid culture, but a very low concentration of *all-trans*-retinoic acid inhibits their proliferation in soft agar. HL-60 promyelocytes seem to go through one differentiation step to become myelocytes and metamyelocytes in the presence of retinoids. The metamyelocytes can no longer proliferate and therefore would be unable to form a colony of progeny cells.

Breitman: The results from the assays that you use are not consistent with those obtained in liquid culture. The growth of KG-1 is not inhibited in liquid culture by retinoic acid at the concentrations that inhibit colony formation in your cloning assay (H. Hemmi & T.R. Breitman, unpublished work 1983).

Koeffler: I think that for measuring the ability of a substance to inhibit growth the cloning assay is better than a liquid culture assay. What you are really interested in in leukaemia are the colony-forming cells; that is, the cells that can continue to divide. A liquid culture assay of inhibition of growth is usually not as sensitive; a cell need not divide but can still remain viable for a long time. Many of the substances we study affect cells that are going through the cell cycle.

Sporn: Sensitivity often varies with culture conditions. Many cell lines are more sensitive to retinoid inhibition of soft-agar growth than they are to inhibition of monolayer growth.

Lotan: We have found that several human melanoma lines are insensitive to retinoids when they are cultured in liquid medium, but they are quite sensitive in agar (Lotan et al 1982).

Sherman: Agar culture conditions probably put more stress on a cell to survive at clonal density to begin with. So you may just tip the balance more easily with retinoid in an agar culture than you would in liquid.

REFERENCES

Breitman TR, Collins SJ, Keene BR 1981 Terminal differentiation of human promyelocytic leukemic cells in primary culture in response to retinoic acid. Blood 57:1000-1004

Breitman TR, Keene BR, Hemmi H 1983 Retinoic acid-induced differentiation of fresh human leukaemia cells and the human myelomonocytic leukaemia cell lines, HL-60, U-937, and THP-1. Cancer Surv 2:263-291

Collins SJ, Bodner A, Ting R, Gallo RC 1980 Induction of morphological and functional differentiation of human promyelocyte leukemia cells (HL-60) by compounds which induce differentiation of murine leukemia cells. Int J Cancer 25:213-218

DeRuyter MG, Lambert WE, DeLeenheer AP 1979 Retinoic acid: an endogenous compound of human blood. Unequivocal demonstration of endogenous retinoic acid in normal physiological conditions. Anal Biochem 98:402-409

Ferrero D, Tarrella C, Gallo E, Ruscetti FW, Breitman TR 1982 Terminal differentiation of the human promyelocytic leukemia cell line, HL-60, in the absence of cell proliferation. Cancer Res 42:4421-4426

Flynn PJ, Miller WJ, Weisdorf DJ, Arthur DC, Brunning R, Branda RF 1983 Retinoic acid treatment of acute promyelocytic leukemia: in vitro and in vivo observations. Blood 62:1211-1217

Gold EJ, Mertelsmann RH, Itri LM et al 1983 Phase I clinical trial of 13-*cis*-retinoic acid in myelodysplastic syndromes. Cancer Treat Rep 67:981-986

Lotan R, Lotan D, Kadouri A 1982 Comparison of retinoic acid effects on anchorage-dependent growth, anchorage-independent growth and fibrinolytic activity of neoplastic cells. Exp Cell Res 141:79-86

Nilsson B 1984 Probable *in vivo* induction of differentiation by retinoic acid of promyelocytes in acute promyelocytic leukaemia. Br J Haematol 57:365-371

Olsson I, Olofsson T, Mauritzon N 1981 Characterization of mononuclear blood cell-derived differentiation inducing factors for the human promyelocytic leukemia cell line HL-60. J Natl Cancer Inst 67:1225-1230

Olsson IL, Sarngadharan MG, Breitman TR, Gallo RC 1984 Isolation and characterization of a T lymphocyte-derived differentiation inducing factor for the myeloid leukemic cell line HL-60. Blood 63:510-517

Final general discussion

From basic science to clinical trials

Wald: During this meeting we have been concerned with mechanisms, with understanding the biology of retinoids and of growth in general. The field is extremely wide and complex, but as an epidemiologist I am concerned that our attempts to understand what is going on may be at the expense of a more pragmatic approach. At the moment, the greatest returns in medicine will come from such a pragmatic approach, from acting simply on the basis of rather rough and ready judgements, without understanding exactly why particular substances should be used or how they work. We should choose agents that will probably do no harm in clinical situations but might do some good. What concerns me is that the step from identifying a particular retinoid to putting it into clinical practice seems to take so long. Satisfactory trials are few in number and most are only just beginning. We need a complementary approach; the people who understand the biology and mechanisms should talk more to the clinicians and epidemiologists, but in a simple way that will allow clinical trials to be launched more rapidly and with a greater sense of impetus than they are at present, and with the aim of producing results in two or three years. Perhaps we could discuss whether this should happen and, if so, how we can arrange things to make it happen effectively.

Sporn: If you want to influence a disease process that may take 20 years or more to unfold, I cannot see how you are necessarily going to get any results in two or three years, no matter how many people and how much money you bring to bear on the problem. Retinoids seem to be more useful for the prevention of disease than for therapy, and it takes a long time for results on prevention to unfold. It is not realistic to expect to get meaningful data to assess in only two or three years of a clinical prevention trial. Even in the breast and bladder cancer groups that have been selected for study in Milan and London, the first year or two will be spent getting rid of the data that you don't want.

Hicks: That's an argument for starting trials as soon as possible. The sooner you start, the sooner you will get results.

Wald: For studies like Marian Hicks' study on bladder cancer, once you start recruiting patients you *will* have results in about three years if the trial is big enough.

Sporn: Yes, but some people think that this type of research is gobbling up too much money. A lot of the problems are very practical ones. One must

268

consider how much it costs to do this sort of work and what the other priorities are in research.

Wald: The kind of pragmatic approach I am suggesting is rare and clinical trials are extremely sparse, whereas basic scientific research is much more extensive. I'm not saying that there should be a battle between the two, but there is an enormous imbalance.

Sporn: The people who fund very expensive clinical trials always want to wait for more basic scientific results that show that the drug really works. It is a 'Catch 22' situation.

Wald: That is exactly the sort of thing one should not wait for. Many people think that before one starts testing something in clinical practice one must understand how it works. That is a misconception. Many of the most effective agents in medicine were actually found to work in clinical practice well before their mechanisms of action were understood. In fact, does anyone know how 13-*cis*-retinoic acid works in patients with acne?

Sherman: Although it is very effective in acne, 13-*cis*-retinoic acid, like many other retinoids, has potentially harmful side-effects. One can be pragmatic and hopeful about these drugs, but without doing mechanistic studies one might run a higher risk of doing a disservice to one's patients.

Moon: One of the problems is that clinicians often want to use compounds that have not proved efficacious in animal models. For example, etretinate is not effective in animal models of bladder cancer, but there is talk of using it in a clinical study on the bladder. Although there is some clinical evidence that etretinate might be effective in humans, it is tenuous. The other point is that the animal studies showed that the compound had considerable toxicity. One must take into account not just efficacy, but toxicity aspects as well.

Sporn: I would like to go to patients and tell them that the drug has worked in six different animals in five different laboratories all over the world before asking them to sign an informed consent form. You need some scientific data before it is ethical to ask people to give their consent. You cannot just say 'This is something that might work; why don't you sign up for the trial?' None of these retinoids is going to be totally devoid of toxicity. We live in a climate in which society looks at the introduction of any new chemical substance with scepticism. Although retinoids are relatively safe compared with some other drugs, people still need assurance that you are not going to hurt them while you are trying to help.

Hicks: That does not contradict what Nick Wald is saying. He is not asking you to stop testing retinoids; he wants the efficacy and toxicity data. But he is asking whether we really have to know how retinoids act at a molecular level before we start a clinical trial. If we have a compound that is efficacious in animals and is not so toxic that it is likely to kill the patient, do we really have to wait until we understand the molecular mechanisms before we do a trial?

Sporn: But that is not what is holding up clinical trials.

Moon: The first publication of the effectiveness of N-(4-hydroxy-phenyl)retinamide (HPR) was in 1979, but we are only now getting to the point of starting a clinical trial.

Sporn: The delay has nothing to do with a need to wait until we understand the molecular biology of the drug's action. When we knew seven or eight years ago that HPR was highly active in animal systems and relatively non-toxic, we immediately suggested putting it into a clinical trial to see whether it would do something in dermatological conditions, in patients with bladder papillomas or even in women with a risk of developing premenopausal breast cancer or fibrocystic disease of the breast. The delay in starting trials was due to the constraints on introducing new drugs: legal and ethical requirements, review boards and so on. I'm not very happy with the long delays, but unless we introduce a much greater degree of authoritarianism into society, there is no way of shortening them.

Wald: I'm not suggesting that one should adopt an authoritarian approach. I am concerned that our desire to understand *how* something works should not be satisfied at the expense of the pragmatic approach of seeing whether in fact it *does* work in humans. Clinical groups and pathologists should be encouraged to undertake proper scientific trials. You said that good scientific data are needed before one goes ahead with a trial, and I agree. But there are two types of science: there is basic science, which many of you are doing, and there is clinical science, involving the collection of information on efficacy and toxicity in a controlled way so that we can learn about the drugs in question in the species of interest, i.e. in humans. The fact that we may not know exactly how the drug works should not hold us up from doing a trial, provided that there are reasonable grounds for going ahead. Lack of knowledge about the mechanism is often given as a reason for not going ahead, but I think that is a mistake.

Moon: How would you approach things if I said I had a compound that might be good for bladder cancer and that had relatively little toxicity? Who would we get to produce it, and where would the money come from to synthesize it for a clinical trial? These would be some of the practical problems. If the compound was described in the open literature, nobody would want to take the responsibility for making it.

Wald: These are the kinds of questions that we should address. We need to identify what the problems are in introducing retinoids which we believe would not have high toxicity and would have reasonable efficacy. Then we need to propose practical solutions to overcome those problems. A particular problem that you refer to is that there may be no commercial advantage in producing some of these agents and this will inhibit their development. We therefore should explore methods by which such drugs, for which there are no patents or commercial advantages, can be investigated and if necessary introduced.

Sherman: Several drug companies do have a very strong interest in retinoids and their introduction into the clinic. So the problem is not one of basic scientists having to go to drug companies with retinoids that they think will be effective and having to persuade them to spend money in order to market the drugs. The reality is that there are many safeguards that must be considered and hurdles that have to be passed before you can introduce a new drug into the clinic.

Wald: It is surprising how rarely clinical investigations of the right kind are conducted. This is true not only for retinoids but also for other drugs. There are many small, poor studies. It is quite common for industry to encourage small studies and it is of general concern that there is a tendency for those that are positive to get published and for those that are negative to remain unpublished. This, unfortunately, leads to the conclusion that ineffective regimens are useful when in fact they are not. Small studies can also mean that a quite useful therapeutic result may be completely overlooked. We need to choose about three retinoids that at the moment look the best for a particular site such as the bladder, and get together with colleagues at the National Cancer Institute in the USA or the Medical Research Council in Great Britain to discuss clinical trials. We can decide for how many years the trial should run, how many patients should be included and where they are going to come from. If there is no commercial impetus behind a particular retinoid, we should pay a pharmaceutical company to produce it. If we can all work together effectively to address these problems, there is no real reason why these things should take 10 years.

Hicks: What you are really asking for is a complete change in mental attitude. In general, people who regard themselves as basic scientists are not aiming at a clinical trial as an end-goal of their research. They are not really interested in getting together with people like epidemiologists to apply their research to clinical medicine.

Sporn: There is a tremendous resistance to preventive medicine. Another problem in America is that practising physicians do not want to try new drugs that are not yet 'accepted' by society because they may face litigation for malpractice. The problems in getting a trial off the ground have to do with society and economics rather than the fact that we don't know the molecular mechanisms of action of retinoids. If we are fortunate, and some really good results come out of the London bladder cancer trial, the Milan breast cancer trial, or dermatology trials with retinoids that might have fewer side-effects than 13-*cis*-retinoic acid, then the situation will be changed.

Moon: We have got to educate physicians about the basic science of retinoids because many of them assume that all retinoids are the same. They may expect to get a response with retinyl acetate or retinol in cancer patients because they have read that HPR or 13-*cis*-retinoic acid gives a response in animals. If they then get negative results, they may say that retinoids don't work in cancer patients.

Sporn: We have had this problem with a previous bladder cancer trial in America. The retinoid tested was 13-*cis*-retinoic acid, but academic urologists felt that it was worthless.

Wald: This is a valid point. If you choose the wrong drug and the results are negative, those results may be applied to the whole group of drugs and unjustifiably inhibit the development of potentially effective ones. But I think it is for us to point out the weaknesses of particular studies, how results that relate to one drug need not apply to another similar one, and why we think the drug is wrong for a given disease. If we think a particular clinical trial with a particular drug is of no use, we should treat it on its own merits and criticize it without necessarily suggesting that the whole class of drugs is useless. Retinoids are no different from anything else in medicine; we should be more pragmatic and scientific, and less philosophical.

Moon: But attitudes may develop on the basis of a single case-history study. A physician who is treating a cancer patient may read about a certain retinoid that might help; but if he does not have this retinoid, he may try another one that is useless. It would be fine if these cases were kept in context, but I have found that physicians tend to get together to discuss the drugs they have tried. Their experience of failure with a few isolated cases fosters the general view that none of these things work. We need to have good clinical trials set up so that we can make rational decisions on whether a compound is effective or not in humans.

Lotan: I think that one of the problems with clinical trials is that they *are* done in haste. No one waits for the basic scientific data. These studies are performed on patients who have previously received cytotoxic treatment or have not responded to other treatments. Retinoids are cytostatic rather than cytotoxic, so they have no chance of producing an improvement in these patients. If scientists were consulted, they would advise against doing such studies. The results of Gunther Dennert (this volume) and Suzanne Eccles (see p 129) indicate that retinoids work best under conditions where there is some immune response that can be stimulated. One would avoid giving retinoids to patients who have undergone cytoxic therapy because their immune responses might be completely suppressed; there may be nothing to resurrect. In some clinical trials, retinoids have been used inappropriately as anticancer agents, for example in testicular carcinoma, which is the equivalent of embryonal carcinoma. The trial was unsuccessful but I am not surprised. There is a very responsive mouse model, but embryonal carcinoma cell lines from humans are not responsive to retinoids, so I think this trial failed because of a lack of basic scientific information. Clinicians who have done trials should come back to the scientists to ask why they have not succeeded.

Sporn: We are all concerned that the wrong trial, hastily done, will do a lot of harm, so we need to be very judicious and plan things carefully. The trials that are going to be undertaken in London and Milan *are* very well planned, with good coordination between the basic scientists, the clinicians and the epidemiologists, so they should give some informative results.

Dennert: In clinical trials, it might be better to look at the effects of retinoids on certain infections than to look for cancer prevention or cure. There is now pretty good evidence from animal model systems, and some evidence from humans, that retinoids may stimulate the immune system. So it may be possible to induce a better immune response to viral or bacterial infection. These trials would give a relatively quick return in terms of results.

Turton: We tend to think of retinoids as compounds which may be used in the treatment of disorders of proliferation or differentiation, particularly in diseases of the skin such as psoriasis, in cancer, and in rheumatoid arthritis (Sporn & Harris 1981, Harris 1984). But are retinoids being considered as possible therapeutic agents in disease conditions outside those areas broadly covered by the term 'diseases of proliferation or differentiation'? The possible use of retinoids in the treatment of gout has been mentioned.

Sporn: Of course retinoids are used for the acute treatment of things like cystic acne and keratinizing dermatoses, but I'm not aware of any plans to use them for the treatment of non-dermatological conditions other than cancer.

REFERENCES

Dennert G 1985 Immunostimulation by retinoic acid. In: Retinoids, differentiation and disease. Pitman, London (Ciba Found Symp 113) p 117-131
Harris ED 1984 Retinoid therapy for rheumatoid arthritis. Ann Intern Med 100:146-147
Sporn MB, Harris ED 1981 Proliferative diseases. Am J Med 70:1231-1236

Index of contributors

Entries in **bold** *type indicate papers; other entries refer to discussion contributions*

Subject index